AA003789

XLIX International Symposium on Multiparticle Dynamics (ISMD 2019)

EPJ Web of Conferences Volume 235 (2020)

Santa Fe, New Mexico, USA
9 – 13 September 2019

Editors:

Ivan Vitev
Cesar da Silva
Saskia Mioduszewski
Claudia Ratti
Ina Sarcevic
Mark Schlegel

ISBN: 978-1-7138-1440-5

Printed from e-media with permission by:

Curran Associates, Inc.
57 Morehouse Lane
Red Hook, NY 12571

Some format issues inherent in the e-media version may also appear in this print version.

For additional information, please contact EDP Sciences – Web of Conferences
at the address below.

EDP Sciences – Web of Conferences
17, Avenue du Hoggar
Parc d'Activité de Courtabœuf
BP 112
F-91944 Les Ulis Cedex A
France

Phone: +33 (0) 1 69 18 75 75
Fax: +33 (0) 1 69 28 84 91

contact-edps@webofconferences.org

Additional copies of this publication are available from:

Curran Associates, Inc.
57 Morehouse Lane
Red Hook, NY 12571 USA
Phone: 845-758-0400
Fax: 845-758-2633
Email: curran@proceedings.com
Web: www.proceedings.com

TABLE OF CONTENTS

edp sciences

EPJ Web of Conferences

Statement of Peer review

In submitting conference proceedings to EPJ Web of Conferences, I certify to the Publisher that I adhere to the **Policy on Publishing Integrity** of the journal in order to safeguard good scientific practice in publishing.

1. All articles have been subjected to peer review administered by the proceedings editors.
2. Reviews have been conducted by expert referees, who have been requested to provide unbiased and constructive comments aimed, whenever possible, at improving the work.
3. Proceedings editors have taken all reasonable steps to ensure the quality of the materials they publish and their decision to accept or reject a paper for publication has been based only on the merits of the work and the relevance to the journal.

Title, date and place of the conference

XLIX International Symposium on Multiparticle Dynamics
September 9-13, 2019 Santa Fe, NM USA

Proceedings editor(s): Ivan Vitev

Co-editors: Cesar da Silva, Saskia Mioduszewski
Claudia Ratti, Ina Sarcevic, Mark Schlegel

Date and editor's signature May 14, 2020

Sources of multiparticle correlations

– a microscopic perspective

Christian Bierlich[1,2,*]

[1]Department of Astronomy and Theoretical Physics, Lund University, Sölvegatan 14A, Lund, Sweden.
[2]Niels Bohr Institute, University of Copenhagen, Blegdamsvej 17, Copenhagen, Denmark.

Abstract. Multiparticle correlations is a hallmark measurement characterizing the behaviour of the assumed Quark Gluon Plasma in heavy ion collisions. In these proceedings an alternative, microscopic approach is presented, based on interacting strings and multiparton interactions.

1 Multiparton interactions: From pp to AA

In high energy collisions of protons, the presence of multiparton interactions (MPIs) are ubiquitous, and to a large extent successfully modelled by considering individual interactions as almost independent[1] perturbative scatterings. This model was introduced to the PYTHIA event generator [1] by Sjöstrand and van Zijl [2], and takes as a starting point the $2 \to 2$ scattering cross section, with its low $p_\perp$ divergence regularised using a parameter $p_{\perp 0}$:

$$\frac{d\sigma_{2\to 2}}{dp_\perp^2} \propto \frac{\alpha_s^2(p_\perp^2)}{p_\perp^4} \to \frac{\alpha_s^2(p_\perp^2 + p_{\perp 0}^2)}{(p_\perp^2 + p_{\perp 0}^2)^2}. \tag{1}$$

Naively, this adds sub-scatterings, which in turn each draws two strings (for gluonic interactions) from projectile to target, filling the whole available rapidity range with final state particles. As it was already realized in the original paper, such a simplistic treatment fails to describe data. In particular the rise of $\langle p_\perp \rangle$ with N_{ch} cannot be accounted for. The resolution was to colour-reconnect soft partonic scatterings to hard ones, with the guiding principle of minimizing the total potential energy in the string. Such a treatment can correctly reproduce $\langle p_\perp \rangle(N_{ch})$, and it was later shown to also give effects similar to collectivity [3], also reproduced in later, more elaborate models [4]. These collective effects are, however, all short range in rapidity [5]. The challenge of extending the MPI formalism to pA and AA, lies not in scaling the uncorrelated partonic collisions to a collision systems involving collisions of several nucleons, but rather how the latter part should be superimposed. In refs. [6, 7] the Angantyr model, based on the wounded nucleon model [8] and Regge theory, was introduced, where instead of applying colour reconnection, absorptive (i.e. inelastic non–diffractive) nucleon–nucleon collisions are ordered in primary and secondary collisions (see ref. [7] for a full description of the procedure). Primary collisions are treated as normal absorptive pp collisions, and secondary collisions are treated as a single wounded nucleon

*e-mail: christian.bierlich@thep.lu.se

[1]Not entirely independent, as parton density corresponding to the extracted parton is rescaled by a factor $1 - x$, and energy–momentum conservation is obeyed.

which, inspired by the Fritiof model [9], contributes as a single string with a mass distribution $\propto dM^2/M^2$. This drastically alters the number of particles produced in both central and forward regions, and for high energy Pb-Pb collisions both are in good agreement with data, as seen in figure 1 (left) for forward production and (right) for central production.

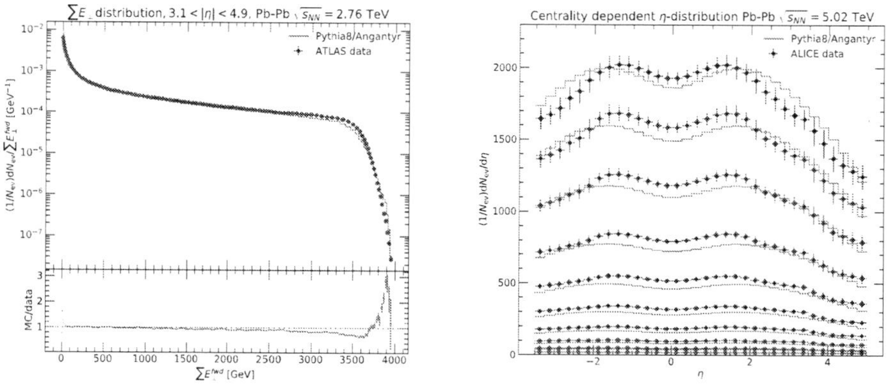

Figure 1. Forward energy deposition (left) and multiplicity as a function of pseudo-rapidity (right) in Pb-Pb collisions shown, compared to simulation by PYTHIA +Angantyr.

The correct description of forward production, as shown in figure 1 (left) is crucial, as it allows for comparison to data in bins of centrality defined in the same way as done in experiment – in particular for pA collisions, the translation from geometric to measured centrality is not straightforward. This task is further eased by the RIVET framework for data comparison [10] (with which the above figures are produced), which has recently been updated to include heavy ion functionality [11], centrality included.

2 String shoving

As mentioned in the previous section, the particle content of the final state, is governed by strings responsible for the partonic final state fragmenting to hadrons. In the following, and in PYTHIA in general, this task is carried out by the Lund model [12–14]. Before hadronizing, strings may interact with each other – especially in AA collisions, or high multiplicity pp, where the density is large – by pushing or "shoving" each other [15, 16] or by forming colour multiplets, so–called ropes [4, 17, 18]. This section is concerned with the former option.

The physical picture of string shoving in transverse space, for a "slice" in rapidity, is sketched in figure 2 (left). On top left there is a piece not overlapping with any other piece, and it will not experience any interaction. Around $b_x = 0$ two pieces are overlapping, and they will push each other in the direction of the arrows. Finally, in the configuration with three string pieces, the two peripheral pieces will be pushed away, while the central piece will stay stationary as it experiences no net force. The actual expression for the force, is derived a) with inspiration from the duality between the QCD vacuum and an electric superconductor [19], and b) lattice calculations of static properties of a string in transverse space. According to lattice calculations, the longitudinal color-electric field of a string, has nearly a Gaussian shape [20]:

$$E_l(x_\perp) = \frac{\Phi}{2\pi R^2} \exp\left(-\frac{x_\perp^2}{2R^2}\right), \tag{2}$$

 https://doi.org/10.1051/epjconf/202023501001

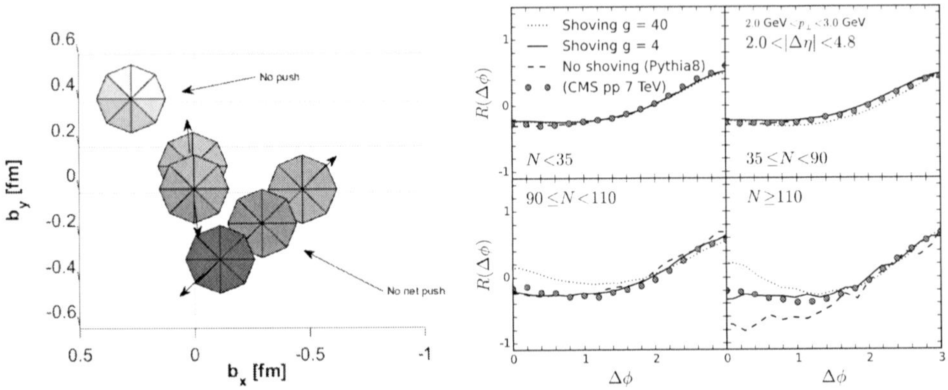

Figure 2. Sketch of the shoving effect (left) showing how, depending on the spatial distribution of strings, there may or may not be an effect, as well as (right) the successful reproduction of the ridge in pp collisions by the shoving model.

where Φ is the flux. From equation (2), the electric field contribution to the string energy density (κ) can be calculated. Assuming that only the electric field contributes to the string energy density, the potential energy of a two-string configuration with a separation $d_\perp$, can be calculated. The force per unit string length, on one string from another, can then be calculated as the gradient of the potential energy, giving:

$$f(d_\perp) = \frac{g\kappa}{R^2} \exp\left(-\frac{d_\perp^2}{4R^2}\right), \tag{3}$$

where g is a free parameter, determining how much of the energy goes into the magnetic current, and breaking the condensate.

Strings are allowed to shove each other from some initial time after the collision, understood as the time it takes for a string to reach its equilibrium transverse size, until the strings hadronize at around $\langle\tau^2\rangle = 2$ fm. In figure 2 (right) it is shown how the shoving mechanism, as implemented in Pythia 8.2, reproduces the ridge as observed by CMS [21].

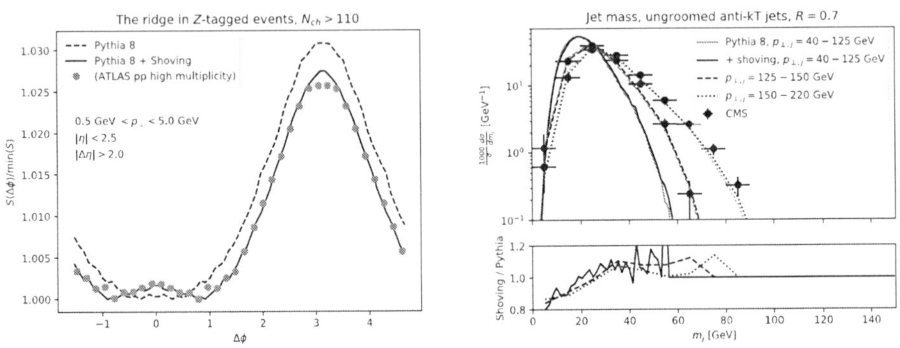

Figure 3. The successful reproduction of the ridge in Z-tagged pp collisions (left) by the shoving model, as well as the effect from shoving on the mass of the recoil jet (right), compared to data.

Since the shoving mechanism is purely based on string interactions, it applies equally well for all types of events, regardless of any type of trigger process. Furthermore, since the string formation is fully integrated in the event generation process, the model allows for full dynamical simulation of kinematical biases present in trigger process events, compared to minimum bias. Consider string interactions in an event with a Z-boson in the di-lepton channel. Since there will not be any string stretched between the leptons, the reconstructed Z will not interact through string interactions, but the rest of the event will. This allows for precision studies of possible jet effects, as the Z gives an unmodified kinematical handle. This possibility was studied in ref. [22], and two key results are shown in figure 3. On the left side, a comparison to data from ATLAS [23], measuring the ridge in Z-tagged events, is shown. It is immediately visible that the ridge effect is reproduced, also around $\Delta\phi = \pi$. Since there is a reasonably hard probe (the Z) present, balancing jets can also be studied. In figure 3 right, the jet mass for anti-k_t jets [24] with $R = 0.7$ in Z-triggered events, is shown with and without shoving enabled, and compared to data from CMS [25]. It is seen that the shoving mechanism affects the jet mass up to about 20% for low mass jets, pointing to a possible source of jet modifications present in pp collisions.

3 The role of the initial state

It is a well known feature of hydrodynamic treatments of heavy ion collisions, that flow coefficients such as v_2, scales with the corresponding eccentricity, ϵ_2. Flow can as such be understood of a response to an initial geometric configuration – the same is true for the string shoving model. In a heavy ion collision such a geometry is usually defined through the Glauber model, possibly plus fluctuations. In small systems, pA and pp collisions, this is more challenging, as sub-nucleonic degrees of freedom dominate the spatial distribution. In order for the shoving model to produce reliable results, a reasonable model for the spatial distribution of strings must be introduced. In order to calculate the transverse distributions of gluons in the colliding nucleons, the Mueller dipole model [26] in a new Monte Carlo implementation [27] is used. The basis of the calculation, is the (leading order) branching probability for a dipole r_{12}, spanned between $\vec{r}_2$ and $\vec{r}_2$ in transverse space, to branch to two dipoles r_{13} and r_{23}, when evolved in rapidity (y):

$$\frac{\mathrm{d}\mathcal{P}}{\mathrm{d}y} = \mathrm{d}^2\vec{r}_3 \frac{N_c\alpha_s}{2\pi^2}\frac{r_{12}^2}{r_{13}^2 r_{23}^2}. \tag{4}$$

Given an initial state to evolve, this allows for the creation of projectile and target Fock states for individual collisions, by the means of a parton shower-like evolution, where equation (4) is modified by a Sudakov form factor:

$$\exp\left(-\frac{N_c\alpha_s}{2\pi^2}\int_{y_{min}}^{y} \mathrm{d}y \int \mathrm{d}^2\vec{r}_3 \frac{r_{12}^2}{r_{13}^2 r_{23}^2}\right).$$

This allows for an evolution strategy ("winner takes it all") where a cascade is ordered in rapidity, with the possible emission in which the lowest rapidity is chosen in each step of the evolution. This process can be iterated until no emissions are allowed without breaking a maximal rapidity, governed by the collision energy. The interaction between projectile and target can then be factorized into dipole–dipole interactions (between dipole r_{12} and r_{34} from projectile and target respectively), given by:

$$\frac{\mathrm{d}\sigma_{dip}}{\mathrm{d}^2\vec{b}} = \frac{\alpha_s^2}{2}\log^2\left(\frac{r_{13}r_{24}}{r_{14}r_{23}}\right) \equiv f_{ij}. \tag{5}$$

Assuming that all dipole–dipole interactions are independent of each other, multiple scatterings exponentiate, resulting in a unitarized scattering amplitude for a single collision:

$$T(\vec{b}) = 1 - \exp\left(-\sum_{ij} f_{ij}\right). \tag{6}$$

The scattering amplitude can be related to cross sections through the optical theorem, fixing all parameters[2], leaving calculations of the proton substructure as real predictions. Interestingly, studies of eccentricities of pp collisions reveals that a full dipole treatment cannot be distinguished from a simple ansatz, where MPIs are distributed according to a 2D Gaussian – but studies of pA collisions can. In figure 4 (left), normalized symmetric cumulants as a function of final state charged multiplicity in pp, pA and AA are shown. Normalized symmetric cumulants signifies the normalized correlation between flow coefficients, and therefore provides a picture of the initial state, given the aforementioned proportionality between eccentricities and flow coefficients, as:

$$\mathrm{NSC}(n, m) = \frac{\langle v_n^2 v_m^2\rangle - \langle v_n^2\rangle\langle v_m^2\rangle}{\langle v_n^2\rangle\langle v_m^2\rangle} \approx \frac{\langle \epsilon_n^2 \epsilon_m^2\rangle - \langle \epsilon_n^2\rangle\langle \epsilon_m^2\rangle}{\langle \epsilon_n^2\rangle\langle \epsilon_m^2\rangle}. \tag{7}$$

Only for pA collisions can the models be qualitatively distinguished, and interestingly, the dipole treatment is the only one predicting a negative NSC(2, 3) in line with ALICE data [28], as opposed to the Gaussian or a normal Glauber treatment. The particular influence on pA collisions can also be seen from the slightly improved description of flow fluctuation measured by CMS [29], shown in figure 4 (right), where the ratio is similarly expected to make the response coefficients cancel each other.

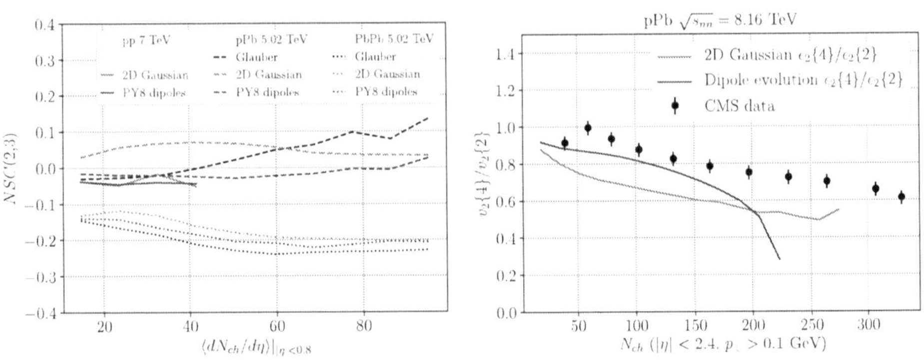

Figure 4. The normalized symmetric cumulant NSC(2, 3) in pp, pA and AA collision by a Glauber treatment, a Gaussian MPI distribution and a Mueller dipole based treatment (left), and a comparison of Gaussian MPI distributions and a dipole based one, compared to data for flow fluctuations in pA collisions (right).

Acknowledgements

I thank the organisers for a pleasant conference. Support from Swedish Research Council, contract number 2017-003, is gratefully acknowledged.

[2]A proton initial state before evolution needs to be modelled, and a model for confinement is also introduced, giving a total of four parameters of the model (including Λ_{QCD}).

References

[1] T. Sjöstrand, S. Ask, J.R. Christiansen, R. Corke, N. Desai, P. Ilten, S. Mrenna, S. Prestel, C.O. Rasmussen, P.Z. Skands, Comput. Phys. Commun. **191**, 159 (2015), 1410.3012

[2] T. Sjostrand, M. van Zijl, Phys. Rev. **D36**, 2019 (1987)

[3] A. Ortiz Velasquez, P. Christiansen, E. Cuautle Flores, I. Maldonado Cervantes, G. Paić, Phys. Rev. Lett. **111**, 042001 (2013), 1303.6326

[4] C. Bierlich, J.R. Christiansen, Phys. Rev. **D92**, 094010 (2015), 1507.02091

[5] C. Bierlich, Nucl. Phys. **A982**, 499 (2019), 1807.05271

[6] C. Bierlich, G. Gustafson, L. Lönnblad, JHEP **10**, 139 (2016), 1607.04434

[7] C. Bierlich, G. Gustafson, L. Lönnblad, H. Shah, JHEP **10**, 134 (2018), 1806.10820

[8] A. Bialas, M. Bleszynski, W. Czyz, Nucl. Phys. **B111**, 461 (1976)

[9] B. Andersson, G. Gustafson, B. Nilsson-Almqvist, Nucl. Phys. **B281**, 289 (1987)

[10] C. Bierlich et al., SciPost Phys. **8**, 026 (2020), 1912.05451

[11] C. Bierlich et al. (2020), 2001.10737

[12] B. Andersson, G. Gustafson, G. Ingelman, T. Sjostrand, Phys. Rept. **97**, 31 (1983)

[13] B. Andersson, G. Gustafson, B. Soderberg, Z. Phys. **C20**, 317 (1983)

[14] T. Sjostrand, Nucl. Phys. **B248**, 469 (1984)

[15] V.A. Abramovsky, E.V. Gedalin, E.G. Gurvich, O.V. Kancheli, JETP Lett. **47**, 337 (1988), [Pisma Zh. Eksp. Teor. Fiz.47,281(1988)]

[16] C. Bierlich, G. Gustafson, L. Lönnblad, Phys. Lett. **B779**, 58 (2018), 1710.09725

[17] T.S. Biro, H.B. Nielsen, J. Knoll, Nucl. Phys. **B245**, 449 (1984)

[18] C. Bierlich, G. Gustafson, L. Lönnblad, A. Tarasov, JHEP **03**, 148 (2015), 1412.6259

[19] M. Baker, J.S. Ball, F. Zachariasen, Phys. Rept. **209**, 73 (1991)

[20] P. Cea, L. Cosmai, F. Cuteri, A. Papa, Phys. Rev. **D89**, 094505 (2014), 1404.1172

[21] V. Khachatryan et al. (CMS), JHEP **09**, 091 (2010), 1009.4122

[22] C. Bierlich, Phys. Lett. **B795**, 194 (2019), 1901.07447

[23] T.A. collaboration (ATLAS) (2017)

[24] M. Cacciari, G.P. Salam, G. Soyez, JHEP **04**, 063 (2008), 0802.1189

[25] S. Chatrchyan et al. (CMS), JHEP **05**, 090 (2013), 1303.4811

[26] A.H. Mueller, Nucl. Phys. **B415**, 373 (1994)

[27] C. Bierlich, C.O. Rasmussen, JHEP **10**, 026 (2019), 1907.12871

[28] S. Acharya et al. (ALICE), Phys. Rev. Lett. **123**, 142301 (2019), 1903.01790

[29] A.M. Sirunyan et al. (CMS), Phys. Rev. **C101**, 014912 (2020), 1904.11519

Overview of jet physics results from ALICE

Filip Krizek[1,*] *for the ALICE Collaboration*

[1]Nuclear Physics Institute of CAS, Husinec-Rez, Czech Republic

Abstract. Recent results from jet shower-shape and substructure analyses done by the ALICE collaboration in central Pb–Pb collisions at $\sqrt{s_{\rm NN}} = 2.76\,\text{TeV}$ and in pp collisions at $\sqrt{s} = 7\,\text{TeV}$ are reviewed. The presented jet shower-shape observables are angularity g and transverse momentum dispersion $p_{\rm T}D$, which were studied for a small resolution parameter $R = 0.2$, track-based jets with a minimum constituent transverse momentum ($p_{\rm T}$) cut-off of $0.15\,\text{GeV}/c$. Jet substructure is explored for track-based anti-$k_{\rm T}$ jets with $R = 0.4$ by means of iterative declustering and grooming techniques, which were used to measure the absolutely-normalized leading subjet momentum fraction $z_{\rm g}$ and the number of hard splittings in the reclustered jet shower, $n_{\rm SD}$. These observables provide complementary information on the jet fragmentation and help to discriminate between different scenarios for medium-induced modifications of the parton shower in heavy-ion collisions due to jet quenching.

1 Introduction

Collisions of ultra-relativistic heavy ions are used to probe strongly interacting matter in the regime of high-energy densities and temperatures. In this regime, color confinement of quarks and gluons in hadrons breaks down and a new state of matter, called Quark-Gluon Plasma, is formed [1]. Properties of this medium can be inferred e.g. based on observed modifications of produced jets, collimated sprays of particles that emerge as a result of fragmentation of highly-virtual partons after an initial hard scattering [2]. Jet-medium interaction proceeds via collisional and radiative processes of the partonic shower inside the medium. As a result, we observe jet quenching [3], which consists in yield suppression of high-$p_{\rm T}$ hadrons and jets with a finite radius R. In addition, the jet-medium interaction also leads to jet substructure modifications which can be quantified by different jet-shape observables or identified as changes in shower evolution when unwinding jet clusterization history.

Jet-shape observables are constructed as functions of jet constituent four-momenta and they are designed to accentuate some characteristic feature of the shower. For instance, angularity helps to quantify a degree of jet collimation [4]. It is defined as

$$g = \sum_{i\in\text{jet}} \frac{p_{{\rm T},i}}{p_{\rm T,jet}} |\Delta R_{{\rm jet},i}|,$$

where $\Delta R_{{\rm jet},i}$ denotes the angle enclosed by the i-th jet constituent and the jet axis. The ratio $p_{{\rm T},i}/p_{\rm T,jet}$ then gives the fraction of jet transverse momentum which is carried by this

*e-mail: krizek@ujf.cas.cz

constituent. Jets with small angularity are more collimated and in the limit of pencil-like jets, the angularity tends to zero.

Another jet-shape observable, called transverse momentum dispersion, quantifies the hardness of the jet fragmentation and it is related to dispersion of the constituent momenta [4],

$$p_{\mathrm{T}}D = \frac{\sqrt{\sum_{i\in \mathrm{jet}} p_{\mathrm{T},i}^2}}{\sum_{i\in \mathrm{jet}} p_{\mathrm{T},i}}.$$

Jets with a large number of soft constituents will have $p_{\mathrm{T}}D$ values closer to 0 whereas jets where just a few constituents carry a large fraction of jet momentum will tend to have $p_{\mathrm{T}}D$ values closer to 1 [4].

Jet substructure measurements analyze history of steps, in which a jet algorithm assembled jet constituents together. This information is accessible via iterative declustering which unwinds the sequence of jet branch mergings. In heavy-ion collisions, jets are usually reconstructed using the anti-k_{T} jet algorithm [5]. This algorithm has reduced susceptibility with respect to the soft underlying event but at the same time, it does not reflect important characteristics of a partonic shower, e.g., the angular ordering. Therefore, anti-k_{T} jet constituents are subsequently once again reclustered with the Cambridge-Aachen algorithm [2]. In order to identify hard splittings within the jet shower, one uses grooming techniques like the Soft Drop (SD) [6]. In the given clusterization step, there are two branches with transverse momenta $p_{\mathrm{T},1}$ and $p_{\mathrm{T},2}$. The SD grooming condition that is evaluated is

$$z = \frac{\min(p_{\mathrm{T},1}, p_{\mathrm{T},2})}{p_{\mathrm{T},1} + p_{\mathrm{T},2}} > z_{\mathrm{cut}}\theta^{\beta},$$

where z_{cut} is a chosen momentum-fraction cutoff, θ is an opening angle of the splitting and β is a parameter. This article reports results for $\beta = 0$ and $z_{\mathrm{cut}} = 0.1$, which allow us to select hard splittings in the jet shower. In the case that a given splitting fails the grooming condition, the harder branch of the splitting is followed and decomposed in the subsequent unwinding step. Two substructure observables will be discussed here. The first one, called z_{g}, measures the momentum fraction of the first splitting that fulfills $z > 0.1$. To leading order, the measurement of z_{g} in vacuum is closely related to the Altarelli-Parisi splitting functions [6]. Theoretical considerations about the in-medium modification of z_{g} can be found in [12–15]. The second observable counts the number of splittings that fulfill $z > 0.1$ when we follow the hardest branch and it will be denoted n_{SD}.

2 Results from jet shape measurements

Recently, the ALICE collaboration reported measurement of angularity and $p_{\mathrm{T}}D$ in Pb–Pb collisions at $\sqrt{s_{\mathrm{NN}}} = 2.76$ TeV and in pp collisions at $\sqrt{s} = 7$ TeV [4]. The analysis was performed for track-based, anti-k_{T} jets with a resolution parameter of $R = 0.2$ and a minimum jet constituent transverse momentum cut-off of 150 MeV/c. The measured spectra corrected for instrumental and underlying event effects are shown in Fig. 1. The upper panels present angularity and $p_{\mathrm{T}}D$ spectra for jets with transverse momentum 40–60 GeV/c measured in pp collisions at $\sqrt{s} = 7$ TeV. The spectra are well reproduced by simulated distributions generated by two different PYTHIA tunes [7]. The bottom panels show the corresponding comparison of jet-shape distributions for pp collisions at $\sqrt{s} = 2.76$ TeV generated by PYTHIA and the fully corrected jet-shape spectra obtained in central Pb–Pb collisions at $\sqrt{s_{\mathrm{NN}}} = 2.76$ TeV (0–10% centrality bin). The comparison suggests that the measured 40–60 GeV/c jets exhibit a shift toward lower angularity and greater $p_{\mathrm{T}}D$ values. Thus it can be seen that jet-medium

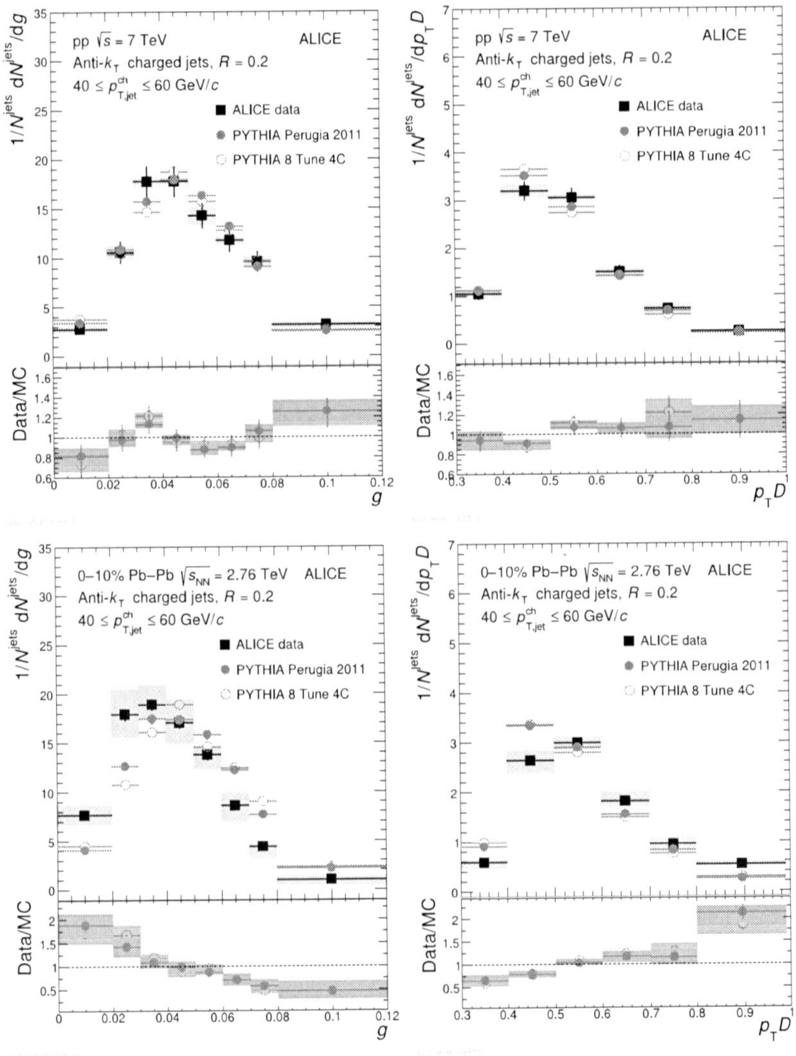

Figure 1. Fully corrected spectra of jet-shape observables as measured by ALICE in pp collisions at $\sqrt{s}$ = 2.76 TeV (top panels) and central Pb–Pb collisions at $\sqrt{s_{NN}}$ = 2.76 TeV (bottom panels) for track-based anti-k_T jets with R = 0.2. Transverse momentum of the jets is constrained to the range 40–60 GeV/c. The left panels show angularity and the right panels present $p_T D$ distributions. Data are compared with predictions of the PYTHIA event generator. See text for further details. Taken from [4].

interaction led to noticeable collimation of these R = 0.2 jets and at the same time resulted also in more asymmetric fragmentation when compared to the vacuum expectation represented here by the PYTHIA simulation. Both features are qualitatively consistent with more quark-like fragmentation [4].

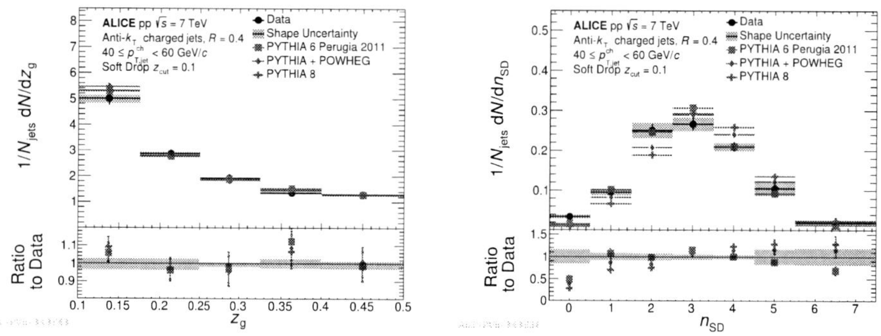

Figure 2. Fully corrected distributions of jet substructure observables z_g and n_{SD} as measured in pp collisions at $\sqrt{s}$ = 7 TeV. Data are compared with PYTHIA event generator calculations. Taken from [8].

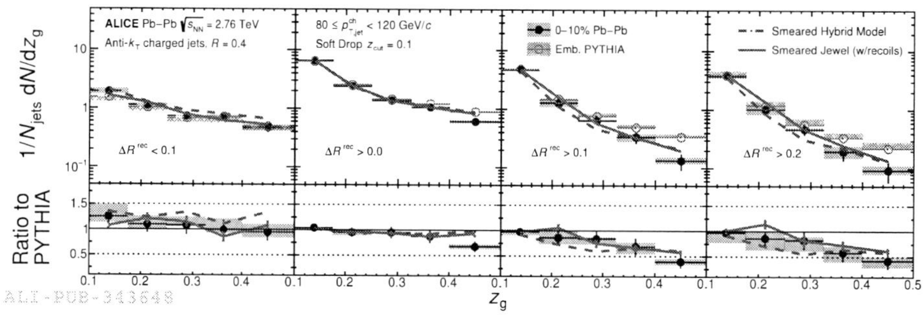

Figure 3. Raw z_g distributions as measured for 80–120 GeV/c track-based jets in Pb–Pb collisions at $\sqrt{s_{NN}}$ = 2.76 TeV (centrality bin 0–10%). The measured data are compared to PYTHIA prediction which was smeared by instrumental effects and embedded into central Pb–Pb collisions and smeared predictions of JEWEL [10] and the Hybrid model [11]. Taken from [8].

3 Results from jet substructure measurements

The same pp and Pb–Pb data were further used to perform jet substructure measurements [8]. The analysis was carried out for track-based, anti-k_T jets with R = 0.4 having a transverse momentum in the range 40–60 GeV/c and 80–120 GeV/c, respectively. Figure 2 presents the fully corrected z_g and n_{SD} distributions measured in pp collisions at $\sqrt{s}$ = 7 TeV. The distributions were compared with the corresponding predictions from PYTHIA simulations. All distributions were normalized by the total number of jets with transverse momentum in the range $p^{ch}_{T.jet} \in (40, 60)$ GeV/c, explicitly including also those jets which did not fulfill the SD condition $z > 0.1$. PYTHIA describes the pp data well except for the the bin $n_{SD} = 0$ where the simulation underestimates the number of jets without a hard splitting.

In case of the central Pb–Pb data, it was not possible to fully correct the raw z_g and n_{SD} distributions for smearing caused by instrumental effects and the underlying event because the unfolding turned out to be unstable due to presence of significant off-diagonal terms in the response matrix. Hence, the measured distributions were compared directly with PYTHIA calculations which were smeared by detector effects and embedded into raw Pb–Pb events.

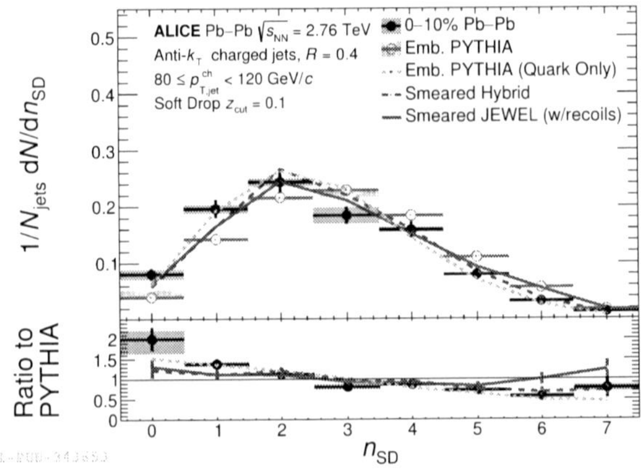

Figure 4. Raw n_{SD} as measured for 80–120 GeV/c track-based jets in central Pb–Pb collisions at $\sqrt{s_{NN}}$ = 2.76 TeV (centrality bin 0–10%). The measured data are compared with PYTHIA predictions which were smeared by instrumental effects and embedded into central Pb–Pb collisions. The PYTHIA calculation was done separately for inclusive-jet and quark-jet fragmentation. The data are further compared with smeared predictions of JEWEL [10] and the Hybrid model [11]. Taken from [8].

The jet transverse momentum range 80–120 GeV/c was chosen to be far away from the region which is dominated by combinatorial background jets. Figure 3 shows the raw measured z_g distributions separately for small and large angle splittings. The two panels on the right show that for Pb–Pb data, splittings with a big opening angle ($\Delta R^{rec} > 0.1$ and $\Delta R^{rec} > 0.2$), get more suppressed as the transverse momentum balance between branches in the splitting gets more symmetric, i.e. when z_g tends to 0.5. On the other hand, this suppression is compensated by a small enhancement of small angle ($\Delta R^{rec} < 0.1$) splittings which have at the same time momentum asymmetry as it can be seen in the first panel on the left. Such behavior is qualitatively expected from the color coherence picture [9] where the medium can resolve splittings with larger opening angle as independent color charges. In addition, from angular ordering arguments, it is expected that splittings with greater opening angles occur earlier and can be thus more affected by the medium. Data are compared with JEWEL [10] and the Hybrid model [11] which both reproduce the suppression.

Figure 4 presents the corresponding raw n_{SD} distributions. In Pb–Pb, it can be seen that the n_{SD} distribution is shifted towards a lower number of splittings that pass the SD condition with respect to the embedded PYTHIA which indicates harder, more quark-like fragmentation in Pb–Pb. Thus, features of the medium induced jet shower modifications which were seen with the angularity and $p_T D$ measurements are apparent also for the jet substructure observables.

4 Summary

Jet-substructure and jet-shape observables complement nuclear modification factor measurements and carry new information about the underlying mechanism of the jet quenching. The measurements presented in this paper analyzed jets having transverse momentum in the intermediate ranges 40–60 GeV/c and 80–120 GeV/c and jet radius $R \leq 0.4$. The data shown suggest that jet-medium interaction tends to shape these jets to have a more collimated jet core and the resulting jet fragmentation resembles the fragmentation of a quark-jet. Similar studies for jets with larger radii are challenging at the moment because of the large underlying event and they require new techniques to be developed.

5 Acknowledgements

This research was funded by the Ministry of Education, Youth, and Sports of the Czech Republic, grant number LTT17018.

References

[1] Stephanov, M.A., arXiv:hep-lat/0701002 (2006).
[2] Salam, G.P., Eur. Phys. J. **C67**, 637–686 (2010).
[3] Majumder, A., Van Leeuwen, M., Prog. Part. Nucl. Phys. **A66**, 41–92 (2011).
[4] Acharya, S. et al. (ALICE Collaboration) JHEP **10**, 139 (2018).
[5] Cacciari, M., Salam, G.P. , and Soyez, G., **JHEP 04**, 063 (2008).
[6] Larkoski, A.J., Marzani,S., Soyez, G. and Thaler, J., JHEP **05**, 146 (2014).
[7] Sjostrand,T.,Mrenna, S., and Skands, P.Z., JHEP **05**, 026 (2006).
[8] ALICE Collaboration, arXiv:1905.02512v1, to be published in PLB.
[9] Casalderrey-Solana, J., Mehtar-Tani, Y., Salgado, C.A., Tywoniuk, K., Phys. Lett. **B725**, 357–360 (2013).
[10] Zapp,K., Ingelman, G., Rathsman,J., Stachel, J., and Wiedemann, U.A., Eur.Phys.J., **C60**, 617–632 (2009).
[11] Casalderrey-Solana, J., Gulhan, D.C., Milhano, J.G., Pablos, D., and Rajagopal, K., Nucl. Phys. **A931**, 487–492 (2014).
[12] Milhano, J.G., Wiedemann, U.A., Zapp, K.C., Phys.Lett. **B779**, 409–413 (2018).
[13] D'Eramo, F., Rajagopal, K., and Yin, Y., JHEP **01**, 172 (2019).
[14] Mehtar-Tani, Y. and Tywoniuk, K., JHEP **04**, 125 (2017).
[15] Chang,N.B., Cao, S. and Qin, G.Y., Phys.Lett. **B781**, 423–432 (2018).

Two-particle correlations on transverse momentum: an untapped resource for studying relativistic heavy-ion collision dynamics

Robert L. Ray[1,⋆] and *Alexander M.* Jentsch[2,⋆⋆]

[1] *Department of Physics, The University of Texas at Austin, Austin, Texas 78712 USA*
[2] *Brookhaven National Laboratory, Upton, New York 11973 USA*

Abstract. Two-particle correlation projections onto two-dimensional transverse momentum coordinates (p_{t1}, p_{t2}) allow access to properties of the relativistic heavy-ion collision system which are complementary to that studied using angular correlations. Examples include the degree of thermal equilibration and the variance of dynamical fluctuations in hard-scattering processes. Results for minimum-bias Au + Au collisions at $\sqrt{s_{NN}} = 200$ GeV are presented, with the structures described by two phenomenological models. The correlations structures and extracted physical quantities are then compared to theoretical predictions. Conclusions from these comparisons regarding global equilibration, fluctuations in soft and semi-hard QCD processes, and the effects of the hot, dense collision medium are presented.

1 Introduction

Two-particle correlation measurements in high-energy, heavy-ion collisions provide access to partonic and hadronic dynamics occurring throughout the spatio-temporal evolution of the hot, dense matter produced in the collisions. In general, these correlations depend on six independent coordinates. However, for most applications only four are relevant, e.g. the relative pseudorapidity and azimuthal angle, $\Delta\eta$ and $\Delta\phi$, and the two-particle transverse momentum coordinates, p_{t1} and p_{t2}. Measurements of correlation distributions on $(\Delta\eta, \Delta\phi)$ are ubiquitous in the heavy-ion literature. Much less attention however, has been given to correlation distributions on transverse momentum. The latter type of measurements have been reported by the NA49 [1, 2], CERES [3], and STAR [4–6] Collaborations. Here, we discuss the features of the nonidentified, two-particle correlation distributions projected on (p_{t1}, p_{t2}) and compare preliminary results with phenomenology and theoretical predictions. Examples of the properties of the heavy-ion collision system which can be deduced by analyzing these correlations are presented and discussed.

2 Analysis method

In Ref. [7] it was shown how to relate two-particle correlation distributions on transverse momentum coordinates to several of the mean-p_t fluctuation quantities in the literature. In the present analysis the

⋆e-mail: ray@physics.utexas.edu
⋆⋆e-mail: alexmichaeljentsch@gmail.com

correlation definition derived from the mean-p_t fluctuation quantity $\Delta\sigma^2_{p_t:n}$, developed and used by the STAR Collaboration [8], was used. The resulting correlation quantity eliminates the statistical bias caused by multiplicity variation within centrality bins. The bias-corrected, event-averaged number of same-event (se) particle pairs and mixed-event (me) pairs, for arbitrary 2D transverse momentum bin k, l, are given by [7]

$$\rho_{\rm se,kl} = \frac{1}{\epsilon}\sum_{j=1}^{\epsilon}(\bar{N}/n_j)n^{\rm se}_{j,kl} \tag{1}$$

$$\rho_{\rm me,kl} = \frac{\bar{N}-1}{\bar{N}\epsilon_{\rm mix}}\sum_{j\neq j'} n_{jk}n_{j'l}, \tag{2}$$

where all charged particle pairs are included, ϵ is the number of collision events in the centrality or multiplicity bin, and index j denotes a specific event. In Eq. (2) indices j and j' denote arbitrary pairs of mixed-events where $\epsilon_{\rm mix}$ is the number of mixed-event permutations included in the multiplicity bin. Also in Eq. (1) $n^{\rm se}_{j,kl}$ is the number of particle pairs from event j in bin k, l. In Eq. (2) n_{jk} is the number of particles in event j and transverse momentum bin k. Quantity $\bar{N}$ is the mean multiplicity within the acceptance defined by $p_t > 0.15$ GeV/c, $|\eta| < 1$ and full 2π in azimuth corresponding to that of the main tracking detector in the STAR experiment [9].

The per-pair correlation quantity is given by

$$\frac{\Delta\rho_{kl}}{\rho_{\rm me,kl}} = \frac{\rho_{\rm se,kl} - \rho_{\rm me,kl}}{\rho_{\rm me,kl}}. \tag{3}$$

Multiplication by a pre-factor, defined in Ref. [10], gives a per final-state particle correlation measure which aligns with standard, normalized covariance measures [11] and which, for example, facilitates binary scaling tests. The prefactor is constructed from Levy-model fits to the measured single-particle p_t spectra. Here, we are interested in testing the centrality dependence of the correlation structures with respect to binary scaling. To facilitate this specific test the prefactor will also include an estimate of the soft-process p_t spectrum. The prefactor is given in Ref. [10]; the final correlation quantity is written for brevity as $\Delta\rho/\sqrt{\rho_{\rm soft}}$. In order to facilitate studies of jet fragment contributions to the correlation structures [12] the correlations are presented on transverse momentum space variable y_t, the *transverse rapidity*, where $y_t = \log[(p_t + m_t)/m_0]$, $m_t = \sqrt{p_t^2 + m_0^2}$ is the transverse mass, and parameter m_0 (pion mass) regulates the divergence as $p_t \to 0$.

Preliminary data from the STAR Collaboration were presented in Refs. [13, 14] for minimum-bias Au+Au collisions at $\sqrt{s_{NN}} = 200$ GeV. These preliminary measurements included all charged-particle pairs but required the relative azimuth of the pairs to be greater than $\pi/2$, i.e. away-side (AS) pairs. Those data were fitted with analytic functions and assigned statistical errors as described in Refs. [10, 14]. The resulting *pseudodata* will be compared to predictions and described with phenomenological models in the following sections.

3 Theoretical predictions

Predictions of HIJING [15] and EPOS [16] are compared with pseudodata for selected centrality bins in figure 1. Both models predict the overall saddle-shape and the peak near $(y_{t1}, y_{t2}) = (3, 3)$ observed in the pseudodata. HIJING combines the LUND [17] model of longitudinal, color-string fragmentation plus PYTHIA [18] which describes semi-hard and hard-scattering jets. In its simplest application HIJING provides a null hypothesis in which heavy-ion collisions are assumed to be a superposition

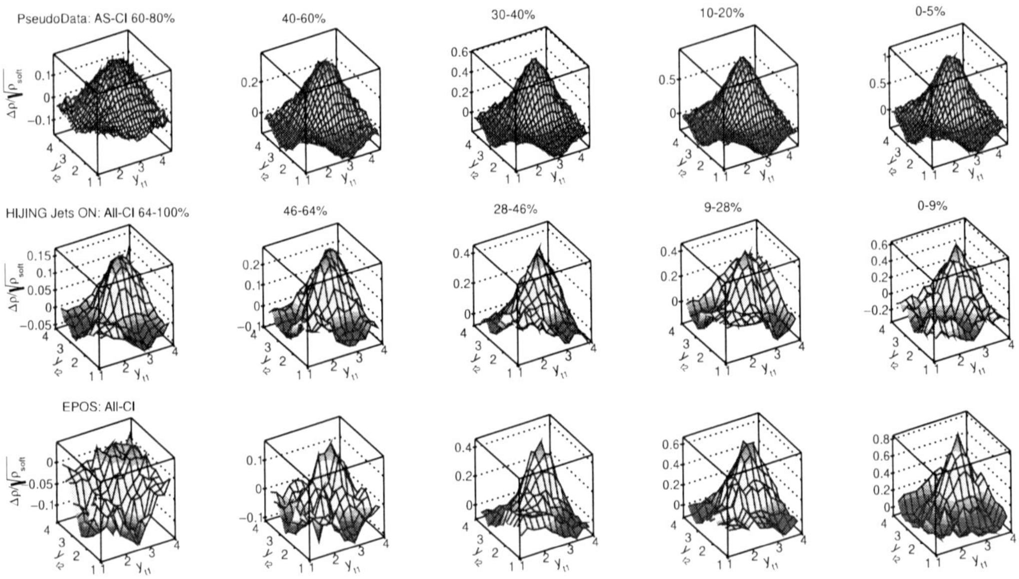

Figure 1. Theoretical model predictions compared with correlation pseudodata $\Delta\rho/\sqrt{\rho_{\rm soft}}$ on coordinates (y_{t1}, y_{t2}) for Au+Au collisions at $\sqrt{s_{NN}} = 200$ GeV using all charged particle pairs (charge-independent, CI) and relative azimuthal angles greater than $\pi/2$. Pseudodata are shown in the upper row of panels for centrality cross section fractions 60-80%, 40-60%, 30-40%, 10-20% and 0-5% from left-to-right, respectively. The next two rows show, respectively, HIJING model predictions for 200 GeV Au+Au collisions with jets-on, and EPOS predictions. Centralities for the HIJING and EPOS predictions are shown in each row from left-to-right for the coarser cross section fractions 64-100%, 46-64%, 28-46%, 9-28% and 0-9%, respectively.

of independent nucleon + nucleon collisions. In EPOS, fluctuating initial conditions are described in a multiple scattering framework using soft and hard Pomerons. The collision volume is separated into "core" and "corona" domains based on the local energy density and transverse momentum of the interacting partons. Evolution of the core is described using (3+1)D viscous hydrodynamics until the hadronization stage, after which UrQMD [19] describes the final-stage hadronic re-scattering and reactions.

For the HIJING predictions the correlation peak near $(y_{t1}, y_{t2}) = (3, 3)$ increases monotonically with centrality, in general agreement with the trend of the pseudodata. The predicted peak amplitudes for the three more-peripheral bins agree well with the pseudodata, but the predictions are smaller than the pseudodata for the two more-central bins. This result may indicate that the event-wise fluctuation mechanisms in HIJING are realistic for peripheral collisions, with respect to this type of correlation, but fail to describe the most-central correlations. HIJING predictions, without jets included (not shown), completely fail to achieve the observed amplitudes of the (3,3) correlation peak.

The EPOS predictions include both the saddle-shape and a strong (3,3) peak. Except for the 64-100% bin, the peak amplitudes monotonically increase with centrality similar to the pseudodata. The EPOS predicted peak amplitudes for the more-central collisions are in general agreement with the pseudodata. In the two more-peripheral bins the EPOS predicted peak amplitudes are much smaller than either the HIJING jets-on predictions or the pseudodata. The EPOS prediction for the most-peripheral bin is quite dissimilar from the pseudodata. These results suggest that for peripheral collisions the hard-

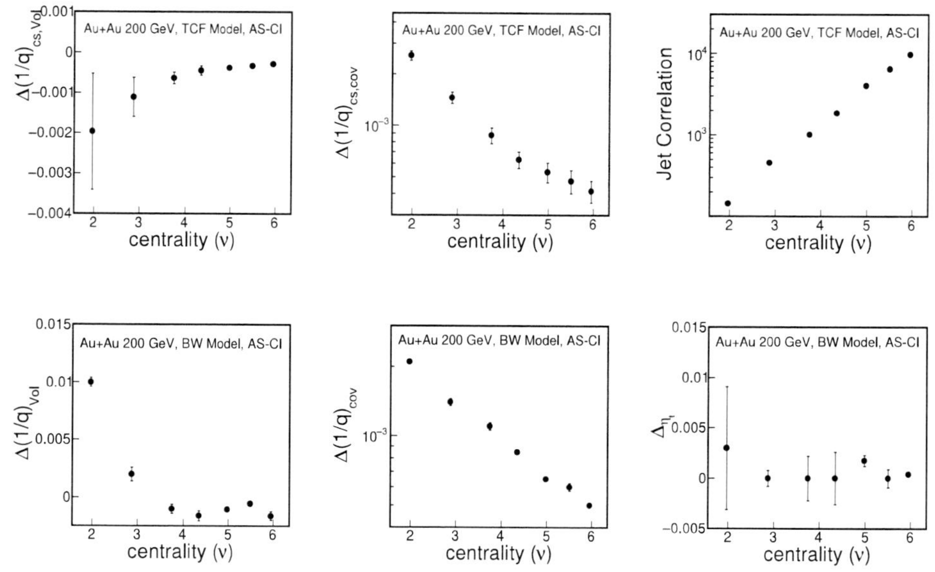

Figure 2. (Upper row): TCF model fit parameters $\Delta(1/q)_{\rm cs,Vol}$, $\Delta(1/q)_{\rm cs,cov}$ and jet correlation (covariance) quantity from left-to-right, respectively, as a function of collision centrality measure $\nu = {\rm N_{bin}}/({\rm N_{part}}/2)$, where $\rm N_{part}$ is the number of participating nucleons. (Lower row): Blast-wave fit parameters $\Delta(1/q)_{\rm Vol}$, $\Delta(1/q)_{\rm cov}$ and Δ_{η_t} from left-to-right as a function of collision centrality.

scattering contribution is under-estimated by EPOS, while for the more-central collisions, the strong medium-interactions in the model are capable of generating realistic charge-independent correlations.

4 Phenomenology

In order to facilitate the interpretation of these new correlation structures and to offer insight into the possible deficiencies in theoretical approaches, two phenomenological models were developed in Ref. [10]. One is based on the two-component multiplicity production model of Kharzeev, Nardi and Levin [20] and the other is based on the blast-wave [21]. In both models, key elements which determine the particle distributions, e.g. jet energies or freeze-out temperatures, are allowed to fluctuate. All details of the models are given in Ref. [10].

The two-component fragmentation (TCF) model includes color-strings (color flux tubes) and minijets (i.e. jets with no artificial lower energy cut-off). Color-string fragmentation is assumed to produce a Maxwell-Boltzmann p_t distribution with slope parameter $\beta_{\rm cs}$. The semi-hard scattering, minijet production is described as the convolution integral of the product of a universal jet fragment distribution [12] times a QCD power-law probability distribution, including a soft-QCD cut-off. Correlations among the final-state particles are generated via correlated $\beta_{\rm cs}$ values between pairs of particles within a collision event and via correlated minijet energies. The correlation pseudodata were fit by adjusting the overall width and covariance of the two-particle $(\beta_{\rm cs_1}, \beta_{\rm cs_2})$ probability distribution, the variance of the event-wise fluctuating semi-hard scattering multiplicity, and the covariance strength of the two-particle, minijet energy distribution. Three of the fit parameters are shown in the upper row of panels in figure 2 where $\Delta(1/q)_{\rm cs,Vol} = (1/2)(\sigma^2_{\beta_\Sigma} + \sigma^2_{\beta_\Delta} - 2\sigma^2_{\beta})/\bar{\beta}^2$ and $\Delta(1/q)_{\rm cs,cov} = (1/2)(\sigma^2_{\beta_\Sigma} - \sigma^2_{\beta_\Delta})/\bar{\beta}^2$. In these definitions σ^2_β is the variance of the fluctuating, single-particle color-string slope parameters whose mean is $\bar{\beta}$. Parameters $\sigma^2_{\beta_{\Sigma,\Delta}}$ are the variances of the 2D $(\beta_{\rm cs_1},\beta_{\rm cs_2})$ distribution along the $\beta_{\Sigma,\Delta} = \beta_{\rm cs_1} \pm \beta_{\rm cs_2}$ directions.

The major features and centrality trends of the data are well-described. The soft-component contributions and the semi-hard scattering multiplicity variance mainly affect the saddle-shape and its amplitude, while the covariance of the two-particle jet energy distribution mainly affects the (3,3) peak amplitude. The centrality dependence of the latter two, semi-hard scattering quantities were compared with binary collision ($N_{\rm bin}$) scaling. The trend of the semi-hard scattering variance is approximately $\sigma^2_{\rm hard} = (0.213 \pm 0.092)N_{\rm bin}^{1.156\pm0.071}$ and $\sigma^2_{\rm hard} = (0.66 \pm 0.31)\bar{N}_{\rm hard}^{1.12\pm0.09}$ where $\bar{N}_{\rm hard}$ is the semi-hard component mean multiplicity. The latter tests the Poisson limit for fluctuations in the semi-hard component. This variance approximately follows binary scaling and approaches the Poisson limit for more-central collisions; a validation of the TCF approach since this variance is expected to be approximately Poisson. The jet covariance scales as $(0.390 \pm 0.085)N_{\rm bin}^{1.451\pm0.035}$ which exceeds binary scaling, indicating the presence of significant interactions within the medium.

The blast-wave (BW) model describes particle emission from an expanding, hydrodynamic-like source with freeze-out temperature $(1/\beta)$ and transverse flow velocity assuming a Maxwell-Boltzmann source distribution [21]. In the model the freeze-out temperature is allowed to fluctuate both locally and globally from event-to-event as described by a two-particle (β_1,β_2) probability distribution analogous to the preceding 2D distribution for color-strings. The transverse flow velocity, or the corresponding rapidity, is similarly allowed to fluctuate and is parametrized by a peaked, two-particle probability distribution. The correlation data were fit by freely adjusting the overall width and variance of the (β_1,β_2) distribution and the co-variance of the transverse flow rapidity distribution. The fit parameters are shown in the lower row of panels in figure 2 where $\Delta(1/q)_{\rm Vol}$ and $\Delta(1/q)_{\rm cov}$ are defined in the same way as the corresponding TCF parameters, and Δ_{η_t} controls the covariance of the two-particle transverse flow rapidity. The correlation pseudodata are well described where the required, relative changes in the overall width and covariance of the (β_1,β_2) distribution are only of order 10^{-3}, and the covariance in the transverse flow is consistent with zero, i.e. no evidence of correlated transverse flow. The two-particle inverse temperature distribution required to fit the correlation pseudodata is strongly inconsistent with an ensemble of globally equilibrated collisions. The latter would be characterized with maximum, relative covariance.

The HIJING and EPOS predictions were also fit with the TCF and BW phenomenologies, to gain insight into those aspects of the theoretical models which might be deficient. Based on comparisons of the fits to the pseudodata and to the theoretical models, we find that EPOS collisions may be too hot, and for peripheral collisions may either be under-estimating hard-scattering processes or over-estimating the degree of dissipation in the medium (core). Similarly, while HIJING predictions, with jets included, appear reasonable for peripheral collisions, the color-string fluctuations are, in general, not sufficiently correlated, and the jet-like correlations for more-central collisions produce only about half the required amplitude.

5 Conclusions

Analysis of two-particle, pair-number correlation distributions on 2D transverse momentum, or transverse rapidity, provides access to additional properties of the relativistic heavy-ion collision system. These include studies of the degree of thermal equilibration within the hydrodynamic picture, or the variance in hard-scattering processes within the fragmentation approach. In combination with 2D angular correlations, these relatively under-appreciated correlation measurements provide significant, new constraints on theoretical models. Phenomenology, as shown here, helps bridge the gap between data and theory by enhancing the interpretive value of these measurements and suggesting what upgrades may be needed in the theoretical approaches. A significant quantity of new correlation measurements on transverse rapidity are being prepared by the STAR Collaboration for publication in the near future.

Acknowledgements

The authors would like to thank Professor Thomas Trainor of the Univ. of Washington for many informative discussions relevant to this work and Professor Rainer Fries of Texas A&M University for discussions related to the blast-wave model. This research was supported in part by the Office of Science of the U. S. Department of Energy under Grants No. DE-FG02-94ER40845 and No. DE-SC0013391.

References

[1] J. G. Reid, Nucl. Phys. A **698**, 611c (2002).
[2] S. V. Afanasiev *et al.* (NA49 Collaboration), Nucl. Phys. A **715**, 55c (2003); T. Anticic *et al.* (NA49 Collaboration), Phys. Rev. C **70**, 034902 (2004).
[3] D. Adamová *et al.* (CERES Collaboration), Nucl. Phys. A **811**, 179 (2008).
[4] J. G. Reid, Ph.D. thesis, University of Washington, 2002 (unpublished); arXiv:nucl-ex/0302001.
[5] J. Adams *et al.* (STAR Collaboration), J. Phys. G: Nucl. Part. Phys. **34**, 799 (2007).
[6] T. A. Trainor and D. J. Prindle, Phys. Rev. D **93**, 014031 (2016).
[7] R. L. Ray and P. Bhattarai, Phys. Rev. C **94**, 064902 (2016).
[8] J. Adams *et al.* (STAR Collaboration), Phys. Rev. C **71**, 064906 (2005).
[9] M. Anderson *et al.*, Nucl. Istrum. Meth. Phys. Research A **499**, 659 (2003).
[10] R. L. Ray and A. Jentsch, Phys. Rev. C **99**, 024911 (2019).
[11] J. L. Rodgers and W. A. Nicewander, Am. Stat. **42**, 59 (1988); B. S. Everitt and A. Skrondal, *The Cambridge Dictionary of Statistics*, 4th ed. (Cambridge Univ. Press, Cambridge, 2010), p. 107.
[12] T. A. Trainor, Phys. Rev. C **80**, 044901 (2009); T. A. Trainor and D. T. Kettler, Phys. Rev. D **74**, 034012 (2006).
[13] E. W. Oldag (STAR Collaboration), J. Phys.: Conf. Ser. **446**, 012023 (2013).
[14] E. W. Oldag, Ph.D. thesis, The University of Texas at Austin, 2013 (unpublished), `https://drupal.star.bnl.gov/ STAR/files/oldag_dissertation_20132.pdf`.
[15] X.-N. Wang, M. Gyulassy, Phys. Rev. D **44**, 3501 (1991).
[16] K. Werner, Nucl. Phys. B (Proc. Suppl.) **175-176**, 81 (2008).
[17] B. Andersson, G. Gustafson, G. Ingelman and T. Sjöstrand, Phys. Rep. **97**, 31 (1983).
[18] T. Sjöstrand and M. van Zijl, Phys. Rev. D **36**, 2019 (1987).
[19] H. Sorge, H. Stöcker, W. Greiner, Nucl. Phys. A **498**, 567 (1989); Ann. Phys. (N.Y.) **192**, 266 (1989).
[20] D. Kharzeev and M. Nardi, Phys. Lett. B **507**, 121 (2001); D. Kharzeev, E. Levin and M. Nardi, Nucl. Phys. A **730**, 448 (2004).
[21] E. Schnedermann, J. Sollfrank, U. Heinz, Phys. Rev. C **48**, 2462 (1993).

Studying baryon production using two-particle angular correlations

Małgorzata Anna Janik[1,*]

[1]Faculty of Phyics, Warsaw University of Technology, ul. Koszykowa 75, 00-662 Warsaw, Poland

Abstract.
Latest measurements of $\Delta\eta\Delta\varphi$ correlations of identified particles show differences in particle production between baryons and mesons. The correlation functions for mesons exhibit the expected peak dominated by effects of mini-jet fragmentation and are reproduced well by general purpose Monte Carlo generators. For baryon pairs (where both particles have the same baryon number) a surprising near-side anti-correlation structure is observed instead of a peak, implying that two such particles are rarely produced with similar momentum. These results present a challenge to the contemporary models and there is no definite theoretical explanation of the observation. In this proceedings an overview of the latest baryon correlation measurements yielding startling results are presented.

1 Introduction

Two-particle angular correlations are a robust tool that enable the exploration of the underlying physics phenomena of particle production in collisions of both protons and heavy ions by studying the distributions of angles in $\Delta\eta\Delta\varphi$ space (where $\Delta\eta$ is the pseudorapidity difference and $\Delta\varphi$ is the azimuthal angle difference between two particles). These correlations open up the possibility to study a number of mechanisms simultaneously. Many phenomena, including mini-jets, elliptic flow, Bose-Einstein correlations, resonance decays, conservation laws, etc. are sources of correlations. Each one produces a characteristic distribution in $\Delta\eta\Delta\varphi$ space and contributes to the final shape of the correlation function. As an example, the clear peak in $(\Delta\eta, \Delta\varphi) = (0, 0)$ for pairs of identical mesons originates from jets and Bose-Einstein correlations. Similar peaks observed for correlations of pairs of kaons with opposite charges is a result of resonances (primarily $\phi \to K^+K^-$) and (also) jets.

The influence of jets on the shape of the correlation function, and hence, the peak in $(\Delta\eta, \Delta\varphi) = (0, 0)$ is expected for all particle pairs. It is indeed observed for all studied meson pairs, as well as for baryon–antibaryon pairs. However, the experimental results [1, 2] show that this peak is not observed for pairs of identical baryons.

*e-mail: majanik@cern.ch

2 ALICE results

The studies performed by the ALICE Collaboration were performed on proton–proton collisions with energy of $\sqrt{s} = 7$ TeV, as well as energy of $\sqrt{s} = 13$ TeV. They were carried out for pions, kaons, protons, and Λ hyperons, separately for particle–particle and particle–antiparticle pairs. Particles selected for the analysis had pseudorapidity within $|\eta| < 0.8$ and transverse momentum $p_T < 2.5$ GeV/c. The particle identification was performed using the TPC and TOF detectors, reaching 99% purity for pions and protons, and 96% purity for kaons. Λ hyperons were reconstructed using their decay topology $\Lambda(\overline{\Lambda}) \rightarrow p\pi^-(\overline{p}\pi^+)$, with purity of more than 95% [1].

2.1 Definition of the correlation function

Experimental correlation function, as reported by ALICE, can be defined as

$$C(\Delta\eta, \Delta\varphi) = \frac{S(\Delta\eta, \Delta\varphi)}{B(\Delta\eta, \Delta\varphi)}. \quad (1)$$

$S(\Delta\eta, \Delta\varphi)$ is a correlated pairs distribution, while $B(\Delta\eta, \Delta\varphi)$ is a reference distribution, obtained using the event mixing technique. Both S as well as B are normalized to the number of pairs in each distribution, therefore reported correlation function is a ratio of probabilities (the chance of observation of two particles in respect to the chance of observation of each of those particles separately). The details of measurement are explained in [1].

2.2 Results

Figure 1 shows correlations for the like-sign pairs of pions, kaons, and protons obtained for $\sqrt{s} = 13$ TeV pp collisions[1] measured by the ALICE experiment.

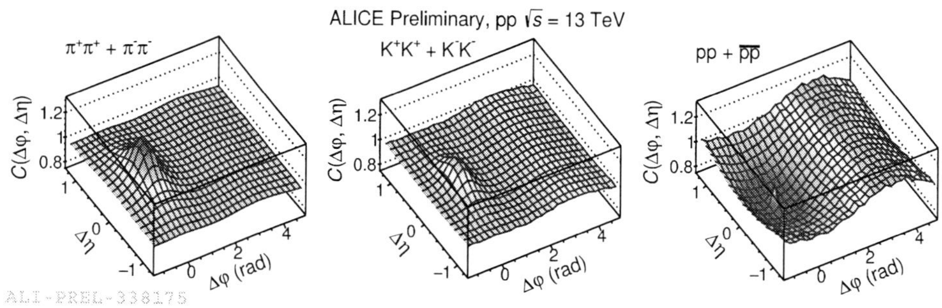

Figure 1. Correlation functions of like-sign pions, kaons, and protons from $\sqrt{s} = 13$ TeV pp collisions measured by ALICE.

Results show clear distinct behavior between mesons and baryons: mesons (pions and kaons, left and middle panels) show characteristic near-side peak in $(\Delta\eta, \Delta\varphi) = (0, 0)$, while for the pairs of identical protons (right panel) we can observe an anti-correlation. Moreover, in Figure 2 $\Delta\eta$ integrated projections of the two-dimensional correlation functions are shown for both unlike-sign (top row) and like-sign (bottom row) pairs of pions, kaons and protons,

[1] All results for $\sqrt{s} = 7$ TeV pp collisions, including correlation functions for pairs of Λ hyperons, can be found in [1].

together with PYTHIA8 [3] model simulations. The majority of Monte Carlo generators[2] do not describe the anti-correlation shape visible fo baryon–baryon pairs, as well as do not quantitatively describe baryon–antibaryon pairs. The most promising are results from the AMPT generator, see [4, 5].

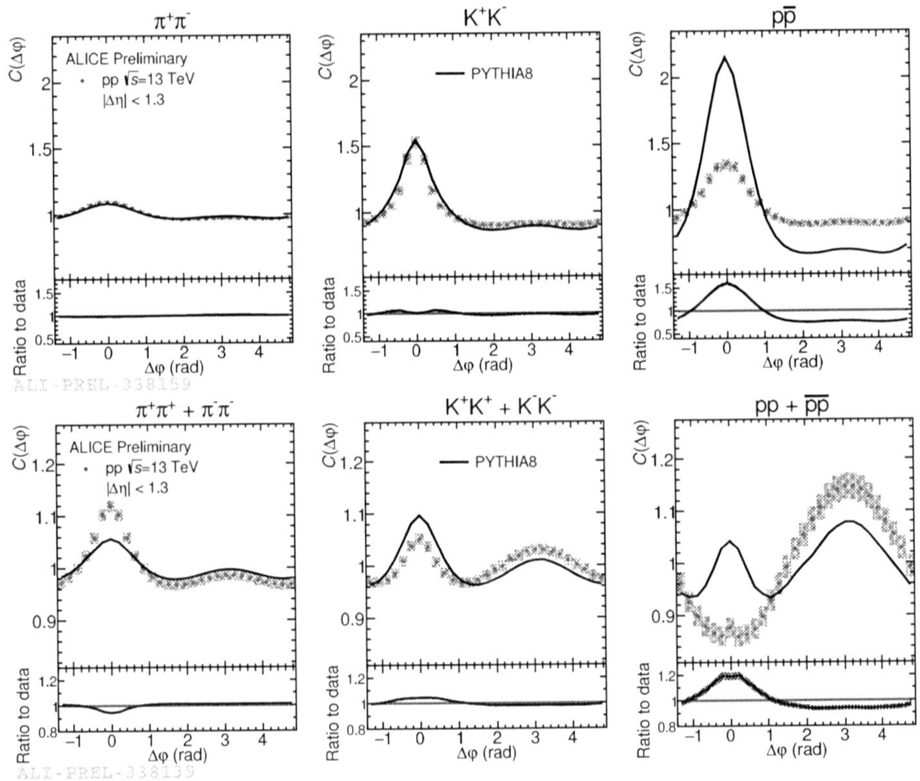

Figure 2. $\Delta\eta$ integrated projection of correlation functions for pairs of pions, kaons, and protons obtained from pp collisions at $\sqrt{s}$ = 13 TeV from ALICE as well as PYTHIA8 Monte Carlo model. Top panels show unlike-sign particles, while bottom panels – like-sign particles. Statistical (bars) and systematic (boxes) uncertainties are plotted.

[2]Comparisons with other models can be found in [1].

3 STAR results

Measurements performed by the STAR experiment were done on $Au - Au$ collisions at $\sqrt{s_{\rm NN}}$ = 7.7–200 GeV. Analysis was performed for pions, kaons, and protons (separately for different charge combinations) as well as for different centrality ranges.

3.1 Correlation function definition

Results of the STAR Collaboration were reported as R_2 quantity, which is defined as follows:

$$R_2(\Delta y, \Delta\varphi) = \frac{\langle n\rangle^2}{\langle n(n-1)\rangle} \frac{\rho_2(\Delta y, \Delta\varphi)}{\rho_1(y_1, \varphi_1)\rho_1(y_2, \varphi_2)} - 1, \quad (2)$$

where $\Delta y = y_1 - y_2$ is the difference between rapidity of two particles, n stands for multiplicity of the given particle type in the collision, while ρ_2 and ρ_1 are the multiplicity density distributions of pairs of particles of single particles, respectively, normalized to the number of events.

3.2 Results

In Figure 3 distributions for the like-sign pairs of protons are presented for two collision energies: $\sqrt{s_{\rm NN}}$ = 62.4 GeV and $\sqrt{s_{\rm NN}}$ = 200 GeV (more results can be found in Ref. [2]). For both distributions previously discussed anti-correlation at $(\Delta y, \Delta\varphi) = (0, 0)$ is clearly visible. The same distribution shape was observed for all studied energies.

Similar anti-correlation is visible also for baryon–antibaryon pairs [2] in heavy-ion collisions, however, of much smaller range; that anti-correlation comes from matter–antimatter annihilation and was observed also in the results of femtoscopy [6] measurements as well as Monte Carlo models [2].

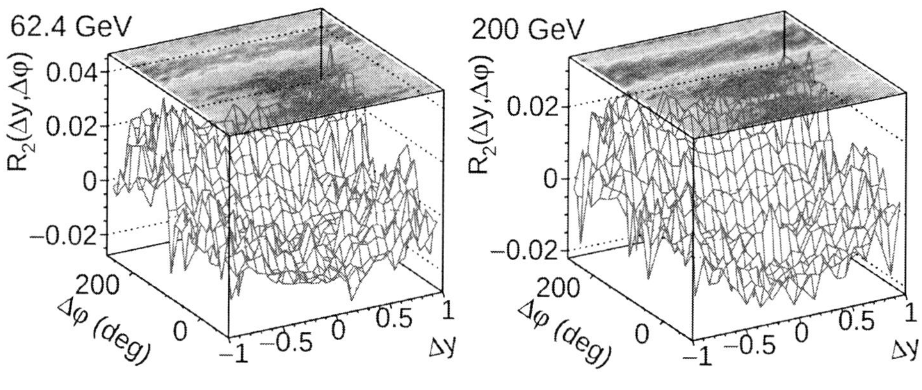

Figure 3. Correlation functions of like-sign protons from Au–Au collisions at $\sqrt{s_{NN}}$ = 62.4 and 200 TeV measured by STAR experiment. Figure from [2].

4 Discussion

Different reasons for the observed shape correlation functions were studied. Possible physics effects that could influence baryon–baryon correlation functions were analyzed:

- The influence of the Coulomb interaction is negligible: the shape of the correlation functions is very similar for proton–proton and lambda–lambda (Λ hyperons are neutral) pairs, regardless of the electric charge of particles [1].
- Fermi-Dirac quantum statistics effects cannot be the sole reason of the observed anti-correlation. The anti-correlation can be observed for pairs of pp, $\Lambda\Lambda$ (identical particles), and pΛ (non-identical particle pairs, that are not influenced by quantum statistics effects) [1].
- The influence of the momentum of particles was studied; regardless of the momentum the anti-correlation was still observed [1].
- The dependence on the collision energy was studied as well: for both lower energies [2] as well as LHC energies ($\sqrt{s} = 7$ TeV [1] and $\sqrt{s} = 13$ TeV) anti-correlation is visible.
- The local conservation of baryon number is not the only reason of the anti-correlation: all studied Monte Carlo models included this effect in the calculations, however, they were unable to reproduce the experimental results [7].

5 Summary

Experimental results, which are not well reproduced by current theoretical models, suggest that the production mechanism of protons – one of the most common particles in the Universe – is not well understood. The surprising effect of anti-correlation was observed first for $\sqrt{s} = 7$ TeV pp collisions by ALICE, but now we know as well, that the same behavior of like-sign baryons is observed for higher, as well as lower energies, and also for different systems.

Acknowledgments

This work was supported by the Polish National Science Centre under decisions no. UMO-2016/22/M/ST2/00176, no. UMO-2017/27/B/ST2/01947 and Polish Ministry of Science and Higher Education.

References

[1] J. Adam et al. (ALICE), Eur. Phys. J. **C77**, 569 (2017), 1612.08975
[2] J. Adam et al. (STAR) (2019), 1906.09204
[3] P. Skands, S. Carrazza, J. Rojo, Eur. Phys. J. **C74**, 3024 (2014), 1404.5630
[4] L.Y. Zhang, J.H. Chen, Z.W. Lin, Y.G. Ma, S. Zhang, Phys. Rev. **C98**, 034912 (2018), 1808.10641
[5] L.Y. Zhang, J.H. Chen, Z.W. Lin, Y.G. Ma, S. Zhang, Phys. Rev. **C99**, 054904 (2019), 1904.08603
[6] S. Acharya et al. (ALICE), Phys. Lett. **B802**, 135223 (2020), 1903.06149
[7] T. Sjöstrand, Nucl. Phys. **A982**, 43 (2019), 1808.03117

 https://doi.org/10.1051/epjconf/202023501005

Self-similarity in Fluctuations and Clustering of Multiplicities**

Maciej Rybczyński[1,⋆] and *Zbigniew* Włodarczyk[1]

[1] *Institute of Physics, Jan Kochanowski University, 25-406 Kielce, Poland*

Abstract. In this talk we summarize the main results of our recent publication [1], where we describe the observed by the CERN NA49 experiment a non-monotonic behavior of the fluctuations in multiplicity as a function of centrality of the collision using the concept of correlations between produced particles promoting cluster formation. A new element is a use of log-normal distribution for particles in a cluster which emerges naturally when one considers particle production based on bivariate branching scenario [2].

1 Introduction

Fluctuations of physical observables in collisions of ions have become in recent years one of the main topics of interest, as they can provide some important signals for the formation of Quark-Gluon Plasma. The possibility of inclusion of dynamical fluctuations in the study of the energy of phase transition and the search for the critical point of strongly interacting matter has become a motivation for extensive program of fluctuations analyzes at the SPS, LHC as well as at RHIC. The NA49 experiment found a non-monotonic behavior of the fluctuations in multiplicity [3] as a function of centrality of the collision at the highest SPS energy. This intriguing result may be the first sign of the presence of the critical point. Therefore, the NA49 efforts will be continued in the NA61/SHINE experiment [4].

Although there are some models trying to describe the non-monotonic behavior of the scaled variance of multiplicity distribution as a function of collision centrality expressed by the number of nucleons participants for the NA49 result measured in the centre-of-mass rapidity range $1.1 < y_\pi < 2.6$ (rapidity calculated assuming that all particles have pion mass), up to now there is no commonly accepted explanation of this phenomenon. In this talk we discuss multiplicity fluctuations of charged particles produced in nuclear collisions measured event-by-event by the NA49 experiment. In our recent work [1] the observed non-monotonic behaviour of the scaled variance of multiplicity distribution as a function of collision centrality (such effect is not observed in a widely used string-hadronic models of nuclear collisions, see [3] for more details) was described using the concept of correlations between produced particles promoting cluster formation. Here we have extended our analysis for description of the very intriguing result observed recently by the NA61/SHINE experiment at CERN SPS [5]. Namely, the scaled variance of multiplicity distributions of charged particles measured in broad rapidity range ($0 < y_\pi < y_{max}$) is significantly higher in inelastic proton+proton interactions than in the 1% most central Pb+Pb collisions measured by NA49 experiment at the same energy per nucleon. This result is in qualitative disagreement with the predictions of the wounded nucleon

**Presented by M. Rybczyński

⋆e-mail: maciej.rybczynski@ujk.edu.pl

model [6] assuming that particle production in nucleon+nucleon and nucleus+nucleus collisions is an incoherent superposition of particle production from wounded nucleons. The proper description allow for verification of models of dynamics of particle production and to put relevant constraints on the elementary entropy deposition in the early phase.

2 Data on multiplicity fluctuations

The NA49 and NA61/SHINE experiments located at CERN SPS analyzed multiplicity fluctuations of charged particles produced in proton+proton, Be+Be, Ar+Sc and Pb+Pb collisions [3, 7, 8]. Both experiments used scaled variance of multiplicity distribution,

$$\omega(N) = \frac{Var(N)}{\langle N\rangle} = \frac{\langle N^2\rangle - \langle N\rangle^2}{\langle N\rangle} \tag{1}$$

as a measure of multiplicity fluctuations. The NA49 Collaboration published data on multiplicity fluctuations in Pb+Pb reactions as a function of collision centrality [3]. Unexpectedly, the measured scaled variance show very non-trivial centrality dependence. It is close to unity at completely central collisions but it manifests a prominent discrepancy from unity at peripheral interactions. The measurement has been performed at the collision center of mass energy $\sqrt{s_{NN}} = 17.3$ GeV for particles produced in forward hemisphere in the restricted rapidity interval $1.1 < y_\pi < 2.6$ in the center of mass frame. The azimuthal acceptance has also been limited, and about 17% of all produced charged particles have been used in the analysis [3]. Later on NA49 and NA61/SHINE experiments registered multiplicity distributions of negatively charged particles produced in proton+proton and the most central (1%) Be+Be, Ar+Sc and Pb+Pb collisions at the same center of mass energy, but emitted to the full forward hemisphere, $y_\pi > 0$ [7, 8].

We note a substantial difference in the values of scaled variances of multiplicity distributions of charged particles produced in proton+proton and most central Pb+Pb collisions. The scaled variance is significantly larger for inelastic proton+proton interactions than for the 1% most central Pb+Pb collisions. This difference was extensively discussed in Ref. [5] within the WNM and the statistical model (SM) [9] of particle production. The NA61/SHINE and NA49 results clearly contradict predictions of both WNM and SM models [5].

3 Clustering of multiplicities

Usually the following bivariate branching process scenario of multiparticle production is considered:

a) In collision of relativistic ions quark-gluon pairs are created.

b) Each quark or gluon can emit a gluon which can convert into a new quark-antiquark pair.

c) Each of the created quarks may again emit a gluon and so on.

Such evolution of the above described cascading process may be treated as a pure birth process. In each vertex there is a certain probability $P(N)$ to produce N particles. Since the probability $P(N)$ is the same in each vertex such process is self-similar [2]. The bivariate branching process has a multiplicative nature, i.e. the number of particles in a given generation is proportional to the number of particles in the previous one.

$$N_i = (1 + \epsilon_i)\,N_{i-1} \tag{2}$$

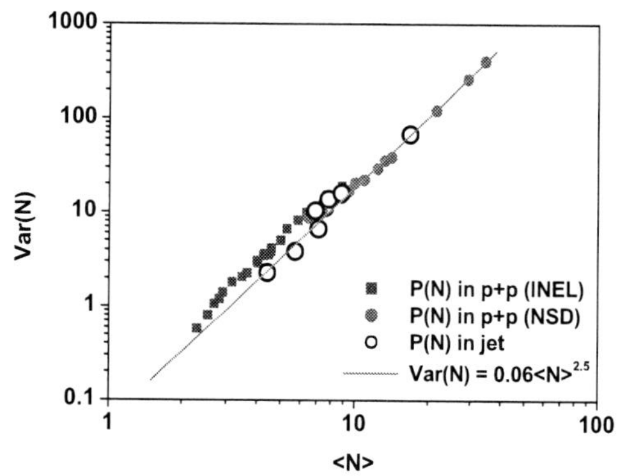

Figure 1. Variance of charged particles multiplicity distribution as a function of average charged multiplicity. By squares and circles we indicate data on particle production in proton+proton collisions: squares (inelastic data) are from the compilation for beam energy 3.7-303 GeV presented in [10], full circles (non-single diffractive data) are from the compilation [11]. Open circles are from data on particle production in jets [12–14]. A line shows our fit to the data.

with ϵ being a random factor. In other words, the final number of particles is a product of a number of random factors:

$$N = N_0 \prod_i (1 + \epsilon_i) . \tag{3}$$

Taking logarithms of both sides we see that right hand side of the above equation is a sum of a random components:

$$\ln (N/N_0) = \sum_i \epsilon'_i . \tag{4}$$

where $\epsilon'_i = \ln(1 + \epsilon_i) > 0$ is a random number. Such a sum should be distributed normally due to the Central Limit Theorem. Hence, also a left hand side, i.e. a logarithm of multiplicity should be distributed normally. In other words, multiplicity should be distributed log-normally [2]:

$$P(N) = \frac{1}{\sqrt{2\pi}\sigma} \cdot \frac{1}{N} \exp\left(-\frac{[\ln N - \mu]^2}{2\sigma^2} \right) . \tag{5}$$

Formally we get the log-normal (LND) distribution in the limit of a large number of steps. Thus log-normal distribution is continuous. In the case of use of LND for the multiplicities we should apply:

$$P_N = \int_N^{N+1} P(N)\, dN. \tag{6}$$

Figure 1 presents a compilation of values of variances of charged particle multiplicity distributions as a function of average charged multiplicity. Such dependence may be well fitted by a simple formula:

$$Var(N) = 0.06 \cdot \langle N \rangle^{2.5} . \tag{7}$$

The rough formula (7) asserts Taylor's law, $Var(N) = a \cdot \langle N \rangle^b$ with exponent $b > 2$. Such behaviour corresponds a geometrical random walk (as opposed to the ordinary additive random walk) if multiplicity density at each step grows on average (super-critical model) [15]. Comparing multiplicity fluctuations in jets and in minimum bias proton+proton interactions one observes a kind of self-similarity

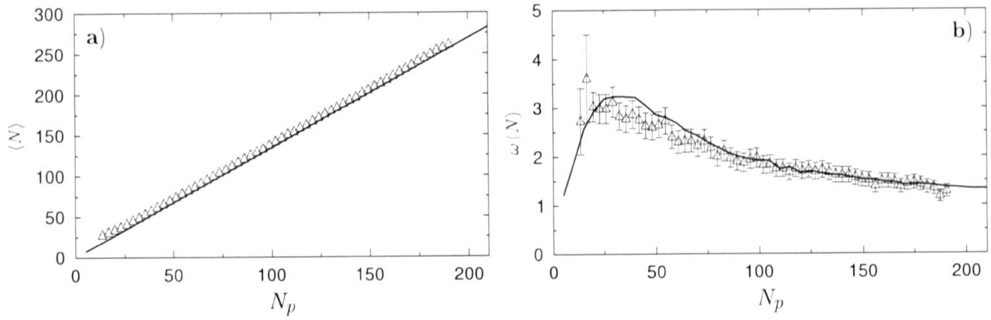

Figure 2. Average number of all charged particles (panel a)) and scaled variance of all charged multiplicity distribution (panel b)) of particles produced in Pb+Pb collisions plotted as a function of number of nucleons from projectile nucleus which participate in the collision. By triangles we indicate data of the NA49 experiment [3].

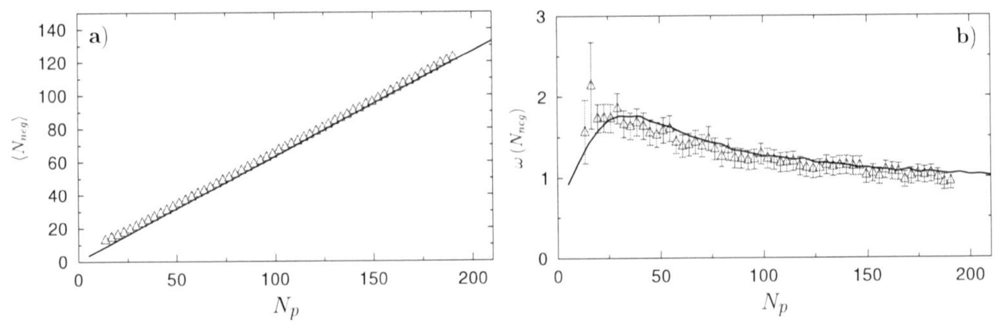

Figure 3. The same as in Figure 2 but for negatively charged particles.

of the multiparticle production processes [16]. Regardless of the amount of the available energy, the variance is the same power function of the average multiplicity.

To describe the NA49 data the particle clusterization method discussed in [1] was used. The modification we made for the purpose of this presentation is the following. The multiplicity in a given cluster is calculated according to LND with its variance dependent on $\langle N \rangle$, according to Eq. (7).

4 Results

For the simulation of nucleus-nucleus collisions in the framework of Glauber Monte Carlo, we have used an accordingly adopted GLISSANDO package [17]. The parameters of the particle production were adjusted by fitting of experimental data. The resultant centrality dependencies of average all charged multiplicities and corresponding scaled variances of multiplicity distributions are presented in Figure 2. To include experimental acceptance we accepted a fraction of 17% of generated particles. See [1] for detailed discussion of acceptance. Figure 3 presents similar results to those shown in Figure 2 but for negatively charged particles. To obtain corresponding fits we had to adjust only one parameter: the average multiplicity in proton+proton collisions. For the case of negatively charged particles $\langle N_{ch} \rangle = 3.6$. The only difference between scaled variances for negatively and all charged particles is the number of accepted particles. The two-particle correlation function, $\langle v_2 \rangle = (\omega(N) - 1) / \langle N \rangle$ [18] which is not dependent on the experimental acceptance, is roughly the

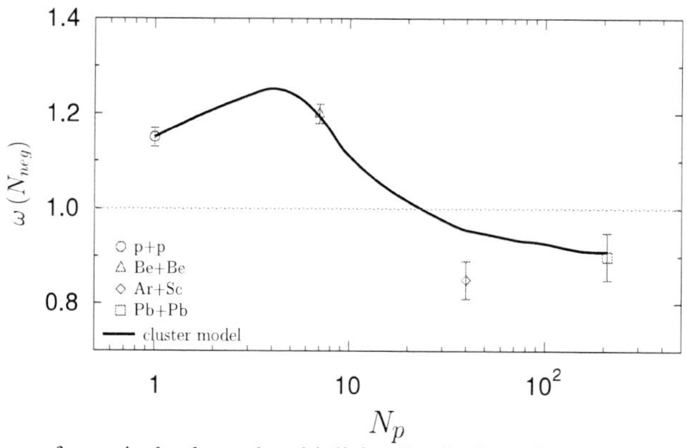

Figure 4. Scaled variance of negatively charged multiplicity distribution of particles produced in proton+proton and the most central (1%) Be+Be, Ar+Sc and Pb+Pb collisions, and emitted to the forward hemisphere, $y_\pi > 0$ plotted as a function of number of nucleons from projectile nucleus which participate in the collision. Symbols present data of the NA49 and NA61/SHINE experiments [7, 8]. With the line we show results obtained using our model.

same in both cases. Using similar considerations we have obtained the values for the scaled variance of negatively charged multiplicity distribution produced in proton+proton and the most central (1%) Be+Be, Ar+Sc and Pb+Pb collisions, and emitted to the forward hemisphere, $y_\pi > 0$ [7, 8], see Figure 4. Our model describe most of the NA61/SHINE data (except Ar+Sc collisions) quite reasonable. The observed difference for the colliding system as regards size being comparable to the cluster's size can be caused by a too rough approximation of the cluster size distribution adopted in the model. We note that the previously published results based on the negative binomial distribution gives very similar results [1].

The presence of self-similar cluster phase in the mechanism of particle production, caused by correlations between produced particles, in the natural way allows for reproduction of the effect of the substantially higher value of scaled variance of multiplicity distribution of charged particles produced in proton+proton interactions in comparison to the most central Pb+Pb collisions.

5 Concluding remarks

In the present study we used the concept of clusterization in the mechanism of multiparticle production for the description of multiplicity fluctuations observed in relativistic ion collisions at CERN SPS. Our conclusions are as follows:

a) The observed non-monotonic behaviour of the scaled variance of multiplicity distribution as a function of collision centrality can be fully explained by the correlations between produced particles promoting cluster formation Such an effect is not present in a commonly used string-hadronic models of nuclear collisions.

b) The success of the approach lies in treating of all particles included in the cluster-phase as a single system with variance given by Eq. (7) rather than in the details of the probability distributions.

c) The presence of self-similar cluster phase in mechanism of particle production in natural way describes substantially higher value of scaled variance of multiplicity distribution of charged particles produced in proton+proton interactions in comparison to the most central Pb+Pb collisions.

d) Multiplicity clustering (being some kind of implementation of the core-corona model [19, 20]) with self-similar fluctuations (power-law dependence given by Eq. (7) obeys the scaling relationship $Var(N; \lambda\langle N\rangle) = \lambda^{2.5} Var(N; \langle N\rangle))$ provides new insights on non-monotonic behaviour of multiplicity fluctuations.

M.R. was supported by the Polish National Science Centre (NCN) grant 2016/23/B/ST2/00692.

References

[1] M. Rybczynski and Z. Włodarczyk, Eur. Phys. J. A **56**, no. 1, 28 (2020) doi:10.1140/epja/s10050-020-00030-1 [arXiv:1904.01366 [hep-ph]].

[2] R. Szwed, G. Wrochna and A. K. Wroblewski, Mod. Phys. Lett. A **5**, 1851 (1990). doi:10.1142/S0217732390002110

[3] C. Alt *et al.* [NA49 Collaboration], Phys. Rev. C **75**, 064904 (2007) doi:10.1103/PhysRevC.75.064904 [nucl-ex/0612010].

[4] K. Grebieszkow [NA61/SHINE Collaboration], Central Eur. J. Phys. **10**, 1333 (2012) doi:10.2478/s11534-012-0059-2 [arXiv:1201.0742 [nucl-ex]].

[5] A. Aduszkiewicz *et al.* [NA61/SHINE Collaboration], Eur. Phys. J. C **76**, no. 11, 635 (2016) doi:10.1140/epjc/s10052-016-4450-9 [arXiv:1510.00163 [hep-ex]].

[6] A. Bialas, M. Bleszynski and W. Czyz, Nucl. Phys. B **111**, 461 (1976). doi:10.1016/0550-3213(76)90329-1

[7] A. Motornenko, K. Grebieszkow, E. Bratkovskaya, M. I. Gorenstein, M. Bleicher and K. Werner, J. Phys. G **45**, no. 11, 115104 (2018) doi:10.1088/1361-6471/aae149 [arXiv:1711.07789 [nucl-th]].

[8] K. Grebieszkow [NA61/SHINE Collaboration], PoS EPS **-HEP2017**, 167 (2017) doi:10.22323/1.314.0167 [arXiv:1709.10397 [nucl-ex]].

[9] E. Fermi, Prog. Theor. Phys. **5**, 570 (1950). doi:10.1143/PTP.5.570

[10] A. Wroblewski, Acta Phys. Polon. B **4**, 857 (1973).

[11] C. Geich-Gimbel, Int. J. Mod. Phys. A **4**, 1527 (1989). doi:10.1142/S0217751X89000662

[12] G. Aad *et al.* [ATLAS Collaboration], Eur. Phys. J. C **76**, no. 6, 322 (2016) doi:10.1140/epjc/s10052-016-4126-5 [arXiv:1602.00988 [hep-ex]].

[13] G. Aad *et al.* [ATLAS Collaboration], New J. Phys. **13**, 053033 (2011) doi:10.1088/1367-2630/13/5/053033 [arXiv:1012.5104 [hep-ex]].

[14] G. Aad *et al.* [ATLAS Collaboration], Phys. Rev. D **84**, 054001 (2011) doi:10.1103/PhysRevD.84.054001 [arXiv:1107.3311 [hep-ex]].

[15] J. E. Cohen, M. Xu M, W. S. F Schuster, Proc. R. Soc. B **280**, 20122955 (2013) doi:10.1098/rspb.2012.2955

[16] G. Wilk and Z. Wlodarczyk, Phys. Lett. B **727**, 163 (2013) doi:10.1016/j.physletb.2013.10.007 [arXiv:1310.0671 [hep-ph]].

[17] P. Bożek, W. Broniowski, M. Rybczynski and G. Stefanek, Comput. Phys. Commun. **245**, 106850 (2019) doi:10.1016/j.cpc.2019.07.014 [arXiv:1901.04484 [nucl-th]].

[18] M. Rybczynski and Z. Wlodarczyk, J. Phys. Conf. Ser. **5**, 238 (2005) doi:10.1088/1742-6596/5/1/022 [nucl-th/0408023].

[19] K. Werner, Phys. Rev. Lett. **98**, 152301 (2007) doi:10.1103/PhysRevLett.98.152301 [arXiv:0704.1270 [nucl-th]].

[20] M. Petrovici, I. Berceanu, A. Pop, M. Târzilă and C. Andrei, Phys. Rev. C **96**, no. 1, 014908 (2017) doi:10.1103/PhysRevC.96.014908 [arXiv:1703.05805 [nucl-th]].

Thermal photons at PHENIX experiment

Veronica Canoa Roman[1,2,⋆]

[1] *for the PHENIX Collaboration*
[2] *Stony Brook University*

Abstract. Direct photons are a unique probe to study the properties of the medium created in heavy ion collisions. In particular low p_T direct photons are of great importance since one expects that they are predominantly of thermal origin. In A+A systems PHENIX has observed a large yield of low p_T direct photon that are emitted with a significant azimuthal anisotropy with respect to the reaction plane (v2). The mechanism responsible for the large yield and large v2 is not understood yet. Following recent evidence for collective behavior of charged particle production from small systems like p+A, d+Au, and 3He+Au, PHENIX has made systematic measurements of direct photons with different collision energies and system configurations. It has been found that the low p_T direct photon yield $dN_\gamma/d\eta$ is proportional to $(dN_{ch}/d\eta)^\alpha$. This behavior holds for beam energies measured both at RHIC and at the LHC in large-on-large systems, while data from small systems suggest an onset of QGP formation at low $dN_{ch}/d\eta$. In this talk, I will report recent measurements of thermal photon and related observables.

1 Introduction

Direct photons are one of the probes to study the properties of a hot and dense QCD medium. Since photons do not interact strongly in heavy ion collisions, they can carry primordial information about the collision from the time of their production. Photons are very abundant in heavy ion collisions. However, the measurement of direct photons is challenging becuase most of them come from hadronic decays, mainly π^0 and η. The so-called direct photons can be determined experimentally by subtracting from the total yield the expected fraction coming from hadron decays. The primary photon contributor in the low p_T region below 3GeV/c is the thermal photons from the medium that keep the thermodynamic information from when they are generated. The contribution of these photons to the yield have an exponential shape, in analogy to the black body radiation. Meanwhile the contribution of the photon coming from initial states, usually call hard photons, have a shape following $1/p_T^n$. In contrast to the thermal photons, the contribuition of these photons start to be dominant at high p_T (5 GeV).
At the PHENIX experiment, it was measured not only the direct photon yield but also the flow in Au-Au collisions at 200 GeV. It was found a large yield and a large v_2. It has brought challenges to the theoretical models. Because a large yield implies early emission when the temperature is higher, but a large v_2 implies late emission when the flow is fully developed. The details can be found in [1].

⋆e-mail: vcanoaroman@gmail.com

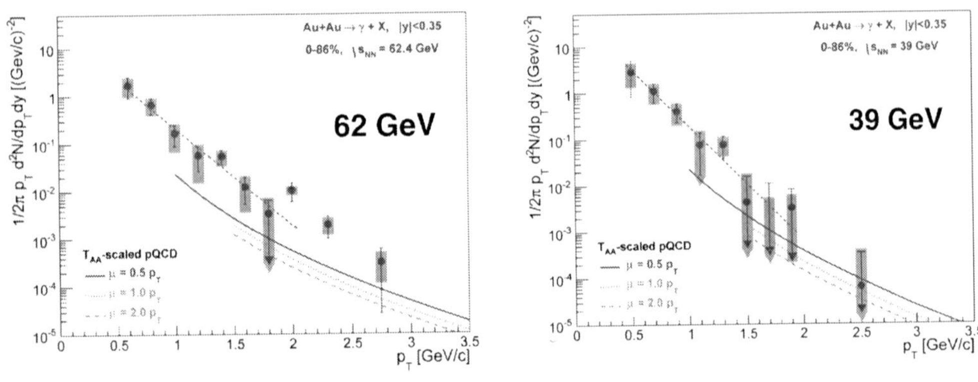

Figure 1. Low p_T direct photon yield for Au+Au collisions at $\sqrt{s_{NN}}$ = 62.4 GeV (left) and Au+Au collisions at $\sqrt{s_{NN}}$ = 39 GeV (right).

2 Direct photons in large systems

The measurement of the direct photon yield in Au+Au collisions at 200 GeV is well established [2]. Recently PHENIX accomplished low momentum direct photon measurements in Au+Au at 62.4 GeV and 39 GeV with the external conversion method [5]. In the external conversion method, the photons are measured through their conversions to electron-positron pairs at the HBD in the PHENIX detector system, and the fraction of direct photons is determined after tagging photons from neutral pion decays. Comparing the data with scaled pQCD calculations to Ncoll, we can find a clear excess over the scaled p+p yield in the low p_T region, as it has been seen in A+Au at 200 GeV. The results can be seen in Figure 1.
Cu+Cu collisions at 200 GeV were also explored [3] via the internal conversion method. A clear direct photon signal excess was found at low p_T region, when we compare with the scaled p+p yield. The direct photon yield for Cu+Cu minimum bias (MB) can be seen in the left panel of Figure 2. If we compare data from Au+Au 40%–60% centrality which have a similar Npart as the Cu+Cu 0%-40% centrality data, where the Au+Au points are scaled by the Npart ratio (66.4/56.0), we can find they are consistent within uncertainties, see Figure 2 in the right panel.

3 Direct photons in small systems

In p+p and p+A collisions at 200 GeV collision, PHENIX has recently measured low momentum direct photons production with the external conversion method. While the p+Au MB direct photon signal is consistent with Ncoll p+p scale, the most central p+Au shows a hint of excess over the binary-scaled baseline, as seen in Figure 3. This non-zero excess yield in central p+Au collisions is compatible with the production of QGP droplets in small central systems. The p+Au result in MB is compatible with a previous measurement in d+Au MB, where zero-excess was found [4].

4 Direct photons scaling

In order to compare the different collision systems and energies, and gain insight into the mechanism of the direct photon production, the yield can be integrated. It can be represented using the number

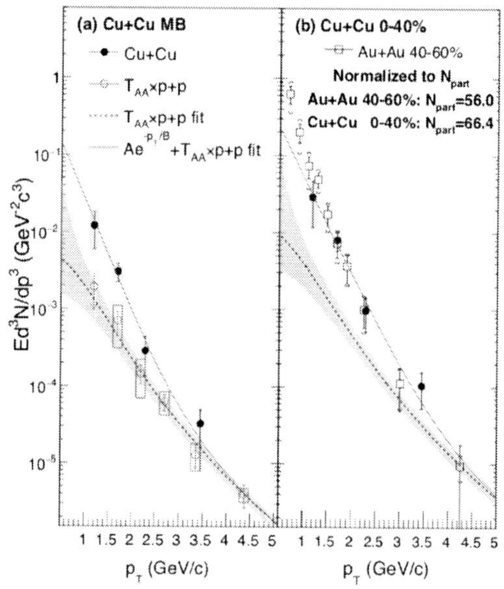

Figure 2. Low p_T direct photon yield for Cu+Cu MB case at $\sqrt{s_{NN}}$ = 200 GeV (left), and Cu+Cu 0%-40% centrality (black circles) and Au+Au at $\sqrt{s_{NN}}$ = 200 GeV in similar number of participants centrality(open squares) (right).

of participants, N_{part}, or the number of binary collisions, N_{coll}. However, this is not useful to compare data at different energies. We can use instead charged particle multiplicity $(dN_{ch}/d\eta)$. It was found a interesting scaling behavior between Ncoll and $(dN_{ch}/d\eta)$ shown in Figure 4. The N_{coll} scales like $(dN_{ch}/d\eta)^{\alpha}$, for all the energies, with $\alpha = 1.25$ [5]. In order to quantify the direct photon spectra, and have a picture with different collision systems and energies together, we can integrate the invariant yield over some p_T threshold value. Figure 5 shows the integrate direct photon spectra for $p_T > 1$ GeV/c (left) and $p_T > 5$ GeV/c (right).

In the plot, we put together all the data for direct photons at PHENIX and also data for ALICE [6]. We can observe a Universal scaling behavior of $dN_{\gamma}/d\eta$ with $(dN_{ch}/d\eta)^{1.25}$, independent of the system, energy or centrality at low p_T direct photons for large systems. In the low multiplicity region one can see the gradually increasing trend of the integrated yield of the small systems, which seems to intersect with the trend from the large systems. The integrated yield at high p_T, $p_T > 5$ GeV/c, follows the p+p fit scaling by the number of collisions for Au+Au data as expected, where the dominant photons are the ones that come from hard scattering.

5 Conclusions

The PHENIX collaboration has performance measurement at low momentum direct photon in Au+Au collisions at 200 GeV, 62 GeV and 39 GeV, Cu+Cu collisions at 200 GeV, and p+Au at 200 GeV. When we put all the data from the large systems together, it was found a surprising scaling behavior of direct photons in large systems. At low p_T the direct photons integrated yield follows a universal scaling as a function of the charged particle multiplicity $(dN_{ch}/d\eta)^{\alpha}$, with $\alpha = 1.25$. The scaling may be an indication that direct photons are produced during the transition from QGP to the hadron

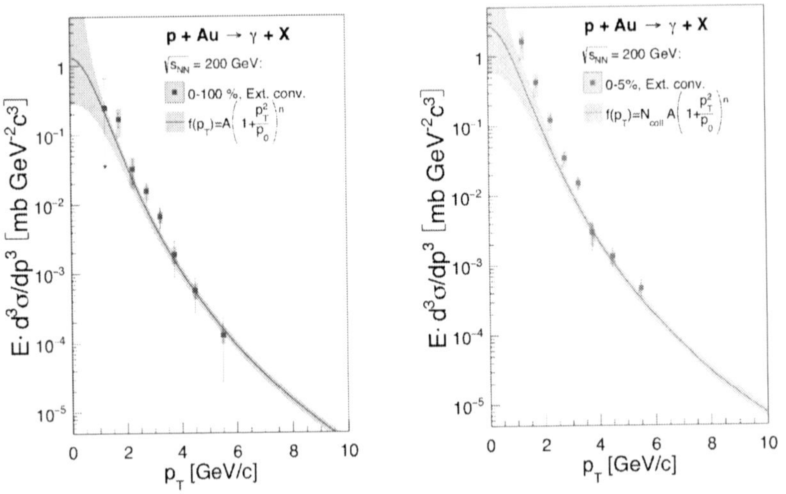

Figure 3. Low p_T direct photon yield in p+Au MB(left) and p+Au central collision(right) at $\sqrt{s_{NN}}$ = 200 GeV.

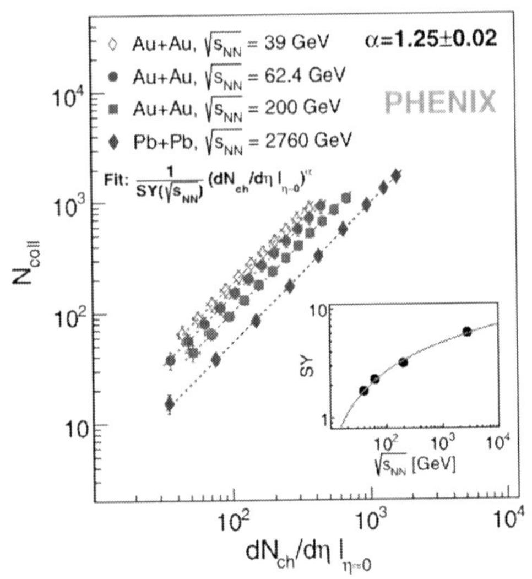

Figure 4. N_{coll} vs. charged-particles multiplicity $dN_{ch}/d\eta$ for four beam energies. More details in [5].

gas. The non-zero excess over the scaled p+p yield in central p+Au collisions can indicate a possible formation of QGP droplets in the small systems.

References

[1] A. Adare et al. (PHENIX Collaboration), Phys.Rev. C **94**, 064901 (2016).

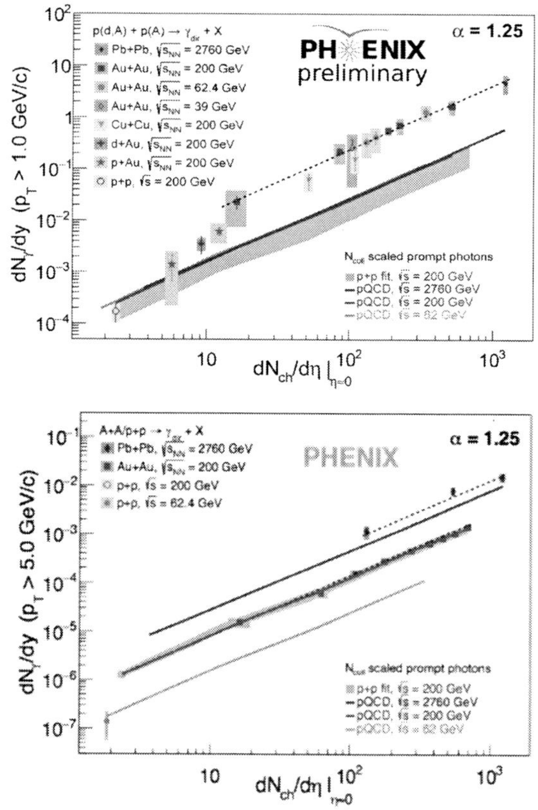

Figure 5. Integrated direct photon yield for different collision systems and energies vs $dN_{ch}/d\eta$, integrated in $p_T > 1$ GeV/c(top) and $p_T > 5$ GeV/c (bottom)[5].

[2] A. Adare et al. (PHENIX Collaboration), Phys.Rev. C **91**, 064904 (2015).
[3] A. Adare et al. (PHENIX Collaboration), Phys.Rev. C **98**, 054902 (2018).
[4] A. Adare et al. (PHENIX Collaboration), Phys.Rev. C **87**, 054907 (2013).
[5] A. Adare et al. (PHENIX Collaboration), Phys.Rev.Lett. **123**, 022301 (2019).
[6] J. Adam et al. (ALICE Collaboration), Phys.Lett.B **754**, 235 (2016).

Dielectrons and Direct Virutal Photons at RHIC-STAR

Chi Yang[1,⋆]

[1]*Shandong University, Qingdao, China 266237*

Abstract. Electromagnetic probes are unique probes of the hot and dense medium created in relativistic heavy-ion collisions due to their minimal interactions with the partonic and hadronic medium. They can be produced at all stages of a collision, thus they provide unique ways to study the medium properties of the whole collision evolution. We present the dielectron invariant mass spectrum at $\sqrt{s_{NN}}$ =27, 39, and 62.4 GeV. Comparing to hadronic cocktail simulation, significant excesses of dielectron invariant mass spectrum at low mass are observed. The direct virtual photon invariant yields derived from the low-mass e^+e^- continuum in Au+Au collisions at $\sqrt{s_{NN}}$ = 200 GeV is presented and the excesses from thermal photons in low transverse momentum (p_T) have been observed. Model calculations can simultaneously describe both dielectron low mass excesses and direct virtual photon low p_T excesses.

1 Introduction

Electromagnetic probes such as dielectrons and direct virutal photons are unique probes in the study of the fundamental properties of the hot and dense medium created in ultra-relativistic heavy-ion collisions [1]. Both dielectrons and direct virtual photons are produced in the whole evolution of the system and escape with minimum interaction with the strongly interacting medium. Thus, they provide information about the various stages of the system during the evolution.

In dielectron low mass region (LMR, $M_{ee} < M_\phi$), the mass spectra of ρ mesons are modified due to their interaction with the medium which is related to the chiral symmetry restoration. At the Relativistic Heavy Ion Collider (RHIC), both STAR and PHENIX Collaborations observed that dilepton mass spectra in Au+Au collisions at $\sqrt{s_{NN}}$ = 200 GeV show a significant excess in the LMR when compared to the known hadronic sources [2, 3]. Theorectical predictions with many-body approach and a transport model can well describe published STAR measurements of the excess yield in the LMR at $\sqrt{s_{NN}}$ = 200 GeV [4, 5]. The RHIC Beam Energy Scan (BES) program [6] provides a unique opportunity to systematically test these calculations as a function of the initial collisions energies.

Direct photon stands for all the produced photons except those from hadron decays in the last stage of the collisions. Measurements at the RHIC [7] and the LHC [8] have shown that the production of the high p_T direct photons in heavy-ion collisions is consistent with the $p + p$ result scaled by the nuclear overlap function T_{AA} for $p_T > 5$ GeV/c, which indicates that the dominant contribution is from hard processes in high p_T. At $1 < p_T < 3$ GeV/c, thermal contributions from the hadronic

⋆e-mail: chiyang@email.sdu.edu.cn

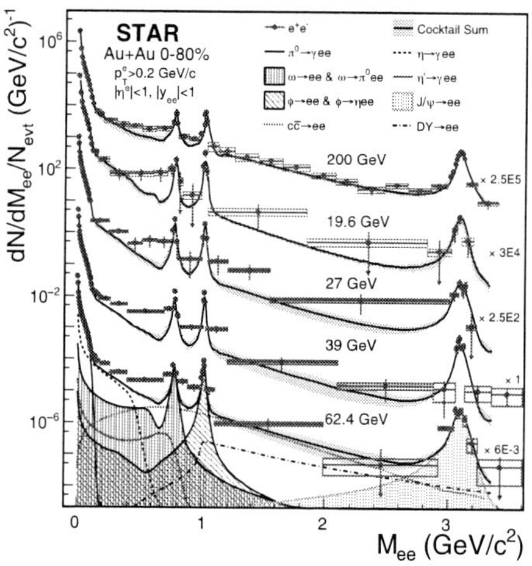

Figure 1. Dielectron invariant mass spectra within the STAR acceptance from $\sqrt{s_{NN}}$ = 19.6, 27, 39, 62.4, and 200 GeV 0-80% most-central Au+Au collisions [10]. The black solid curves represent the hadronic cocktail. The measurements and cocktail predictions are scaled by factors for better presentation.

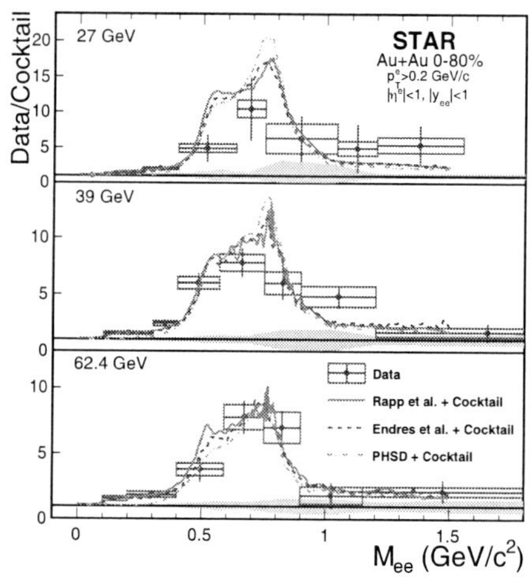

Figure 2. The ratio of the invariant mass spectra to the cocktail with the ω and ϕ yields removed from both the data and cocktail. Model calculations by Rapp *et al.*, Endres *et al.*, and PHSD were separately added to the reference cocktail and compared to the reference cocktail as shown with the curves [10].

medium and Quark-Gluon-Plasma (QGP) play a major role [9]. At 3< p_T <5 GeV/c, the interaction of high energy partons with the QGP has been predicted to contribute a major part of the direct photon production [9].

In this talk, we will present recent dielectron and direct virtual photon measurements from STAR. The measurements of dielectron production in Au+Au collisions with collisions at 27, 39, and 62.4 GeV are presented [10]. The direct virtual photon invariant yields in the transverse momentum ranges $1 < p_T < 3$ GeV/c and $5 < p_T < 10$ GeV/c at mid-rapidity derived from the low-mass e^+e^- invariant mass continuum in 0-80% minimum-bias Au+Au collisions at $\sqrt{s_{NN}}$ = 200 GeV are presented [11]. Comparisons to model calculations with thermal contributions from the hadronic medium and QGP are discussed.

2 Dielectron Results from STAR BES Program

The dielectron invariant mass spectrum measured within STAR acceptance (p_T^e > 0.2 GeV/c, $|\eta^e|$ <1, and $|y_{ee}|$<1) for 0-80% most-central Au+Au collisions at $\sqrt{s_{NN}}$ = 27, 39, and 62.4 GeV are shown in Fig. 1 [10]. The back solid curves represent the hadronic cocktail without the vacuum ρ−meson which contribution is expected to be strongly modified in the medium. Results from 0-80% most-central Au+Au colllisions at $\sqrt{s_{NN}}$ = 19.6 and 200 GeV are shown for a systematic study of the initial collisions energy dependence. A significant of yield excess are shown in the LMR relative to hadronic cocktail in all energies. To estimate this yield excess, a ratio of the present data to the hadronic cocktail with yields from ω and ϕ subtracted from both the data and cocktail were shown in Fig. 2. The open boxes and gray bands represent the experimental systematic uncertainties and cocktail simulation uncertainties separately. A clear enhancement is observed in the LMR relative to the hadronic cocktail for each of the three collision energies. Different model calculations within the

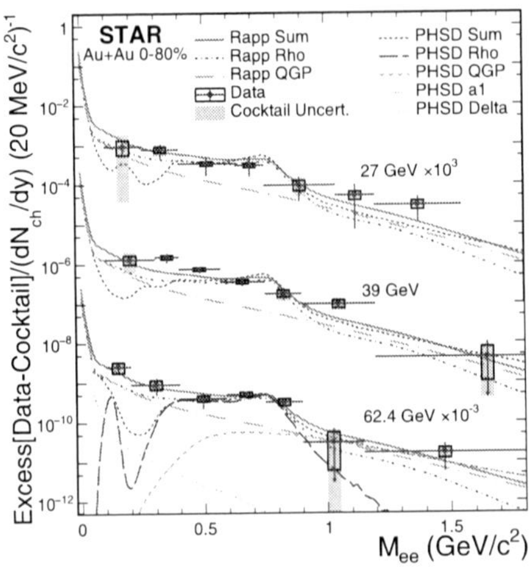

Figure 3. STAR acceptance-corrected dielectron excess invariant mass spectra normalized by dN_{ch}/dy in minimum-bias Au + Au collisions at $\sqrt{s_{NN}}$ = 27, 39, and 62.4 GeV. Model calculations containing a broadened ρ spectral function in hadronic medium (Rapp Rho) and QGP thermal radiation (Rapp QGP) are also shown.

STAR acceptance are added to hadronic cocktails to compare with measurements. Rapp *et al.* model is an effective many-body calculation for vector mesons in a QGP where the spectral function of ρ is broadened primarily due to interactions with baryons and meson in a hadron gas. Endres *et al.* model is a coarse-grained transport approach that includes the ρ spectral function. PHSD is a microscopic transport model which includes the collisional broadening of the ρ meson.

The dielectron excess invariant mass spectra, corrected for detector acceptance, are shown in Figure 3 for minimum-bias Au+Au collisions at $\sqrt{s_{NN}}$ = 27, 39, and 62.4 GeV. The spectra are normalized by mid-rapidity dN_{ch}/dy to cancel out the volume effect, and compared with model calculations. The model calculations can describe the acceptance-corrected excess invariant mass spectra within uncertainty at different collisions energy range. Model calculations from Rapp *et al.* in Figure 3 containing a broadened ρ spectral function in hadronic medium (Rapp Rho) and QGP thermal radiation (Rapp QGP) are also shown. The PHSD model calculations in Figure 3 have contributions from the ρ meson (PHSD Rho), QGP (PHSD QGP), Dalitz decays of the a1 (PHSD a1), and Δ resonances (PHSD Delta). Both models are compared with the excess yield at each energy. Calculations by Rapp *et al.* have an uncertainty on the order of 15%, and PHSD model calculations have an uncertainty on the order of 30% . Within uncertainties, the model calculations are found to reproduce the acceptance-corrected excess in Au+Au collisions at each of the collision energies.

3 Direct Virtual Photon Results in Au+Au Collisions at STAR

Figure 4 shows centrality dependence of the invariant yields of the direct virtual photons in Au+Au collisions at $\sqrt{s_{NN}}$ =200 GeV. The $p + p$ result from PHENIX Collaboration is parameterized by a power-law function [12]. The parameterized distribution is then scaled by the nuclear overlap function T_{AA}, and compared to the Au+Au results in different centralities, as shown in the solid curves. For 1< p_T < 3 GeV/c, the Au+Au results are higher than the T_{AA} scaled $p + p$ results, while at p_T > 6 GeV/c the Au+Au yield is consistent with the scaled $p + p$ expectation. However, one should notice

https://doi.org/10.1051/epjconf/202023502002

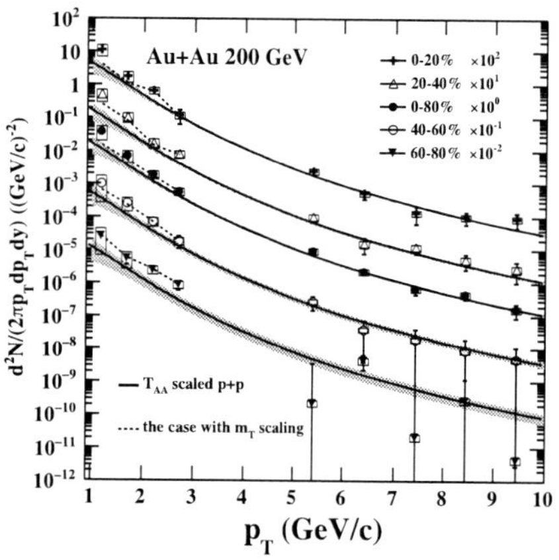

Figure 4. (Color online) Centrality dependence of the direct virtual photon invariant yields as a function of p_T in Au+Au collisions at $\sqrt{s_{NN}}$ =200 GeV. The solid curves represent a power-law fit to PHENIX 200 GeV p+ p results, scaled by T_{AA}. The bands on the curves represent the uncertainties in the parameterization and in T_{AA}. The error bars and boxes represent the statistical and systematic uncertainties, respectively.

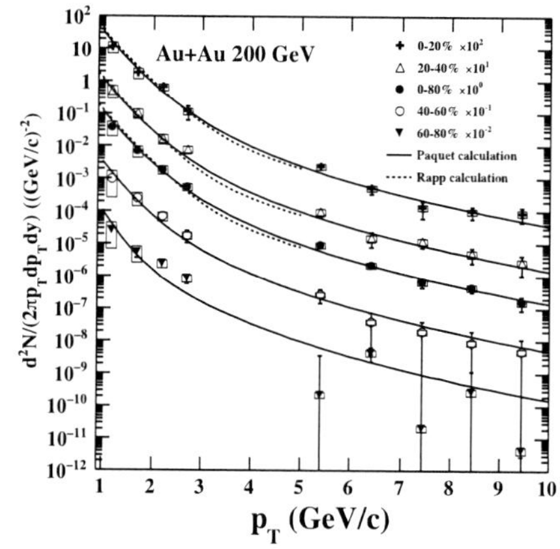

Figure 5. The direct virtual photon invariant yields as a function of p_T in Au+Au collisions at $\sqrt{s_{NN}}$ = 200 GeV compared to model predictions from Rapp et al. and Paquet et al. . The statistical and systematic uncertainties are shown by the bars and boxes, respectively.

that due to larger uncertainties for data points at $1 < p_T < 2$ GeV/c in 40-60% and 60-80% Au+Au collisions, the measurements at this range are also consistent with the scaled $p + p$ expectations.

A comparison between STAR Au+Au data and model calculations from Rapp *et al.* and Paquet *et al.* is shown in Figure 5. Both models include the contributions from QGP thermal radiation, in-medium ρ meson and other mesonic interactions in the hadronic gas, and primordial contributions from the initial hard parton scattering. In Rapp's model, an elliptic thermal fireball evolution is employed for the bulk medium. The sum of the thermal medium and primordial contributions for the former case is shown in Figure 5. In the model from Paquet *et al.*, a (2+1)-D hydrodynamic evolution is employed for the bulk medium. The comparison between the model and data shows that in the p_T range 1-3 GeV/c the dominant source is the thermal radiation, while the initial hard-parton scattering becomes dominant in p_T range 5-6 GeV/c. The comparison shows agreement between both model calculations and our measurements within uncertainties for all the other centralities except 60-80% centrality which includes most peripheral collisions, where hydrodynamic calculations might not be applicable.

4 Summary

Electromagnetic probes provide unique ways to study the hot and dense medium over the whole evolution. Dielectron invariant mass spectrum in the LMR are observed statistically significant excesses when comparing the data to hadronic cocktails that do not include vacuum ρ decay contributions in Au+Au collisions at $\sqrt{s_{NN}}$ =27, 39, and 62.4 GeV. The model calculations with ρ broadening scenario can describes the excess in LMR at all RHIC energies in BES program. The low p_T excesses of the direct virtual photon invariant yields compared to $p + p$ references are observed in Au+Au collisions at STAR. Model calculations can simultaneously describe both dielectron and direct virtual photon results.

5 Acknowledgement

The work is partly supported by the Shandong Provincial Natural Science Foundation, China (Grant Number: ZR2019MA005) and the Program of Qilu Young Scholars of Shandong University.

References

[1] G. David, R. Rapp and Z. Xu, Phys. Rept. 462, 176 (2008).
[2] L. Adamczyk, et al., [STAR Collaboration], Phys. Rev. Lett. 113, 022301 (2014); Phys. Rev. C 92, 024912 (2015).
[3] A. Adare, et al., [PHENIX Collaboration], Phys. Rev. C 93, 014904 (2016).
[4] R. Rapp, Adv. High Energy Phys. 2013, 1 (2013).
[5] O. Linnyk, W. Cassing, J. Manninen, E. L. Bratkovskaya, and C. M. Ko, Phys. Rev. C 85, 024910 (2012).
[6] Experimental Diagram & Study of Search the QCD for the Phase Critical Point, STAR Note SN0493 (2009), URL: https://drupal.star.bnl.gov/STAR/starnotes/public/sn0493.
[7] S. Afanasiev, et al., [PHENIX Collaboration], Phys. Rev. Lett. 109, 152302 (2012).
[8] J. Adam, et al., [ALICE Collaboration], Phys. Lett. B 754, 235 (2016).
[9] S. Turbide, C. Gale, E. Frodermann, and U. Heinz, Phys. Rev. C 77, 024909 (2008).
[10] J. Adam, et al., [STAR Collaboration], arXiv:1810.10159.
[11] L. Adamczyk, et al., [STAR Collaboration], Phys. Lett. B 770, 451 (2017).
[12] S.S. Adler, et al., [PHENIX Collaboration], Phys. Rev. Lett. 98, 012002 (2007) .

Transverse Momentum Resummation for Z boson Plus Jet Production at the LHC

Peng Sun[1,⋆], *Bin* Yan[2,3,⋆⋆], *C.-P.* Yan[3,⋆⋆⋆], and *Feng* Yuan[4,⋆⋆⋆⋆]

[1] *Department of Physics and Institute of Theoretical Physics, Nanjing Normal University, Nanjing, Jiangsu, 210023, China*
[2] *Theoretical Division, Group T-2, MS B283, Los Alamos National Laboratory, P.O. Box 1663, Los Alamos, NM 87545, USA*
[3] *Department of Physics and Astronomy, Michigan State University, East Lansing, MI 48824, USA*
[4] *Nuclear Science Division, Lawrence Berkeley National Laboratory, Berkeley, CA 94720, USA*

Abstract. We study the transverse momentum resummation effects for the Z boson plus jet associated production at the LHC within the transverse momentum dependent (TMD) factorization formalism. The large logarithm $\ln(Q^2/q_\perp^2)$ with $Q \gg q_\perp$ has been resummed to Next-to-Leading Logarithm (NLL) accuracy. We compare our numerical results to the CMS data by reweighting method. It shows that our prediction agree with the data very well.

1 Introduction

Precisely study the Z boson and jet associated production at the Large Hadron Collider (LHC) is important for us to test the standard model (SM) and search the possible new physics effects. In order to give a robust prediction and compare it to the experimental data, the higher order QCD correction is necessary. The next-to-next-to-leading order (NNLO) QCD correction has been finished in Refs. [1–3] and it shows that the theoretical uncertainties could be reduced to percent-level. However, for the observables which are sensitive to the soft and collinear radiations (e.g. the total transverse momentum of Z boson and leading jet system), we should use the resummation technique to improve the theoretical prediction, not by the fixed-order calculation. Here, we use the transverse momentum dependent (TMD) factorization formalism to discuss the kinematical distributions of Z boson plus one jet [4]. A similar work based on the soft-collinear effective theory can be found in Ref. [5].

2 TMD resummation

TMD resummation formula can be written as [6]:

$$\frac{d^5\sigma}{dy_Z dy_J dP_{J\perp}^2 d^2\vec{q}_\perp} = \sum_{ab}\left[\int \frac{d^2\vec{b}_\perp}{(2\pi)^2} e^{-i\vec{q}_\perp\cdot\vec{b}_\perp} W_{ab\to ZJ}(x_1, x_2, b_\perp) + Y_{ab\to ZJ}\right], \quad (1)$$

⋆e-mail: pengsun@msu.edu
⋆⋆e-mail: binyan@lanl.gov
⋆⋆⋆e-mail: yuan@pa.msu.edu
⋆⋆⋆⋆e-mail: fyuan@lbl.gov

where y_Z and y_J denote the rapidity of the Z boson and the leading jet; $P_{J\perp}(P_{Z\perp})$ and $\vec{q}_\perp = \vec{P}_{Z\perp} + \vec{P}_{J\perp}$ are the leading jet (Z boson) transverse momentum and the imbalance transverse momentum of the Z boson and the jet system. The all order resummation effects have included in $W_{ab\to ZJ}$. $Y_{ab\to ZJ}$ is the remainder function between fixed order and asymptotic result. x_1 and x_2 are the momentum fractions of the incoming hadrons carried by the two incoming partons,

$$x_{1,2} = \frac{\sqrt{m_Z^2 + P_{Z\perp}^2}\,e^{\pm y_Z} + \sqrt{P_{J\perp}^2}\,e^{\pm y_J}}{\sqrt{S}}. \tag{2}$$

The $W_{ab\to ZJ}$ can be further written as,

$$\begin{aligned} W_{ab\to ZJ}(x_1, x_2, b) &= x_1 f_a(x_1, \mu_F = b_0/b_\perp) x_2 f_b(x_2, \mu_F = b_0/b_\perp) \\ &\times H_{ab\to ZJ}(s, \mu_{\rm res}, \mu_R) e^{-S_{\rm Sud}(s,\mu_{\rm res},b_\perp)} e^{-\mathcal{F}_{NP}}, \end{aligned} \tag{3}$$

Here μ_F, μ_R and $\mu_{\rm res}$ are the factorization, renormalization and resummation scales.

The Sudakov form factor can be expressed as,

$$S_{\rm Sud} = \int_{b_0^2/b_\perp^2}^{\mu_{\rm res}^2} \frac{d\mu^2}{\mu^2}\left[\ln\left(\frac{s}{\mu^2}\right) A + B_1 + B_2 + D \ln\frac{1}{R^2}\right], \tag{4}$$

where R is the jet cone size. The coefficients A, $B_{1,2}$ and D can be expanded perturbatively in α_s. The NLO results have show in Ref. [4]. Factor $\mathcal{F}_{NP}$ denotes the contribution from non-perturbative effects [7]. The hard function $H_{ab\to ZJ}$ is depending on the initial state and the full NLO results can be found in Ref. [4].

3 Phenomenology

We apply the resummation formula in Eq. 1 to calculate the $q_\perp$ spectrum and ϕ angle distribution between Z boson and leading jet. The anti-k_t jet algorithm with cone size $R = 0.4$ is used in our numerical calculation. In order to compare our theoretical prediction to data, we apply a reweighting procedure to estimate the effect from imposing kinematic cuts on the leptons from Z boson decay. The differential cross section after we including the kinematic cuts can be written as,

$$\left.\frac{d\sigma}{dq_\perp}\right|_{\rm decay} = \left.\frac{d\sigma}{dq_\perp}\right|_{\rm stable,Z} \times \kappa(m_{\ell^+\ell^-}, y_{\ell^+\ell^-}, p_T^{\ell^+\ell^-}), \tag{5}$$

where $\kappa(m_{\ell^+\ell^-}, p_T^{\ell^+\ell^-}, y_{\ell^+\ell^-})$ is the reweighting factor which depends on lepton pair invariant mass ($m_{\ell^+\ell^-}$), transverse momentum ($p_T^{\ell^+\ell^-}$) and rapidity ($y_{\ell^+\ell^-}$). We find that the reweighting factor could approximate as a constant when $|y_Z| < 1.5$. Therefore, we have,

$$\left.\frac{d\sigma}{\sigma dq_\perp}\right|_{\rm decay} \simeq \left(\left.\frac{d\sigma}{\sigma dq_\perp}\right|_{\rm stable,Z}\right), \text{ for } |y_{\ell^+\ell^-}| < 1.5. \tag{6}$$

We calculate the normalized $q_\perp$ distribution of Z boson plus one jet production at the $\sqrt{s} = 13$ TeV with CT14NNLO PDF [8] in Fig. 1. The ϕ angle distribution at the $\sqrt{s} = 8$ TeV is shown in Fig. 2. The blue and red bands represent the experimental and scale uncertainty respectively. It is clear that our theoretical prediction agree well with the experimental data.

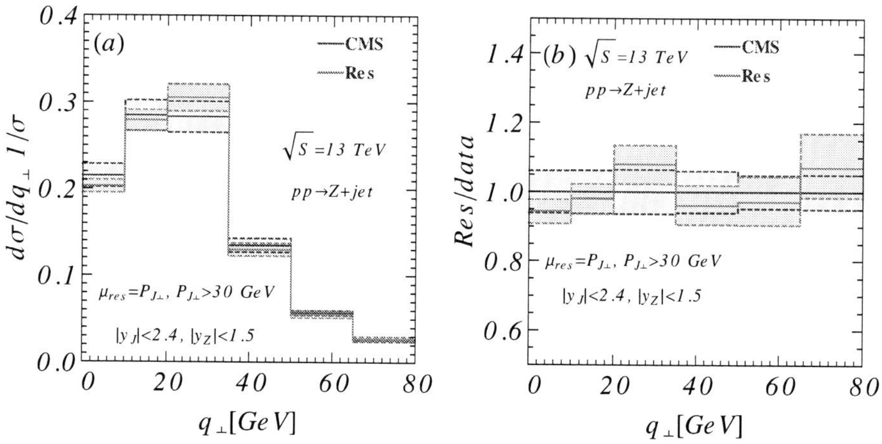

Figure 1. (a) The normalized $q_\perp$ distribution of the Z boson plus one jet system, produced at the $\sqrt{S}$ = 13 TeV LHC with $|y_J| < 2.4$ and $P_{J\perp} > 30$ GeV. The blue and red bands represent the CMS experimental uncertainty [9] and the resummation calculation (Res) scale uncertainty, respectively. (b) The ratio of resummation prediction to CMS data as a function of $q_\perp$.

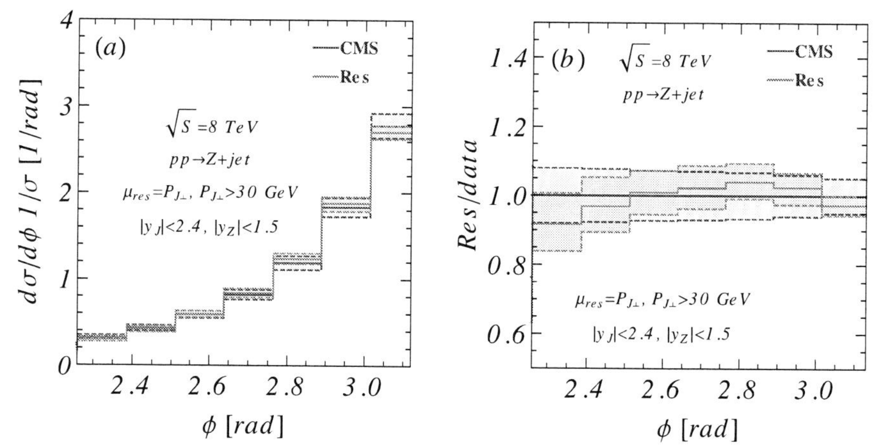

Figure 2. The normalized distribution of ϕ, the azimuthal angle between the final state jet and Z boson measured in the laboratory frame, for $pp \to Z + jet$ production at the $\sqrt{S}$ = 8 TeV LHC with $|y_J| < 2.4$, $|y_Z| < 1.5$ and $P_{J\perp} > 30$ GeV. The blue and red bands represent the CMS experimental error [10] and resummation scale uncertainty respectively.

4 Summary

In this work, we discuss the transverse momentum resummation effects for the Z boson and jet associated production. The large logarithms of $\ln(Q^2/q_\perp^2)$ has be resuumed to NLL accuracy. In order to compare to data directly, the reweighting method is used in our analysis. It shows that our resummation calculation can describe well the CMS data, both the transverse momentum distribution and azimuthal angle of Z boson plus jet system.

Acknowledgment: This work is partially supported by the U.S. Department of Energy, Office of Science, Office of Nuclear Physics, under contract number DE-AC02-05CH11231, and by the U.S. National Science Foundation under Grant No. PHY-1719914. C.-P. Yuan is also grateful for the support from the Wu-Ki Tung endowed chair in particle physics.

References

[1] R. Boughezal, J. M. Campbell, R. K. Ellis, C. Focke, W. T. Giele, X. Liu and F. Petriello, Phys. Rev. Lett. **116**, no. 15, 152001 (2016) doi:10.1103/PhysRevLett.116.152001 [arXiv:1512.01291 [hep-ph]].

[2] A. Gehrmann-De Ridder, T. Gehrmann, E. W. N. Glover, A. Huss and T. A. Morgan, Phys. Rev. Lett. **117**, no. 2, 022001 (2016) doi:10.1103/PhysRevLett.117.022001 [arXiv:1507.02850 [hep-ph]].

[3] R. Boughezal, X. Liu and F. Petriello, Phys. Rev. D **94**, no. 7, 074015 (2016) doi:10.1103/PhysRevD.94.074015 [arXiv:1602.08140 [hep-ph]].

[4] P. Sun, B. Yan, C.-P. Yuan and F. Yuan, Phys. Rev. D **100**, no. 5, 054032 (2019) doi:10.1103/PhysRevD.100.054032 [arXiv:1810.03804 [hep-ph]].

[5] Y. T. Chien, D. Y. Shao and B. Wu, JHEP **1911**, 025 (2019) doi:10.1007/JHEP11(2019)025 [arXiv:1905.01335 [hep-ph]].

[6] P. Sun, C.-P. Yuan and F. Yuan, Phys. Rev. Lett. **114**, no. 20, 202001 (2015) doi:10.1103/PhysRevLett.114.202001 [arXiv:1409.4121 [hep-ph]].

[7] P. Sun, J. Isaacson, C.-P. Yuan and F. Yuan, Int. J. Mod. Phys. A **33**, no. 11, 1841006 (2018) doi:10.1142/S0217751X18410063 [arXiv:1406.3073 [hep-ph]].

[8] S. Dulat *et al.*, Phys. Rev. D **93**, no. 3, 033006 (2016) doi:10.1103/PhysRevD.93.033006 [arXiv:1506.07443 [hep-ph]].

[9] A. M. Sirunyan *et al.* [CMS Collaboration], Eur. Phys. J. C **78**, no. 11, 965 (2018) doi:10.1140/epjc/s10052-018-6373-0 [arXiv:1804.05252 [hep-ex]].

[10] V. Khachatryan *et al.* [CMS Collaboration], JHEP **1704**, 022 (2017) doi:10.1007/JHEP04(2017)022 [arXiv:1611.03844 [hep-ex]].

Searching the QCD critical endpoint with lattice simulations

Szabolcs Borsanyi[1], *Zoltan* Fodor[1,2,3,4], *Matteo* Giordano[2], *Jana N.* Guenther[1,5,⋆], *Kornél* Kapás[2], *Sandor K.* Katz[2], *Kalman K.* Szabó[1,3], *Attila* Pasztor[2], *Israel* Portillo[6], and *Claudia* Ratti[6]

[1] *University of Wuppertal, Department of Physics, Wuppertal D-42097, Germany*
[2] *Eötvös University, Budapest 1117, Hungary*
[3] *Jülich Supercomputing Centre, Jülich D-52425, Germany*
[4] *Physics Department, UCSD, San Diego, CA 92093, USA*
[5] *University of Regensburg, Department of Physics, Regensburg D-93053, Germany*
[6] *Department of Physics, University of Houston, Houston, TX 77204, USA*

Abstract. We discuss the usefulness of various lattice observables especially fluctuations to locate the QCD critical endpoint. We apply different models to interpret our results for the baryon fluctuations up to μ^8 from simulations at imaginary chemical potentials.

1 Introduction

When investigating Quantum Chromodynamics (QCD) an important but challenging goal is the study of the phase diagram. At zero chemical potential lattice QCD predicts a smooth crossover between hadrons and the quark gluon plasma [1–5], taking place in the temperature range $T \simeq 145 - 165$ MeV. Due to the sign problem lattice QCD is unable to study the region with finite chemical potential.

With the advent of the second Beam Energy Scan (BESII) at the Relativistic Heavy Ion Collider (RHIC), scheduled for 2019-2020, there is a renewed interest in the heavy ion community towards the phases of QCD at moderate-to-large densities. A rich theoretical effort is being developed in support of the experimental program; several observables are being calculated, in order to constrain the existence and location of the QCD critical point and to observe it experimentally.

Fluctuations of conserved charges (electric charge Q, baryon number B and strangeness S) are important observables for the finite-density investigations. One possible way to extend lattice results to finite density is to perform Taylor expansions of the thermodynamic observables around chemical potential $\mu_B = 0$ [6–10]: fluctuations of conserved charges are directly related to the Taylor expansion coefficients of such observables. They allow for a comparison between theoretical and experimental results to extract the chemical freeze-out temperature T_f and chemical potential μ_{Bf} as functions of the collision energy [11–14]. The higher order fluctuations are also an important signature for the critical endpoint, as they give access to the correlation length [8, 15, 16].

In this work we apply the method of analytical continuation from imaginary chemical potential [17–21]. It agrees well with the results of the Taylor expansion as shown for the transition temperature [22]. We present results for χ_2^B to χ_6^B in the temperature range 140 MeV $\leq T \leq$ 220 MeV. Several diagonal and non-diagonal fluctuations of conserved charges up to sixth-order are available in our recent paper [23].

⋆e-mail: Jana.Guenther@t-online.de

2 Fluctuations

We present results of an high-precision analysis on a $48^3 \times 12$ lattice. A more detailed description as well as precise information on the lattice set-up can be found in refs. [23, 24]. We use analytical continuation from imaginary chemical potential to determine the χ^B fluctuations at $\mu_B = 0$. We analyze data for eight different values of $\mu_B = i\frac{j\pi}{8}$ with $j \in \{0, 1, 2, 3, 4, 5, 6, 7\}$. In our analysis we use the following ansatz for the pressure:

$$\chi_0^B(\hat{\mu}_B) = \frac{p}{T^4} = c_0 + c_2\hat{\mu}_B^2 + c_4\hat{\mu}_B^4 + c_6\hat{\mu}_B^6 + \frac{4!}{8!}c_4\epsilon_1\hat{\mu}_B^8 + \frac{4!}{10!}c_4\epsilon_2\hat{\mu}_B^{10}, \tag{1}$$

where ϵ_1 and ϵ_2 are drawn randomly from a normal distribution with $\mu = -1.25$ and $\sigma = 2.75$. The values were chosen in a way to allow for χ_8^B to take the value predicted by the hadron resonance gas, as well as the result from the toy model introduced in section 4. From the ansatz we can calculate the derivatives that can be measured on the lattice:

$$\chi_1^B(\hat{\mu}_B) = 2c_2\hat{\mu}_B + 4c_4\hat{\mu}_B^3 + 6c_6\hat{\mu}_B^5 + \frac{4!}{7!}c_4\epsilon_1\hat{\mu}_B^7 + \frac{4!}{9!}c_4\epsilon_2\hat{\mu}_B^9, \tag{2}$$

$$\chi_2^B(\hat{\mu}_B) = 2c_2 + 12c_4\hat{\mu}_B^2 + 30c_6\hat{\mu}_B^4 + \frac{4!}{6!}c_4\epsilon_1\hat{\mu}_B^6 + \frac{4!}{8!}c_4\epsilon_2\hat{\mu}_B^8, \tag{3}$$

$$\chi_3^B(\hat{\mu}_B) = 24c_4\hat{\mu}_B + 120c_6\hat{\mu}_B^3 + \frac{4!}{5!}c_4\epsilon_1\hat{\mu}_B^5 + \frac{4!}{7!}c_4\epsilon_2\hat{\mu}_B^7, \tag{4}$$

$$\chi_4^B(\hat{\mu}_B) = 24c_4 + 360c_6\hat{\mu}_B^2 + c_4\epsilon_1\hat{\mu}_B^4 + \frac{4!}{6!}c_4\epsilon_2\hat{\mu}_B^6. \tag{5}$$

We perform a correlated fit for $\chi_1^B(\hat{\mu}_B)$, $\chi_2^B(\hat{\mu}_B)$, $\chi_3^B(\hat{\mu}_B)$ and $\chi_4^B(\hat{\mu}_B)$ for the different values of μ_B to determine the fitting parameters c_2, c_4 and c_6. From the parameters we can determine $\chi_2^B(0) = 2c_2$, $\chi_4^B(0) = 24c_4$, $\chi_6^B(0) = 720c_6$ and $\chi_8^B(0) = 24c_4\epsilon_1$. The results are shown in figure 1. These equations show the relation between χ_4^B and χ_8^B that are just related by the factor of ϵ_1 (in the same way χ_4^B and χ_{10}^B are related by a factor of ϵ_2). In this way we take into account the influence of higher order corrections to our fit function. We choose 1000 different values for ϵ_1 and ϵ_2 and in addition we include either seven or eight different values of μ_B in our data. All resulting fits are combined in a histogram and weighted with the Akaike information criteria [25], thus allowing to estimate the systematic error. The statistical error is determined by the Jackknife method and both errors are added quadratically to get the combined error shown in the plots.

3 Cumulants

For a comparison with heavy ion collision experiments the cumulants of the net baryon distribution are a useful tool. The first four cumulants are the mean M_B, the variance σ_B^2, the skewness S_B and the kurtosis κ_B. By forming appropriate ratios, we can cancel out explicit volume factors. However the measured distributions themselves may still depend on the volume, which one should take into account, when comparing to experiments.

Heavy ion collisions with lead or gold take place at $\mu_B > 0$, $\langle n_S \rangle = 0$ and $\langle n_Q \rangle = 0.4\langle n_B \rangle$. Since our simulations are done at $\mu_S = \mu_Q = 0$ and $\mu_B \neq 0$ we have to do some calculations to arrive at the same observables that are measured in experiments (see for example [27]). We investigate three different ratios of cumulants and write each as a Taylor expansion:

$$\frac{M_B}{\sigma_B^2} = \frac{\chi_1^B(T, \hat{\mu}_B)}{\chi_2^B(T, \hat{\mu}_B)} = \hat{\mu}_B r_{12}^{B,1} + \hat{\mu}_B^3 r_{12}^{B,3} + \dots \tag{6}$$

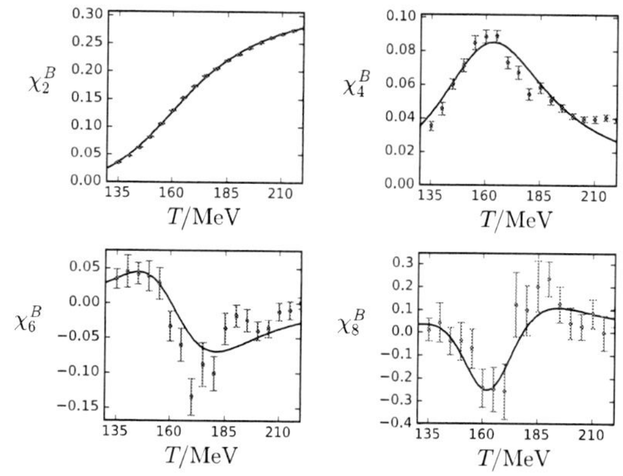

Figure 1: Results for χ_2^B, χ_4^B, χ_6^B and an estimate for χ_8^B on a $N_t = 12$ lattice as functions of the temperature, obtained from the single-temperature analysis (see text). We plot χ_8^B in green to point out that its determination is guided by a prior, which is linked to χ_4^B. The black curve in each panel corresponds to the toy model introduced in [23, 26].

$$\frac{S_B\sigma_B^3}{M_B} = \frac{\chi_3^B(T,\hat{\mu}_B)}{\chi_1^B(T,\hat{\mu}_B)} = r_{31}^{B,0} + \hat{\mu}_B^2 r_{31}^{B,2} + \ldots \tag{7}$$

$$\kappa_B\sigma_B^2 = \frac{\chi_4^B(T,\hat{\mu}_B)}{\chi_2^B(T,\hat{\mu}_B)} = r_{42}^{B,0} + \hat{\mu}_B^2 r_{42}^{B,2} + \ldots \tag{8}$$

The μ_B dependence of the $\chi_i^B(T,\hat{\mu}_B)$ can be again written in terms of the Taylor expansion:

$$\begin{aligned}\chi_{i,j,k}^{BQS}(\hat{\mu}_B) &= \chi_{i,j,k}^{BQS}(0) + \hat{\mu}_B\left[\chi_{i+1,j,k}^{BQS}(0) + q_1\chi_{i,j+1,k}^{BQS}(0) + s_1\chi_{i,j,k+1}^{BQS}(0)\right]\\ &+ \frac{1}{2}\hat{\mu}_B^2\left[\chi_{i+2,j,k}^{BQS}(0) + s_1^2\chi_{i,j+2,k}^{BQS}(0) + q_1^2\chi_{i,j,k+2}^{BQS}(0)\right.\\ &\left.+2q_1s_1\chi_{i,j+1,k+1}^{BQS}(0) + 2s_1\chi_{i+1,j+1,k}^{BQS}(0) + 2q_1\chi_{i+1,j,k+1}^{BQS}(0)\right] + \ldots\end{aligned} \tag{9}$$

with $q_j = \frac{1}{j!}\frac{d^j\hat{\mu}_Q}{(d\hat{\mu}_B)^j}(0)$ and $s_j = \frac{1}{j!}\frac{d^j\hat{\mu}_S}{(d\hat{\mu}_B)^j}(0)$.

We can now use the constraints $\langle n_S\rangle = 0$ and $\langle n_Q\rangle = 0.4\langle n_B\rangle$ which can be rewritten as $\chi_1^Q = 0.4\chi_1^B$ and $\chi_1^S = 0$ to determine the $r_{ij}^{B,k}$ coefficients form the equations 6, 7 and 8. However we now need to know not only the behavior of the χ_i^B but also of derivatives with respect to μ_S and μ_Q. For now our simulations are restricted to ensembles with finite μ_B. Therefore the μ_S and μ_Q derivatives have to be calculated directly and without the support from the fit that we used in the μ_B direction. We calculate various $\chi_{i,j,k}^{B,Q,S}$ with the appropriate values of j and k and all possible values for i so that $i + j + k \leq 4$. For each group of fluctuations with the same j and k we perform a fit analogous to the procedure described in section 2. This is sufficient to determine the first two $r_{ij}^{B,k}$ coefficients for all three observables. The results are shown in figure 2.

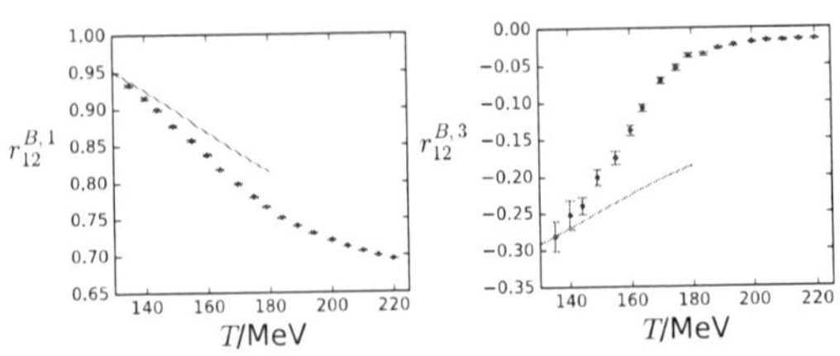

a) $r_{12}^{B,1}$ (left panel) and $r_{12}^{B,3}$ (right panel).

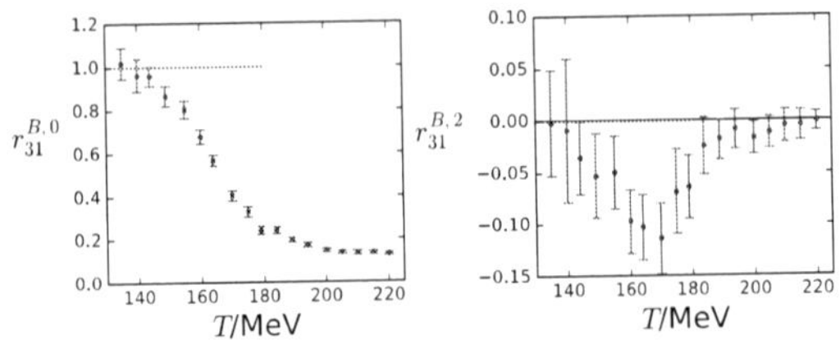

b) $r_{31}^{B,0}$ (left panel) and $r_{31}^{B,2}$ (right panel).

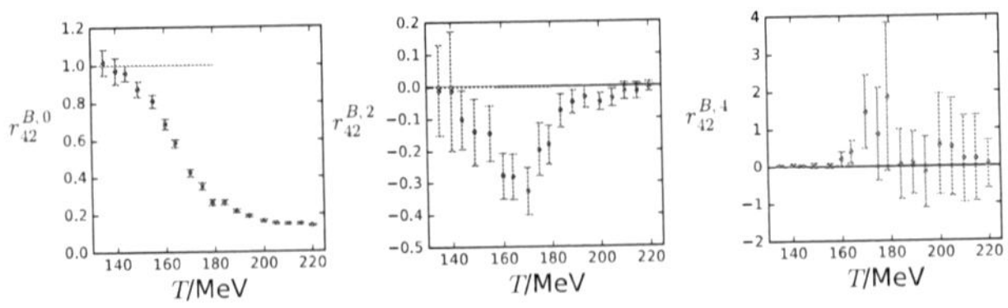

c) $r_{42}^{B,0}$ (left panel) $r_{42}^{B,2}$ (middle panel), $r_{42}^{B,4}$ (right panel). The latter is not obtained independently, but by means of the prior ansatz (see text): for this reason, we plot it in green.

Figure 2: Taylor expansion coefficients as functions of the temperature.

4 Looking for the critical point

To look for the critical endpoint in the QCD phase diagram one can try to calculate the radius of convergence of an expansion in μ_B. Two obvious expansions for this are either the pressure $p(\mu) = p_0 + p_2\hat{\mu}^2 + p_4\hat{\mu}^4 + p_6\hat{\mu}^6 + \ldots$ or the fluctuations that are directly related: $\chi_2^B(\mu) = 2p_2 + 12p_4\hat{\mu}^2 + 30p_6\hat{\mu}^4 + \ldots$. We define

$$r_{2n}^p = \sqrt{\frac{p_{2n}}{p_{2n+2}}} \text{ and } r_{2n}^\chi = \sqrt{\frac{2n(2n-1)}{(2n+1)(2n+2)}} r_{2n}^p. \tag{10}$$

In the limit of $n \to \infty$ if either r_{2n}^p or r_{2n}^χ converge, they converge to the same value, the radius of convergence, which guarantees that there is no criticality within this radius.[1] However, since we only know the fluctuations up to χ_8^B as discussed in the previous section, we will first test this procedure for a toy model in which the critical endpoint is known. We use unimproved staggered fermions on an $N_t = 4$ lattice. For this set up the critical endpoint has been already determined [20, 21, 29]. The results for r_{2n}^p and r_{2n}^χ are shown in the left panel of figure 3. For a temperature where the critical endpoint is close by (right site of the left panel of figure 3), the ratios seem to converge to the correct value. However, as discussed in more detail in ref. [29], due to the structure of χ_6^B there is always a temperature for which the ratios seem to converge, independent of the real value for the critical point. For the $N_t = 12$ data the r_{2n}^χ and the ratios from the hadron resonance gas are shown in the right panel of figure 3. Here the errors are still large.

Instead of investigating a toy model with a known critical endpoint, we can also try to describe the data with a toy model without any critical behavior. If one fits the data for $\chi_1^B/\hat{\mu}_B$ at $\mu_B = 0$ with an

[1] As was shown in [28], the ratio estimator is never convergent in a finite volume, and it is problematic even when using the p_n extrapolated to infinite volume. It will work if one uses infinite-volume Taylor coefficients and the singularity determining the radius of convergence corresponds to a real phase transition in the infinite-volume limit.

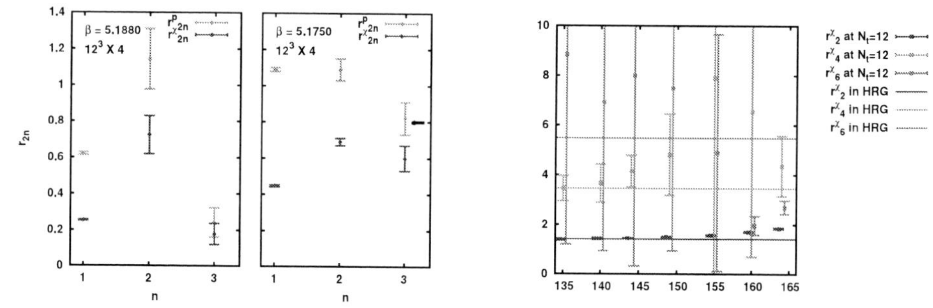

Figure 3: On the left panel: The ratios r^P_{2n} and r^χ_{2n} (eqn. 10) on an $N_t = 4$ lattice. On the very left the temperature is close to the crossover temperature. Next to it the temperature is close to the temperature for the critical endpoint. The black arrow marks the value for the critical endpoint from [21]. On the right panel: The r^χ_{2n} (eqn. 10) ratios for different temperatures [29].

analytic function of T and assumes that any change with respect to the chemical potential is a linear shift of this function, one can determine all fluctuations analytically (more detail on this toy model can be found in ref. [26]). The results of this toy model are shown with black curves in figure 1. They agree well with the data.

Acknowledgements

This project was funded by the DFG grant SFB/TR55. This work was supported by the Hungarian National Research, Development and Innovation Office, NKFIH grants KKP126769 and K113034. An award of computer time was provided by the INCITE program. This research used resources of the Argonne Leadership Computing Facility, which is a DOE Office of Science User Facility supported under Contract DE-AC02-06CH11357. The authors gratefully acknowledge the Gauss Centre for Supercomputing e.V. (www.gauss-centre.eu) for funding this project by providing computing time on the GCS Supercomputer JUQUEEN[30] at Jülich Supercomputing Centre (JSC) as well as on HAZEL-HEN at HLRS Stuttgart, Germany. This material is based upon work supported by the National Science Foundation under grants no. PHY-1654219 and OAC-1531814 and by the U.S. Department of Energy, Office of Science, Office of Nuclear Physics, within the framework of the Beam Energy Scan Theory (BEST) Topical Collaboration. C.R. also acknowledges the support from the Center of Advanced Computing and Data Systems at the University of Houston. A.P. is supported by the János Bolyai Research Scholarship of the Hungarian Academy of Sciences and by the ÚNKP-19-4 New National Excellence Program of the Ministry for Innovation and Technology.

References

[1] Y. Aoki, G. Endrodi, Z. Fodor, S.D. Katz, K.K. Szabo, Nature **443**, 675 (2006), hep-lat/0611014
[2] Y. Aoki, S. Borsanyi, S. Durr, Z. Fodor, S.D. Katz, S. Krieg, K.K. Szabo, JHEP **06**, 088 (2009), 0903.4155
[3] S. Borsanyi, Z. Fodor, C. Hoelbling, S.D. Katz, S. Krieg, C. Ratti, K.K. Szabo (Wuppertal-Budapest), JHEP **09**, 073 (2010), 1005.3508
[4] T. Bhattacharya et al., Phys. Rev. Lett. **113**, 082001 (2014), 1402.5175
[5] A. Bazavov et al., Phys. Rev. **D85**, 054503 (2012), 1111.1710
[6] C.R. Allton, S. Ejiri, S.J. Hands, O. Kaczmarek, F. Karsch, E. Laermann, C. Schmidt, L. Scorzato, Phys. Rev. **D66**, 074507 (2002), hep-lat/0204010
[7] C.R. Allton, M. Doring, S. Ejiri, S.J. Hands, O. Kaczmarek, F. Karsch, E. Laermann, K. Redlich, Phys. Rev. **D71**, 054508 (2005), hep-lat/0501030
[8] R.V. Gavai, S. Gupta, Phys. Rev. **D78**, 114503 (2008), 0806.2233
[9] S. Basak et al. (MILC), PoS **LATTICE2008**, 171 (2008), 0910.0276
[10] O. Kaczmarek, F. Karsch, E. Laermann, C. Miao, S. Mukherjee, P. Petreczky, C. Schmidt, W. Soeldner, W. Unger, Phys. Rev. **D83**, 014504 (2011), 1011.3130
[11] F. Karsch, Central Eur. J. Phys. **10**, 1234 (2012), 1202.4173
[12] A. Bazavov et al., Phys. Rev. Lett. **109**, 192302 (2012), 1208.1220
[13] S. Borsanyi, Z. Fodor, S.D. Katz, S. Krieg, C. Ratti, K.K. Szabo, Phys. Rev. Lett. **111**, 062005 (2013), 1305.5161
[14] S. Borsanyi, Z. Fodor, S.D. Katz, S. Krieg, C. Ratti, K.K. Szabo, Phys. Rev. Lett. **113**, 052301 (2014), 1403.4576
[15] M.A. Stephanov, K. Rajagopal, E.V. Shuryak, Phys. Rev. **D60**, 114028 (1999), hep-ph/9903292
[16] M. Cheng et al., Phys. Rev. **D77**, 014511 (2008), 0710.0354
[17] Z. Fodor, S.D. Katz, Phys. Lett. **B534**, 87 (2002), hep-lat/0104001
[18] P. de Forcrand, O. Philipsen, Nucl. Phys. **B642**, 290 (2002), hep-lat/0205016
[19] M. D'Elia, M.P. Lombardo, Phys. Rev. **D67**, 014505 (2003), hep-lat/0209146
[20] Z. Fodor, S.D. Katz, JHEP **03**, 014 (2002), hep-lat/0106002
[21] Z. Fodor, S.D. Katz, JHEP **04**, 050 (2004), hep-lat/0402006
[22] C. Bonati, M. D'Elia, F. Negro, F. Sanfilippo, K. Zambello (2018), 1805.02960
[23] S. Borsanyi, Z. Fodor, J.N. Guenther, S.K. Katz, K.K. Szabo, A. Pasztor, I. Portillo, C. Ratti, JHEP **10**, 205 (2018), 1805.04445
[24] R. Bellwied, S. Borsanyi, Z. Fodor, S.D. Katz, A. Pasztor, C. Ratti, K.K. Szabo, Phys. Rev. **D92**, 114505 (2015), 1507.04627
[25] H. Akaike, IEEEE **19**, 716 (1974)
[26] S. Borsanyi, Z. Fodor, J.N. Guenther, S.K. Katz, A. Pasztor, I. Portillo, C. Ratti, K.K. Szabó, Nucl. Phys. **A982**, 223 (2019)
[27] L. Adamczyk et al. (STAR), Phys. Rev. Lett. **112**, 032302 (2014), 1309.5681
[28] M. Giordano, A. Pásztor, Phys. Rev. **D99**, 114510 (2019), 1904.01974
[29] Z. Fodor, M. Giordano, J.N. Guenther, K. Kapas, S.D. Katz, A. Pasztor, I. Portillo, C. Ratti, D. Sexty, K.K. Szabo (2018), 1807.09862
[30] Jülich Supercomputing Centre, Journal of large-scale research facilities **A1**, 1 (2015)

Production of exotic charmonium in heavy ion collisions

F. S. Navarra[1,a]

[1] *Instituto de Física, Universidade de São Paulo, Rua do Matão 1371, 05508-090, São Paulo, SP, Brazil*

Abstract. We present a short review of exotic charmonium production in heavy ion collisions. We discuss predictions for the production cross section of several of these states in ultra-peripheral Pb-Pb collisions at the LHC. The experimental study of these processes is feasible and can be used to yield valuable information about the structure of multiquark states. We also address $X(3872)$ production in central Pb-Pb collisions. In particular, we discuss the suppression of $X(3872)$ during the hadron gas phase. Finally, we comment on the very recent CMS data on the $X(3872)$ yield in Pb-Pb collisions.

1 Introduction

During the past decade, the existence of exotic hadrons has been firmly established [1, 2]. These hadrons are not simply made by quark-antiquark or by three quarks. They are genuine multiquark states but we do not know yet how they are organized. Among the proposed configurations, the meson molecule and the tetraquark are the most often discussed. A tetraquark is compact and the interaction between the constituents occurs through color exchange forces whereas a meson molecule is an extended object and its constituents are weakly bound by meson exchange. In principle the new exotic states can be tetraquarks, meson molecules or tetraquark-molecule mixtures. Indeed this mixed approach has led to the best description of the $X(3872)$. In Ref. [3] the mass and strong decay width were very well reproduced assuming that the $X(3872)$ has a $c\bar{c}$ component with a weight of 97 % and a $D\bar{D}^*$ component with 3 % weight. Even if the best description is given by a mixture it is still very important to understand the individual role played by each component.

The production of exotic particles in hadronic colliders is not yet well understood. It has been argued [4, 5] that it is difficult to produce molecules in p-p collisions. In a pure molecular approach the cross section for $X(3872)$ production estimated in [4, 5] is two orders of magnitude smaller than the measured one. However, in [6] the procedure to obtain these estimates was criticized and the molecular picture was reconciled with data. In Ref. [7] it was shown that the best description can be achieved with a charmonium-molecule combination, i.e. $\chi'_{c1} - D\bar{D}^*$, in which the $c\bar{c}$ component is of the order of $28 - 44$ %. An attempt to explain the data in the tetraquark approach was performed in [8].

Relativistic heavy ion collisions offer a new testing ground for the ideas about the structure of exotic charmonium states. In ultra-peripheral collisions the photon flux is extremely high and one can look for states (X) produced through photon-photon fusion ($\gamma\gamma \to X$) in a very clean environment. In central collisions the number of produced charm quarks is very large and so is the expected exotic

[a] e-mail: navarra@if.usp.br

charmonium yield. In the next section we briefly review the existing results on the exotic charmonium production cross section in ultra-peripheral collisions. In the following section we focus on the production of $X(3872)$ in central collisions. We then give a summary of the situation and try to list the next steps in the field.

2 Ultra-peripheral heavy ion collisions

The strong electromagnetic fields generated by these ions allow the production of a meson in photon-photon interactions. The idea of studying exotic meson production in UPCs was pioneered in [9]. Later, in Ref. [10], the same formalism was applied to the production of mesons and exotic states in proton-proton collisions. In [11] these calculations were updated and extended to proton-proton and nucleus-nucleus collisions at LHC energies, focusing on photon-photon production of the heavy exotic charmonium states and including $X(3915)$, $Z(3930)$ and $X(4160)$. All the ingredients of these calculations are fixed with the exception of the two-photon decay width of the exotic state. In principle, tetraquark and molecular configurations would yield quite different numbers for the decay widths, which would yield quite different production cross sections. The two-photon decay width of the exotic states has been calculated in the molecular approach in several works [12]. Unfortunately the theoretical predictions of the tetraquark model are not yet available.

In [11] all the charmonium states were considered for which either a measurement or a theoretical estimate of the decay width is available. Moreover, the two possible assignments, 0^{++} and 2^{++}, for the states $X(3940)$ and $X(4140)$ were considered.

State	Mass	$\Gamma_{\gamma\gamma}^{theor}$(keV)	$\sigma_{b_{min}}$ (μb)		σ_F (μb)	
			2.76 TeV	5.5 TeV	2.76 TeV	5.5 TeV
X(3940), 0^{++}	3943	0.33	4.2	8.2	6.5	11.8
X(3940), 2^{++}	3943	0.27	17.2	33.6	26.5	48.4
X(4140), 0^{++}	4143	0.63	6.5	12.9	10.2	18.7
X(4140), 2^{++}	4143	0.50	26.0	51.2	40.3	74.3
Z(3930), 2^{++}	3922	0.083	5.4	10.5	8.3	15.2
X(4160), 2^{++}	4169	0.363	18.4	36.4	28.6	52.7
$Y_p(3912)$, 2^{++}	3919	0.774	50.5	98.6	77.9	142.2
X(3915), 0^{++}	3919	0.20	2.6	5.1	4.0	7.34

Table 1. Cross sections for exotic meson production in PbPb collisions using the theoretical decay rates published in Ref. [12]. $\sigma_{b_{min}}$ and σ_F refer to different assumptions for the nuclear form factors.

In Table 1 we show the predictions for the cross sections for the production of several exotic mesons in Pb-Pb collisions at $\sqrt{s}$ = 2.76 TeV and $\sqrt{s}$ = 5.5 TeV using two types of nuclear form factor. The cross section is enhanced by a factor Z^4 in Pb-Pb collisions. This is reflected in the results, with the cross sections ranging from a few hundred nb up to a few tens of μb.

In order to illustrate the behavior of the cross section with the collision energy $\sqrt{s}$ we take the state $X(3915)$ as example. Fig. 1a shows the behavior of the production cross section in Pb-Pb collisions with $\sqrt{s}$ from 10 GeV to 100 TeV. As it can be seen, this cross section can reach large values at energies covered by the future colliders such as the FCC (Future Circular Collider) and the CEPC - SPPC (Circular Electron Positron Collider - Super Proton Proton Collider). Fig. 1b, shows the rapidity distribution of the $X(3915)$ in Pb-Pb collisions at $\sqrt{s}$ = 5.5 TeV. The width of the rapidity distributions and the total cross sections are the most useful quantities to be compared with theoretical predictions of particle production in UPCs in relativistic heavy ion collisions.

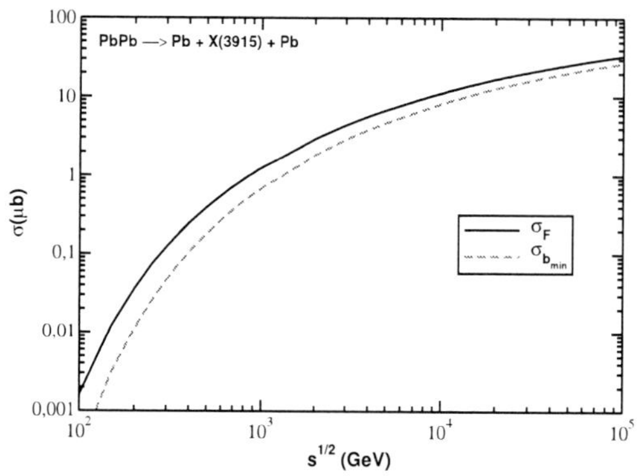

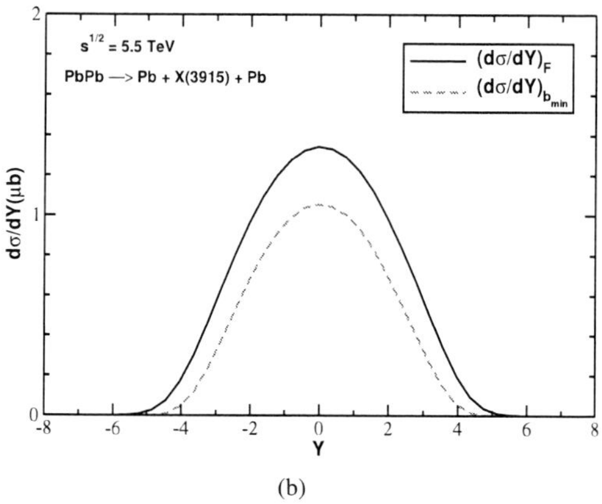

Figure 1. a) Cross section of the process $PbPb \to X(3915)$ as a function of the energy $\sqrt{s}$. b) Rapidity distribution of the produced $X(3915)$ with the same definitions as in a). Figure taken from Ref. [11].

3 Central heavy ion collisions

As pointed out by the EXHIC Collaboration [13], high energy heavy ion collisions offer an interesting scenario to study the production of multiquark states. In these processes, a quite large number of heavy quarks is expected to be produced. Moreover, the formation of quark gluon plasma (QGP) may enhance the production of exotic charmonium states, since the charm quarks are free to move over a large volume and they may coalesce to form bound states at the end of the QGP phase.

One of the main conclusions of Refs. [13] was that, if the production mechanism is coalescence, then the production yield of these exotic hadrons at the moment of their formation strongly reflects their internal structure. In particular it was shown that in the coalescence model the production yield of the $X(3872)$, at RHIC or LHC energies, is almost 20 times bigger for a molecular structure than for a tetraquark configuration. After being produced at the end of the quark gluon plasma phase, the $X(3872)$ interacts with other hadrons during the expansion of the hadronic matter. Therefore, the $X(3872)$ can be destroyed in collisions with the comoving light mesons, such as $X + \pi \to \bar{D} + D$, $X + \pi \to \bar{D}^* + D^*$, but it can also be produced through the inverse reactions, such as $D + \bar{D} \to X + \pi$, $\bar{D}^* + D^* \to \pi + X$. We expect these cross sections to depend on the spatial configuration of the $X(3872)$.

The interactions of the X in a hadronic medium were studied in the framework of $SU(4)$ effective Lagrangians [14]. The authors computed the corresponding production and absorption cross sections, finding that the absorption cross section is two orders of magnitude larger than the production one. In particular, for the largest thermally averaged cross sections they find: $< \sigma v >_{\pi X \to D^* \bar{D}^*} \simeq 30 < \sigma v >_{D^* \bar{D}^* \to \pi X}$. In spite of this difference, the authors of [14] arrived at the intriguing conclusion that the number of X's stays approximately constant during the hadronic phase. Moreover, the terms with anomalous couplings were not included in the calculations. In Ref. [15] the authors showed that the inclusion of the couplings of the $X(3872)$ to charged D's and D^*'s and those of the anomalous vertices, $\pi D^* \bar{D}^*$ and $XD^* \bar{D}^*$, increases the cross sections by more than one order of magnitude. These anomalous vertices also give rise to new reaction channels, namely, $\bar{D} + D^* \to \pi + X$ and $\pi + X \to \bar{D} + D^*$. The changes that the above mentioned contributions can produce in the X abundance (and in its time evolution) were evaluated in Ref. [16] and they are shown in Fig. 2. In the figure the solid line represents the result obtained in [14]. The dashed line shows the result when one includes the couplings of the $X(3872)$ to charged D's and D^*'s. The light-shaded band shows the results obtained with the further inclusion of the diagrams containing the anomalous vertices. The band reflects the uncertainty in the $X\bar{D}^*D^*$ coupling constant, which is $g_{X\bar{D}^*D^*} = 12.5 \pm 1.0$.

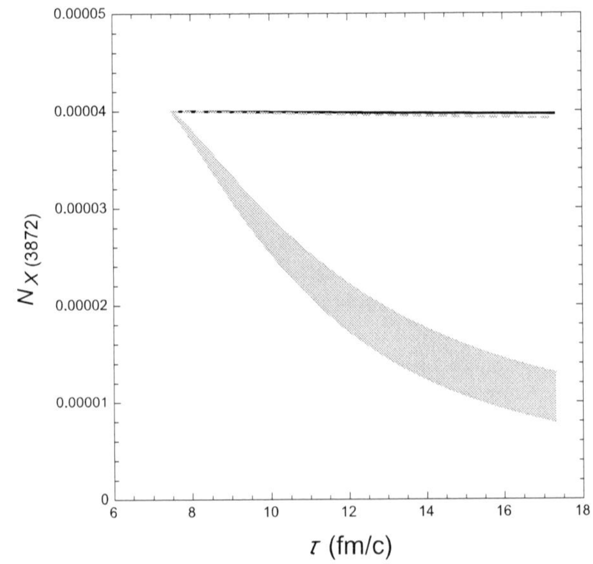

Figure 2. Time evolution of the $X(3872)$ abundance as a function of the proper time τ in central Au-Au collisions at $\sqrt{s_{NN}} = 200$ GeV. The solid line, the dashed line and the light-shaded region represent the results obtained considering only the neutral D's and D^*'s, adding the contribution from charged D's and D^*'s and including contributions from the anomalous vertices respectively. Figure taken from Ref. [16].

As can be seen, without the inclusion of the anomalous coupling terms, the abundance of X remains basically constant. Including the coupling of the X to charged D's and D^*'s does not bring any important change in the time evolution of the X abundance. On the other hand, the inclusion of the anomalous coupling terms modifies the behavior of the $X(3872)$ yield, producing a fast decrease of the X abundance with time.

If the $X(3872)$ would be a tetraquark state it would be formed at the end of the QGP phase and would be massively destroyed during the hadron gas phase. If the $X(3872)$ would be a molecular state, it would be formed *with an abundance 20 times larger* at the end of the QGP phase. Moreover it would be also formed by *meson coalescence at the end of the hadronic phase*. According to [14], at this time the average number of X's, considering it as a $D\bar{D}^*$ molecule, is $N_{X(mol)} \approx 7.8 \times 10^{-4}$, which is about 80 times larger than yield of a tetraquark state at the end of the hadronic phase (see Fig. 2). We can then conclude that the QGP contribution for the $X(3872)$ production (as a tetraquark state and from quark coalescence) after being suppressed during the hadronic phase, becomes insignificant at the end of the hadronic phase. On the other hand, if the $X(3872)$ is a molecular state it will be produced by hadron coalescence at the end of the hadronic phase.

In Ref. [16] it was argued that if the $X(3872)$ were observed in a heavy ion collision it must have been produced at the end of the hadronic phase and, hence, it must be a molecular state. Apparently this seems to be the case!

Very recently, the CMS Collaboration measured the ratio of $1.1 \pm 0.51(\text{stat.}) \pm 0.53(\text{syst.})$ for high - p_T prompt $X(3872)$ and $\psi(2S)$ produced in Pb+Pb collisions at $\sqrt{s_{NN}} = 5.02$ TeV [17]. The central value is about a factor of 10 larger than that in p + p collisions. These data suggest that, not only the $X(3872)$ is observed in heavy ion collisions but its production rate is larger than expected. If this exciting new experimental information is confirmed, it will motivate a refinement of the existing calculations. If the existing theoretical results are confirmed, we will be led to the conclusion that the $X(3872)$ is a meson molecule or, at least, has a molecular component.

4 Future Perspectives

The recent measurement of the $X(3872)$ by CMS, will give great impulse to the search for exotic charmonium. Hopefully this will motivate the experimental searches for other exotic charmonium states both in central and ultra-peripheral collisions. From the theoretical side, it is clear that there are some urgent tasks. The first one is to conclude the study of $X(3872)$ production in proton-proton collisions. We need a reliable calculation of the $X(3872)$ production cross section in the tetraquark approach, we need a final assessment about the existing calculations with the molecular approach and also a confirmation of the results obtained with the mixture approach. We need a calculation of the two-photon decay width of the 0^{++} and 2^{++} charmonium exotic states in the tetraquark approach. With them we will be able to compare the preditions of both approaches for the production cross sections in UPCs. Finally, we need to subject the coalescence model to a rigorous scrutiny and confirm its predictions for the $X(3872)$ (and also for the other exotic states) abundancies in central heavy ion collisions. The near future looks very busy and exciting for the field.

Acknowledgments

This work has been supported by CNPq and FAPESP-Brazil.

References

[1] For a recent review: R. M. Albuquerque, J. M. Dias, K. P. Khemchandani, A. Martinez Torres, F. S. Navarra, M. Nielsen, C. M. Zanetti, J. Phys. G **46**, 093002 (2019).

[2] M. Nielsen and F. S. Navarra, Mod. Phys. Lett. A **29**, 1430005 (2014); M. Nielsen, F. S. Navarra and S. H. Lee, Phys. Rept. **497**, 41 (2010); F. S. Navarra, M. Nielsen and S. H. Lee, Phys. Lett. B **649**, 166 (2007).

[3] R. D. Matheus, F. S. Navarra, M. Nielsen and C. M. Zanetti, Phys. Rev. D **80**, 056002 (2009).

[4] C. Bignamini, B. Grinstein, F. Piccinini, A. D. Polosa and C. Sabelli, Phys. Rev. Lett. **103**, 162001 (2009).

[5] A. Esposito, A. L. Guerrieri, L. Maiani, F. Piccinini, A. Pilloni, A. D. Polosa and V. Riquer, Phys. Rev. D **92**, 034028 (2015).

[6] E. Braaten, L. P. He and K. Ingles, Phys. Rev. D **100**, no. 9, 094024 (2019)

[7] C. Meng, H. Han and K. T. Chao, Phys. Rev. D **96**, 074014 (2017).

[8] F. Carvalho, E. R. Cazaroto, V. P. Gonçalves and F. S. Navarra, Phys. Rev. D **93**, 034004 (2016).

[9] C.A. Bertulani, Phys. Rev. C **79**, 047901 (2009).

[10] V. P. Goncalves, D. T. Da Silva and W. K. Sauter, Phys. Rev. C **87**, 028201 (2013).

[11] B. D. Moreira, C. A. Bertulani, V. P. Goncalves and F. S. Navarra, Phys. Rev. D **94**, 094024 (2016).

https://doi.org/10.1051/epjconf/202023502005

[12] T. Branz, T. Gutsche and V. E. Lyubovitskij, Phys. Rev. D **80**, 054019 (2009); Phys. Rev. D **82**, 054010 (2010); T. Branz, R. Molina and E. Oset, Phys. Rev. D **83**, 114015 (2011).

[13] S. Cho, et al., Prog. Part. Nucl. Phys. **95**, 279 (2017).

[14] S. Cho and S. H. Lee, Phys. Rev. C **88**, 054901 (2013).

[15] A. Martinez Torres, K. P. Khemchandani, F. S. Navarra, M. Nielsen, L. M. Abreu, Phys. Rev. D **90**, 114023 (2014), [Erratum: Phys. Rev.D **93**, 059902 (2016)].

[16] L. M. Abreu, K. P. Khemchandani, A. Martinez Torres, F. S. Navarra, M. Nielsen, Phys. Lett. B **761**, 303 (2016).

[17] Z. Tang, arXiv:2002.1079

QGP droplet formation in small asymmetric collision systems

Thomas A. Trainor[1]*

[1]University of Washington, Seattle, USA

Abstract. The journal Nature recently published a letter titled "Creating small circular, elliptical, and triangular droplets of quark-gluon plasma" [1]. The basis for that claim is a combination of measured Fourier amplitudes v_2 and v_3 from collision systems p-Au, d-Au and h-Au (helion h is the nucleus of atom ^{3}He), Glauber Monte Carlo estimates of initial-state transverse collision geometries for those systems and hydrodynamic Monte Carlo descriptions of the v_n data. Apparent correspondence between hydrodynamic model v_n trends and data trends is interpreted as confirmation of "collectivity" occurring in the small collision systems, further interpreted to indicate QGP formation. QGP formation in small systems runs counter to pre-RHIC theoretical assumptions that QGP formation should require large collision systems (e.g. central A-A collisions). There is currently available a broad context of experimental data from p-p, A-A and p-Pb collisions at the RHIC and LHC against which the validity of the Nature letter claims may be evaluated. This talk provides a summary of such results and their implications.
[1] Nature Phys. 15, no. 3, 214 (2019).

1 Introduction

A recent letter published in the journal Nature reports observation of "short-lived QGP droplets" in 200 GeV p-Au, d-Au and h-Au (h representing the helion nucleus of ^{3}He) [1]. The claimed observation is based on a combination of Glauber Monte Carlo estimates of initial conditions (IC), hydrodynamic theory evolution from the IC to the hadronic final state and comparison of theory results with measurements of azimuth Fourier components $v_2(p_t)$ and $v_3(p_t)$. The letter concludes: "...hydrodynamical models which include QGP formation provide a simultaneous and quantitative description of the data in three systems." The overall argument is based on five critical assumptions that have been challenged in Ref. [2]. In this presentation I confront the five assumptions with evidence from a broad array of published data that contradict them. The context for comparison is the two-component (soft + hard) model (TCM) of hadron production in high-energy nuclear collisions. A third component inferred from measurements of a *nonjet* (NJ) azimuth quadrupole structure is also introduced.

2 Is a perfect liquid formed in A-A collisions?

Arguments in favor are based on azimuthal asymmetries (v_2, v_3 etc.) and jet quenching (R_{AA} high-p_t suppression) as common manifestations of a dense and flowing QCD medium or QGP

*e-mail: ttrainor99@gmail.com

in A-A collisions. Reference [1] argues by analogy that if some or all of those phenomena are observed in small asymmetric collision systems then QGP may be formed there as well.

Figure 1 (left) shows identified-hadron (PID) $v_2(p_t)$ for pions, kaons and Lambdas from 200 GeV Au-Au collisions. The conventional wisdom is that *mass ordering* at lower p_t indicates the presence of radial flow. However, a simple sequence of transformations takes those data from the first to the second panel [3]. Resulting *quadrupole spectra* for three species plotted in the boost frame fall on the same Lévy distribution with slope parameter $T = 92$ MeV. The transformations assume a fixed boost $\Delta y_{t0} = 0.6$ from lab to boost frame. That result falsifies the notion that most hadrons experience Hubble expansion of a bulk medium. The small minority of hadrons "carrying" the NJ quadrupole have a very different spectrum structure and experience a fixed boost corresponding to an expanding thin shell.

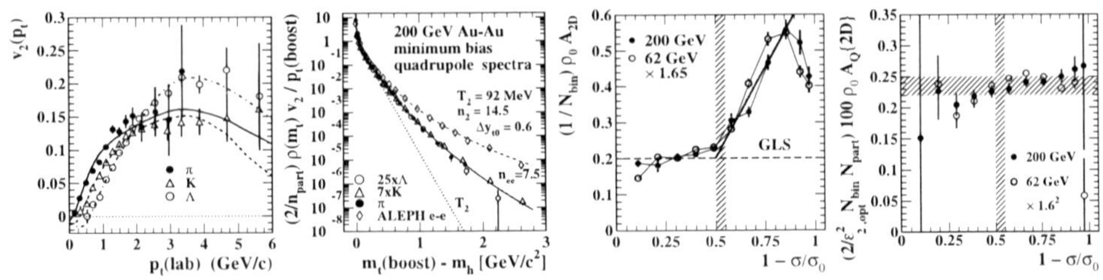

Figure 1. Left: PID $v_2(p_t)$ data on p_t(lab) and corresponding quadrupole spectra on m_t(boost) from 200 GeV Au-Au collisions [3]. Right: Centrality trends for jet quenching and v_2 from 200 GeV Au-Au [4].

Figure 1 (right) shows a comparison (based on 200 GeV Au-Au 2D angular correlations) between jet modification (third) exhibiting a *sharp transition* near 50% centrality (hatched band) and the NJ quadrupole trend (fourth) showing no deviation from a fixed trend from peripheral to central collisions [4]. If there were a common dense QCD medium or QGP, with onset near the sharp transition as suggested by jet modification, it is reasonable to expect that elliptic flow as manifested by v_2 data should show a corresponding marked change in its centrality trend, but no correspondence is observed. In other analysis the jet contribution to p_t spectra is described accurately by pQCD for all Au-Au centrality, albeit fragmentation functions are modified in more-central collisions [5, 6]. That combination, consistent with other evidence, casts strong doubts on formation of a "perfect liquid" in A-A collisions.

3 Is a Monte Carlo Glauber valid for x-A collision geometry?

The hydrodynamic model utilized in Ref. [1] requires estimation of the IC (i.e. the initial transverse geometry) for small asymmetric x-A collision systems. It is assumed that the Glauber Monte Carlo (MC) provides meaningful estimates of the IC for such systems. That assumption can be strongly questioned based on analysis of ensemble-mean $\bar{p}_t$ data from 5 TeV p-Pb collisions.

Figure 2 (left) compares MC Glauber estimates to TCM estimates for nucleon participant-pair number $N_{part}/2$ (first) and ensemble-mean $\bar{p}_t$ (second) vs mean charge density $\bar{\rho}_0 = n_{ch}/\Delta\eta$ from 5 TeV p-Pb collisions [7, 8]. MC Glauber estimates are based on the assumption that p-N collisions within p-Pb collisions are on average equivalent to isolated non-single-diffractive (NSD) p-p collisions. In contrast, TCM estimates are based on an analysis of $\bar{p}_t$ data which indicates that p-N mean multiplicities vary strongly with p-Pb centrality. The MC Glauber estimates (solid points) dramatically fail to describe $\bar{p}_t$ data (open squares) in the second panel whereas the TCM description (solid curve) describes the data accurately.

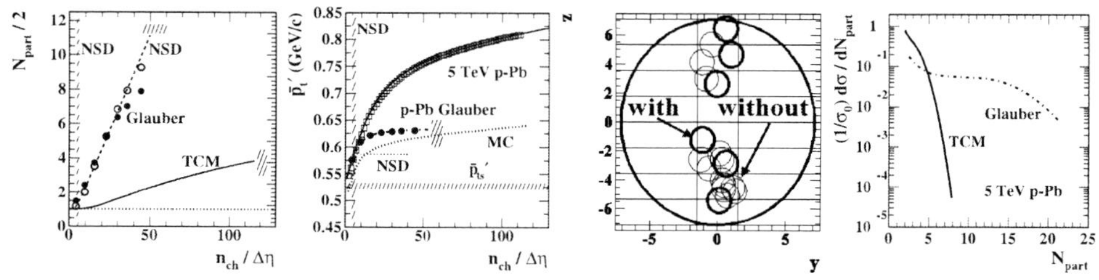

Figure 2. TCM vs Glauber geometry for 5 TeV p-Pb collisions. Left: Estimates of $N_{part}/2$ and $\bar{p}_t$ vs mean charge density $\bar{\rho}_0 = n_{ch}/\Delta\eta$ [7, 8]. Right: Monte Carlo Glauber simulation of p-Pb collisions [9].

Figure 2 (right) provides an explanation for the discrepancy. The third panel shows a Glauber MC simulation of a central p-Pb collision. The total number of small circles denotes the number of geometric *encounters* of the projectile proton with target nucleons. The bold circles denote actual p-N *collisions* allowed by an "exclusivity" condition: the projectile proton can interact with only *one nucleon at a time* [9]. The MC Glauber *with* exclusivity (labeled TCM in fourth panel) is consistent with $\bar{p}_t$ data and other aspects of p-Pb data. The MC Glauber *without* exclusivity produces severely biased IC estimates for x-A collisions.

4 Do A-B p_t spectra provide evidence for radial flow?

Radial flow is conventionally inferred from p_t spectra by qualitative observations of increasing spectrum "hardening" (increased slope parameter) with increasing collision centrality and hadron mass, and by quantitative application to spectra of a blast-wave fit model with parameters T_{kin} and $\bar{\beta}_t$, the latter interpreted as a quantitative measure of radial flow.

Figure 3 (left) shows a TCM analysis of proton spectra from 200 GeV Au-Au collisions [6]. The first panel shows the full spectra while the second panel shows the extracted spectrum hard components (complete minimum-bias jet contribution). The spectra show no evidence for radial flow (boost or translation of the spectrum soft component on y_t leading to *suppression* at lower y_t). The hard components (second panel) indicate suppression at higher y_t (consistent with R_{AA} and "jet quenching") but substantial *enhancement* near $y_t \approx 3.5$ compared to the TCM references (dotted). The hard components below $y_t = 3$ show no change with centrality. These Au-Au proton spectra thus provide no evidence for radial flow.

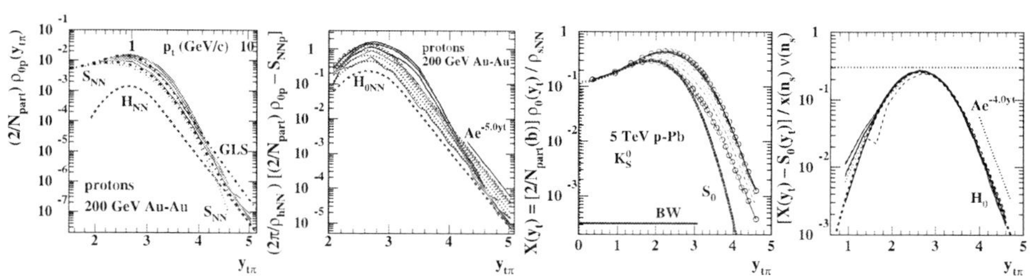

Figure 3. Left: Proton y_t (transverse rapidity $y_{t\pi} = \ln[(p_t + m_{t\pi})/m_\pi]$) spectrum TCM for 200 GeV Au-Au collisions [6]. Right: K_S^0 y_t spectrum TCM for 5 TeV p-Pb collisions [7].

Figure 3 (right) shows K_S^0 spectra from 5 TeV p-Pb collisions extending down to zero momentum (full spectra and jet-related hard components respectively) [7]. If significant radial flow played a role there should be substantial suppression at lower p_t as a consequence of spectra boosted (translated to the right) on y_t. These spectra show no evidence for radial flow. The spectra above 0.5 GeV/c ($y_t = 2$) are dominated by the minimum-bias jet contribution, and the jet contribution (fourth) shows no evidence for jet modification (changes in shape).

5 Do Fourier amplitudes v_n measure flows?

Fourier amplitudes v_n inferred from two-particle angular correlations can be derived from direct model fits to full 2D angular correlations or from Fourier fits to 1D azimuth projections of 2D angular correlations. In the latter case there is the possibility that certain jet-related structures when projected will contribute substantially to inferred v_n as a "nonflow" bias.

Figure 4 (left) shows 2D angular correlations from high-multiplicity 200 GeV p-p collisions (first) and an inferred NJ quadrupole amplitude vs multiplicity soft component $\bar{\rho}_s$ (second) [5]. The quadrupole amplitude ($\propto$ number of correlated pairs) increases $\propto \bar{\rho}_s^3$ accurately over a thousand-fold amplitude increase. Referring to participant low-x gluons with $N_{part} \propto \bar{\rho}_s$ and $N_{bin} \propto \bar{\rho}_s^2$ (dijet production) the p-p quadrupole then varies as $\propto N_{part}N_{bin}$.

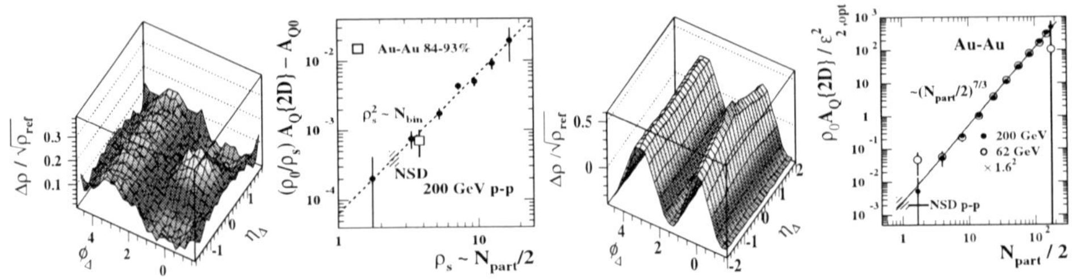

Figure 4. Left: NJ quadrupole n_{ch} trend for 200 GeV p-p collisions [5]. Right: NJ quadrupole centrality trend for 200 GeV Au-Au collisions [10, 11].

Figure 4 (right) shows 2D angular correlations from mid-central 200 GeV Au-Au collisions, dominated by the NJ quadrupole, and the centrality trend for the quadrupole amplitude (divided by ϵ_{opt}^2) vs *nucleon* participant pairs $N_{part}/2$ [10, 11]. The solid line is consistent with the amplitude trend $\propto N_{part}N_{bin}\epsilon_{opt}^2$. In both cases quadrupole amplitudes are inferred by model fits to 2D angular correlations that exclude any significant jet contribution. The simple p-p and Au-Au trends appear to be closely related, and there is no evidence for response to high densities and QGP formation. The p-p $\bar{\rho}_s^3$ trend suggests that the quadrupole arises from a three-gluon interaction and is not a manifestation of hydrodynamic expansion (flows).

Figure 5 (left) shows $v_2\{\text{EP}\}$ (event-plane method) data (open circles) from more-central 200 GeV Au-Au collisions compared with $v_2\{\text{SS}\}$ derived from the same-side 2D jet peak (solid triangles) [11]. $v_2\{\text{2D}\}$ NJ quadrupole data (solid points) are also inferred from 2D model fits. In more-central Au-Au collisions $v_2\{\text{EP}\}$ data are dominated by or entirely determined by jets – i.e. as a Fourier component of the SS 2D jet peak projected onto 1D azimuth.

Figure 5 (right) shows quadrupole amplitude $B_Q\{\text{2D}\}$ derived from $v_2\{\text{2D}\}$ data for 200 GeV Au-Au collisions vs transverse rapidity y_t (third) and quadrupole spectra in the lab frame inferred from those data (fourth, see Fig. 1, left) [11]. The quadrupole spectrum shapes and the source boost $\Delta y_{t0} \approx 0.6$ derived from $v_2\{\text{2D}\}$ data are independent of Au-Au centrality even for peripheral collisions. There is no evidence for the presence of a dense QCD medium.

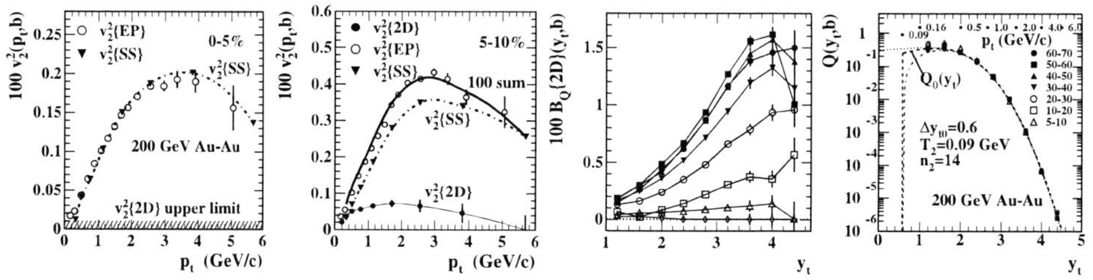

Figure 5. Left: $v_2^2\{\text{EP}\}(p_t)$ trend for 200 GeV Au-Au collisions. Right: $v_2^2\{\text{2D}\}(p_t)$ trend for 200 GeV Au-Au collisions.

6 Do hydrodynamic models have any relation to real A-B collisions?

The apparent success of hydrodynamic models in describing spectra and angular-correlation data from an array of A-B collisions has been viewed as convincing evidence for QGP formation. Hydrodynamic descriptions of x-A data as reported in Ref. [1] are invoked to buttress claims of QGP-droplet formation in those collision systems. However, such interpretations can be questioned.

Figure 6 (left) shows hydrodynamic theory (solid curves) [12] compared to a pion spectrum from 0-5% central 2.76 TeV Pb-Pb collisions (open circles). In the first panel, on linear p_t, the apparent agreement between theory and data at higher p_t is emphasized. However, in the second panel, on logarithmic $y_{t\pi}$, the large discrepancy below $y_t = 2$ ($p_t = 0.5$ GeV/c) falsifies the hydrodynamic model. Strong suppression at lower p_t is expected from a hydrodynamic model (arising from a source boost distribution) but is not observed in data. hydrodynamic theory in effect accommodates the large jet contribution H_{NN} (TCM hard component) peaked near $y_t = 2.7$ ($p_t \approx 1$ GeV/c) [2].

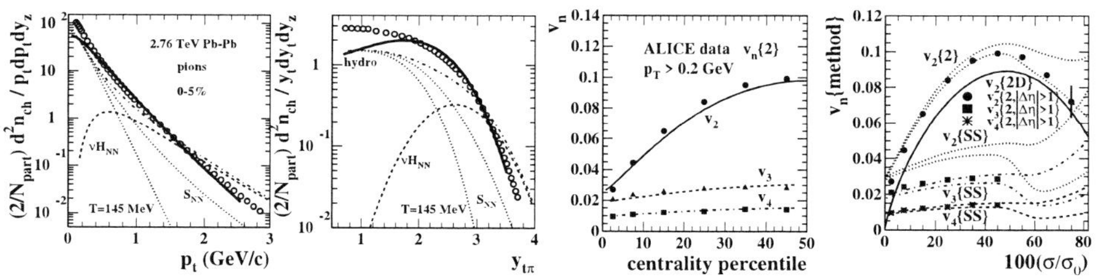

Figure 6. Left: Hydrodynamic theory compared to pion spectrum data from 2.76 TeV Pb-Pb collisions [2, 12]. Right: Hydrodynamic theory compared to $v_2(b)$ centrality data from 2.76 TeV Pb-Pb collisions [12, 13].

Figure 6 (right) shows hydrodynamic theory (curves in third panel) [12] compared to v_n data vs centrality from 2.76 TeV Pb-Pb collisions (points). The hydrodynamic curves appear to describe the data well. However, the fourth panel shows an alternative description of the same ALICE data (albeit with additional points that are missing from the hydrodynamic treatment) [13]. The dotted, dashed and dash-dotted curves in the fourth panel are *predictions* for v_n derived from jet features inferred from model fits to 2D angular correlations [10]. The solid curve represents the NJ quadrupole feature obtained from the same fits. The step up in jet-related trends from peripheral to central collisions (near $\sigma/\sigma_0 \approx 0.5$) corresponds to the "sharp transition" in jet structure occurring in Fig. 1 (third panel) that relates to "jet quenching." Especially for v_3 and v_4 the hydrodynamic model accommodates only jet-related correlation structure. One can conclude from these examples that at least some current hydrodynamic models are sufficiently complex and variable to accommodate data that may have no relation to flows. They are not predictive.

7 Summary

Sections 2-6 respond to five critical assumptions (expressed here in the form of questions) that form the basis for argument in a recent Nature letter introduced to claim formation of QGP droplets in small asymmetric *x*-A collisions. The evidence invoked for QGP formation in small collision systems presents a problem with two possible resolutions: (a) QGP formation is apparently a universal phenomenon in high-energy nuclear collisions, requiring novel theoretical approaches in response, or (b) certain "signatures" conventionally associated with QGP formation in A-A collisions are misinterpreted in any collision system and hydrodynamic descriptions of certain data features do not necessarily correspond to a flow mechanism, may actually be accommodating minimum-bias jet contributions. Experimental evidence presented in this talk responding to the five assumptions calls into question identification of certain data features with QGP formation and challenges hydrodynamic theory descriptions of data that have been interpreted to confirm the presents of flows.

The two-component (soft + hard) model (TCM) of hadron production provides a simple alternative description of hadron production in high-energy nuclear collisions that is consistent with basic QCD and with a broad array of jet measurements. The TCM offers methods and results that enable accurate distinctions among jet contributions (hard) and nonjet contributions (soft) to yields, spectra and two-particle correlations. In *p-p* and *p*-A collisions the TCM provides an exhaustive description of data with no need for exceptional mechanisms (e.g. no flows or jet quenching). In A-A collisions the TCM provides a stable and predictive *reference* against which deviations from linear superposition (e.g. jet modification) can be assessed quantitatively. A third, nonjet quadrupole, component with simple trends in *p-p* and A-A collisions appears to arise from an elementary QCD process (few-gluon interactions).

References

[1] C. Aidala *et al.* (PHENIX Collaboration), Nature Phys. **15**, no. 3, 214 (2019).
[2] T. A. Trainor, arXiv:1905.11585 [hep-ph].
[3] T. A. Trainor, Phys. Rev. C **78**, 064908 (2008)
[4] T. A. Trainor, D. T. Kettler, D. J. Prindle and R. L. Ray, J. Phys. G **42**, no. 2, 025102 (2015).
[5] T. A. Trainor and D. J. Prindle, Phys. Rev. D **93**, 014031 (2016).
[6] T. A. Trainor, Int. J. Mod. Phys. E **17**, 1499 (2008).
[7] T. A. Trainor, arXiv:1812.01151.
[8] T. A. Trainor, arXiv:1801.05862
[9] T. A. Trainor, arXiv:1801.06579.
[10] G. Agakishiev, *et al.* (STAR Collaboration), Phys. Rev. C **86**, 064902 (2012).
[11] D. T. Kettler, D. J. Prindle and T. A. Trainor, Phys. Rev. C **91**, no. 6, 064910 (2015).
[12] C. Gale, S. Jeon, B. Schenke, P. Tribedy and R. Venugopalan, Phys. Rev. Lett. **110**, no. 1, 012302 (2013).
[13] T. A. Trainor, J. Phys. G **40**, 055104 (2013)

EPJ Web of Conferences **235**, 03001 (2020) https://doi.org/10.1051/epjconf/202023503001
ISMD 2019

Constraints of the Low-x Structure of Protons and Nuclei

Thomas Peitzmann[1],*

[1] *Institute for Subatomic Physics, Utrecht University, The Netherlands*

Abstract. I review recent developments in the study of the low-x partonic content of protons and nuclei, with a focus on the latter, as one expects possible deviations from linear QCD evolution to be most pronounced in that case. I give examples of recent theoretical descriptions of HERA measurements with a focus on the role of BFKL evolution. I then concentrate on the status and assumptions of nuclear PDFs and the possibility to use forward particle production at the LHC as further constraint, in particular measurements of open charm and the potential of electromagnetic probes.

1 Introduction

Recent high-energy physics experiments, most notably at the LHC, are providing high precision measurements, which have to be confronted with similarly precise theoretical calculations. For QCD processes, important quantitative ingredients to theoretical calculations, the parton distribution functions, carry considerable uncertainties, which in turn limits the predictive power of theory. In particular, gluon PDFs are only very poorly constrained. Advances on theoretical and experimental sides are needed to improve the status.

In addition, it has been realised that the increasing parton density for small x at moderate Q^2 should become problematic for a description via a purely linear QCD evolution, via DGLAP or BFKL equations. In fact, current frameworks describing initial state parton dynamics rely entirely on DGLAP evolution, so even BFKL has not been well established as a theoretical tool. At high density, partons should start to overlap, leading to recombination processes. The evolution would have to be described by non-linear equations, the JIMWLK equations, or approximately by the BK equations. The corresponding non-linear processes should lead to a saturation of the parton density at small x below a characteristic momentum scale, the *saturation scale* Q_s. As for small x gluons are the most important degree of freedom, this phenomenon is usually termed *gluon saturation*. The transverse gluon density in a nucleus should increase with $A^{1/3}$, so one expects gluon saturation effects to be stronger in heavy nuclei. This is reflected in the expected dependence of the saturation scale on x and A:

$$Q_s^2(x) \approx \frac{\alpha_s}{\pi R^2} xG(x, Q^2) \propto A^{1/3} \cdot x^{-\lambda}, \qquad (1)$$

with $\lambda \approx 0.3$. A potential proof of gluon saturation would be related to the observation of strong nuclear modification of the gluon density and a deviation from linear QCD evolution.

*e-mail: t.peitzmann@uu.nl

2 Proton PDFs

Most of the recent work on proton PDFs is about improved theoretical predictions. As this is not a comprehensive review, I will just give two specific examples, both related to BFKL evolution.

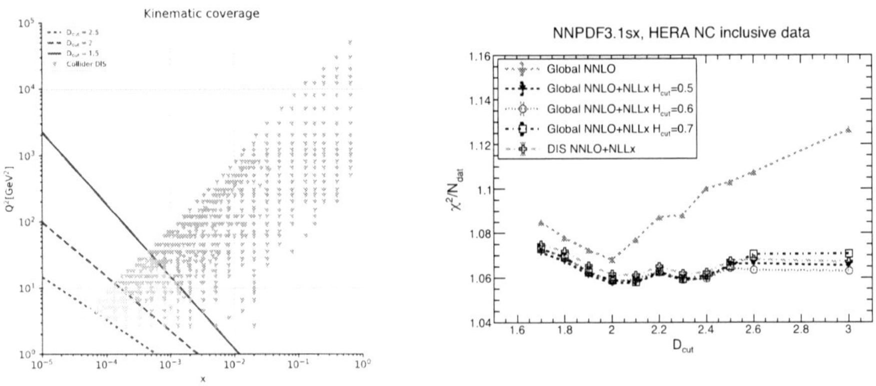

Figure 1. Left: Kinematic coverage of DIS data as a function of x and Q^2 compared to cut values used in the NNPDF fits. Right: χ^2 per degree of freedom for different theory implementations as a function of the value of the kinematic cut used. (From [1].)

In [1] the authors have performed a systematic comparison of different theoretical approaches to the description of DIS data. They have implemented BFKL effects in the NNPDF framework and have performed global fits of HERA data using either fixed-order calculations (NNLO) or calculations using also small-x resummation (NNLO+NLLx). Fits have been performed in different kinematic regions, which is controlled via a parameter D_{cut} as shown in the left panel of Fig. 1. Small values of this parameter exclude a part of the low-x and low-Q^2 region, so that a fixed-order calculation should be sufficient to describe the data, while large values include more of the data where the small-x resummation should become important. The behaviour of the fit quality is displayed in the right panel of Fig. 1, which shows the χ^2 per degree of freedom as a function of D_{cut} for the different calculations. For small values of D_{cut}, all calculations describe the data equally well, i.e. small-x resummation is not really needed, while for large values of of D_{cut} the fixed-order calculation shows a significantly larger χ^2. The behaviour for inclusion of the small x, small Q^2 data strongly suggest that BFKL evolution is necessary to describe the data, and that this particular kinematic region is sensitive to it.

A related analysis is performed in [2] implementing both conventional fixed-order approaches and also approaches including BFKL evolution in the xFitter program, and similar conclusions are found. As an example illustrating their findings, Fig. 2 shows gluon PDFs obtained from different fits using the HERAPDFs and as alternative the NNPDFs. The left panel shows results using NNLO results - here it is apparent that there is a strong tension between the different approaches, particularly at small x. This is remedied when the fits include BFKL evolution as shown in the right panel, where the different approaches are consistent with each other. These findings demonstrate clearly the necessity of incorporating BFKL dynamics also in the context of PDF fits.

No clear sign of saturation has been observed in the behaviour of parton densities in protons. To observe potential deviations from linear evolution one apparently needs additional constraints from measurements at still smaller x, which are currently not available.

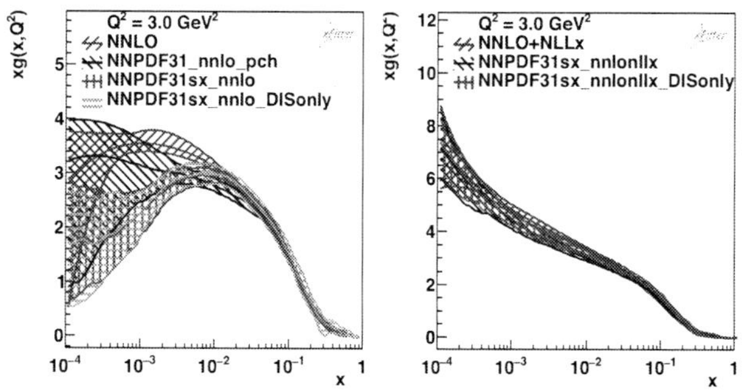

Figure 2. Gluon PDFs resulting from fits of different theoretical descriptions (HERAPDF and NNPDF) to HERA DIS data. Left: Fits using fixed-order calculations. Right: Fits including $1/x$-resummation. (From [2].)

3 Parton distributions in nuclei

Parton distributions are still much less known in nuclei. Very little nuclear DIS data is available, and the kinematic reach is very limited. At the same time, knowledge of the parton densities would be extremely interesting here, as saturation effects should be much more prominent. The current state of theoretical descriptions can be illustrated by the nuclear modification factors of the PDFs, which parameterise the difference of the parton distributions in nucleons inside a nucleus relative to the free nucleon. The left panel of Fig. 3 shows examples of the gluon nuclear modification factor from state-of-the-art nuclear PDF sets.

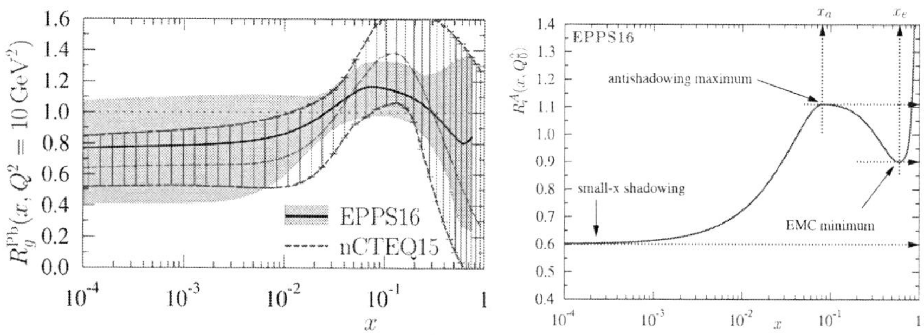

Figure 3. Left: Nuclear modification factors R_g^{Pb} of gluon PDFs in Pb at $Q^2 = 10\,\mathrm{GeV}^2$ as obtained in the EPPS16 and nCTEQ15 frameworks. Right: Analytic shape of R_g^{A} as used in EPPS16. (From [3].)

Both nuclear PDFs show qualitative features that are expected, in particular a so-called antishadowing maximum at moderate x and a value smaller than one for small x, which is usually attributed to shadowing. Both sets show large uncertainties. We are most interested in the features of these distributions at small x, and it is notable that both descriptions show very little to no x dependence of R_g^{Pb} there. This is related to the analytical shapes of functions allowed for the description of R_g^{A}. This is illustrated by the right hand panel of the figure, which displays the parameterisation used in EPPS16 nPDFs. The function levels off to a constant value at small x, which is an invariant feature of the parameterisation used. As the shape of the function is similar for nCTEQ15, we have to assume a similar feature in their parameterisation. The corresponding fit shows smaller uncertainties with usage of the same data, which may be related to a different treatment of flavour dependence or even less flexibility in the shape of the nCTEQ15 parameterisation. Stronger modifications of the PDFs appearing at very small x cannot be described by these functions, which is a limitation that biases the uncertainties obtained. We will revisit this issue again below.

As nuclear DIS data are scarce, one can use particle production at the LHC to constrain the PDFs. In pA collisions one could thus obtain important constraints on nuclear PDFs. Using parton level kinematics at leading order, a measurement of particles with a given transverse momentum p_T at rapidity y in collisions with a CM energy of $\sqrt{s}$ would be sensitive to Bjorken x values as low as:

$$x = \frac{2p_T}{\sqrt{s}} \exp(-y) \tag{2}$$

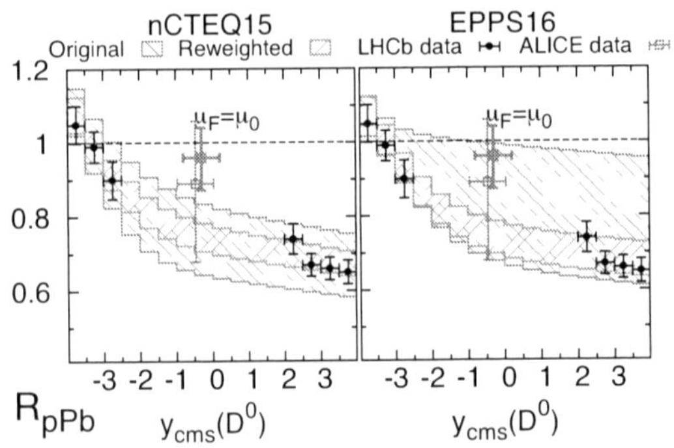

Figure 4. Nuclear modification factor for D^0 production in p–Pb collisions at the LHC as a function of rapidity. Measurements from LHCb and ALICE are compared to refits of nuclear PDF sets nCTEQ15 (left) and EPPS16 (right) for one choice of scales. Bands show the original uncertainties (red) and those after the reweighting including the charm data (grey). The figure is reproduced from [4], in addition, recent data from ALICE [5], which have not been used in the reweighting, are added as blue symbols.

So far, measurements of electroweak probes or jets, which are preferable as the production processes are well understood, are only available at midrapidity and relatively high p_T, so these would constrain rather high values of x only. Forward measurements at not too high p_T have strong advantages. Considering the lack of forward EW probes, open charm measurements are the most promising among existing measurements at the LHC, as charm production should proceed via hard scattering processes. First attempts have been made to include charm production into refits of nPDFs. The results of one such analysis [4] are shown in Fig. 4. Data from LHCb and preliminary data from ALICE have been used in a reweighting procedure, both with nCTEQ15 and EPPS16. At backward rapidity, the nuclear modification factor R_{pPb} is consistent with one, i.e. no modification, while it shows a strong suppression at forward rapidity. The calculations follow nicely the LHCb results, and the inclusion of those data reduces the apparent uncertainty considerably.

Interestingly, R_{pPb} shows very little variation between midrapidity and forward rapidity. While this appeared to be consistent with data in view of the preliminary ALICE results with their large uncertainties, there is some tension between the calculation and the more recent published ALICE results at midrapidity [5], which are added to the figure as blue symbols. The sensitivity range in x of a measurement depends on rapidity (see equation 2), so it seems likely that the lack of rapidity dependence in R_{pPb} can be traced to a lack of x-dependence in the nuclear modification of the PDFs, i.e. R_g^{Pb}. This is a strong hint that the assumptions underlying the functional shapes used in the PDF sets should be revisited.

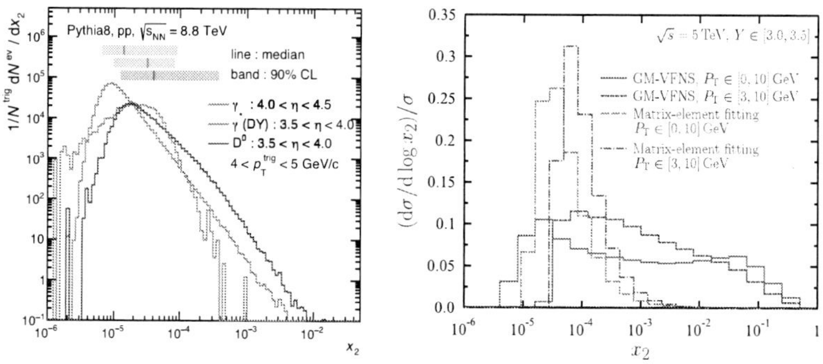

Figure 5. Distributions of x_2 probed in forward production at LHC. Left: Distributions for DY pairs and D^0 mesons in LHCb and for real photons in the proposed forward photon detector (FoCal) in ALICE, from PYTHIA8 calculations. Right: Distributions for D^0 production using different theoretical approximations (from [8]).

When nuclear PDF parameterisations allow for more significant x dependence, the different sensitivity of given probes to specific x-ranges become more important. In this context, it is interesting to compare the sensitivity of different probes - this has been done in Fig. 5. The left panel shows the distributions for different probes of similar p_T as obtained from PYTHIA simulations. All probes are sensitive to values of $x < 10^{-5}$, but the distributions have significant tails towards very large x. For measurements possible within LHCb, DY pairs would have a strong advantage from their x-sensitivity, however, they are sensitive to gluons only at next-to-leading order, and, likely more important, their cross section is extremely small making a measurement very challenging. A measurement of real photons, as it should become possible with the proposed forward calorimeter (FoCal) in ALICE [6, 7], would have an x-sensitivity similar to DY, but would not suffer from those disadvantages.

The usage of charm measurements for PDF fits may suffer from another complication. The authors of [8] have shown that a more realistic study can be done using the variable flavour number scheme (VFNS), and the corresponding x-sensitivity distributions show an even much stronger tail towards large x, as displayed in the right panel of Fig. 5. This suggests, that the sensitivity of charm production to very small x may be much more limited than previously thought. The situation is even more complex, when considering the hints for final state modification of charm distributions, as seen in the nuclear modification [5] and the values of elliptic flow v_2 [9].

Recently, the NNPDF collaboration also provided a nuclear PDF set [10]. The neural network approach allows for high flexibility, as it doesn't use a fixed analytical shape. This leads to even much larger uncertainties of nuclear PDFs, in particular at small x. The left panel of Fig. 6 shows the nuclear modification factor of the gluon density in Pb as a function of x for three different theoretical frameworks, when fitted to the same data [10]. Results for nCTEQ15 and EPPS16 show the same feature as mentioned before, i.e. no x-dependence at low x, and relatively narrow error bands. The fit result from the nNNPDF set shows significantly larger error bands. This demonstrates the impact of larger flexibility of the PDF parameterisation, and it strongly suggests that assumptions entering the nCTEQ15 and EPPS16 frameworks are too restrictive to allow for realistic error estimates.

While it is not clear, whether the recipe used by nNNPDF is fully justified, this seems to be a more promising approach. This has therefore been used in [11] for more recent estimates of the performance of future measurements related to nuclear PDFs as shown in the right panel of Fig. 6. Here the grey band again shows the uncertainties for a fit of nNNPDF to nuclear DIS data, where uncertainties are very large already for $x < 10^{-2}$. Using data of the future EIC facility, the uncertainties would be reduced to the green band, with strong constraints down to $x \approx 10^{-3}$ – still, below this, uncertainties remain very large. A measurement of direct photons at forward rapidities at the LHC with the proposed FoCal detector would provide much stronger constraints still, significantly reducing uncertainties down to values as low as $x \approx 10^{-5}$.

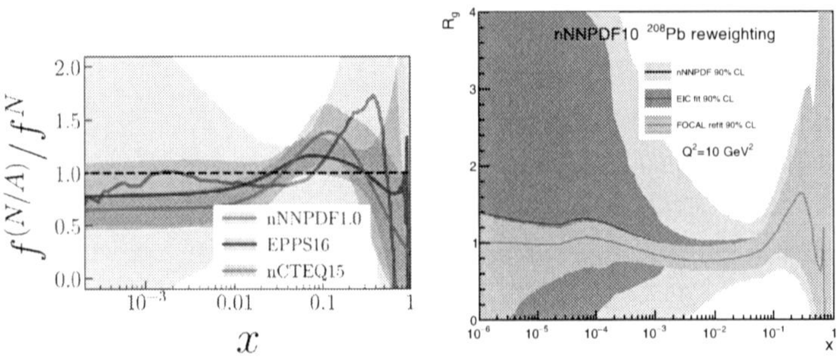

Figure 6. Nuclear modification factor R_g^{Pb} of gluon PDFs for Pb nuclei at a scale of $Q^2 = 10\,\mathrm{GeV}^2$ with their 90% CL uncertainty bands. Left: A comparison of R_g^{Pb} as obtained in fits of different theoretical frameworks to the same data. (From [10].) Right: Results using nNNPDFs with existing DIS data (grey band) and with an additional reweighting with data from future facilities: the EIC (green) and the FoCal detector in ALICE (light red). (From [11]).

4 Conclusion

The small-x region for proton PDFs is used by theorists to demonstrate the necessity of BFKL evolution, however, no proof of non-linear evolution has been observed. Nuclear PDFs with their much larger uncertainties should profit strongly from using LHC data as constraints. However, it is important to revisit the assumptions on small-x behaviour in the theoretical descriptions to allow for realistic estimates of the uncertainty. Forward measurements at LHC, in particular those of real photons, have likely the best potential to provide information on the gluon density at very small x.

References

[1] R.D.Ball et al., Eur. Phys. J. C **78** 321 (2018).
[2] H.Abdolmaleki et al., Eur. Phys. J. C **78** 621 (2018).
[3] K.J.Eskola, P.Paakkinen, H.Paukkunen, and C.A.Salgado, Eur. Phys. J. C **77** 163 (2017).
[4] A.Kusina et al., Phys. Rev. Lett. **121** 052004 (2018).
[5] ALICE Collaboration, submitted to JHEP, preprint arXiv:1906.03425.
[6] T. Peitzmann, PoS HardProbes **2018** 062 (2018), preprint arXiv:1812.07246.
[7] ALICE Collaboration, note ALICE-PUBLIC-2019-005, https://cds.cern.ch/record/2696471.
[8] K.J.Eskola, I.Helenius, P.Paakkinen, and H.Paukkunen, preprint arXiv:1906.02512.
[9] A. M. Sirunyan et al., CMS Collaboration, Phys. Rev. Lett. **121** 082301 (2018).
[10] R.A.Khalek, J.J.Ethier, and J. Rojo, Eur. Phys. J. C **79** 035201 (2019).
[11] M.van Leeuwen, preprint arXiv:1909.05338.

EPJ Web of Conferences **235**, 03002 (2020) https://doi.org/10.1051/epjconf/202023503002
ISMD 2019

Recent results on cold-QCD from RHIC

Qinghua Xu[1,⋆], for the STAR Collaboration

[1] *Institute of Frontier and Interdisciplinary Science & Key Laboratory of Particle Physics and Particle Irradiation of Minstry of Education, Shandong University, Qingdao, Shandong, 266237, China*

Abstract. Polarized proton-proton collisions at the Relativistic Heavy Ion Collider (RHIC) provide unique opportunities to study the spin structure of the nucleon. We will highlight recent results on the nucleon spin structure from the STAR and PHENIX experiments at RHIC: (1) A sizable gluon polarization in the proton is measured with longitudinal double spin asymmetries of jet and hadron production; (2) Longitudinal single spin asymmetries in W boson production improve constraints on the sea quark polarization. The new spin asymmetry results for W boson confirmed the SU(2) flavor asymmetry of the light sea quark polarization in the proton; (3) Transverse spin effects in hadronic systems offer new implications on parton distribution functions in the collinear and transverse momentum dependent frameworks. We will also discuss near term plans for the STAR forward detector upgrade and prospects for proton-proton and proton-ion collisions in the years beyond 2021 at STAR.

1 Introduction to RHIC spin program

In addition to colliding heavy ions, RHIC at Brookhaven National Laboratory is the world's only polarized proton-proton (pp) collider. By colliding high-energy beams of polarized protons, RHIC provides unique opportunities for exploring the spin structure of the proton [1]. In 1988 the EMC experimental results showed that the contribution from quark/antiquark spin to nucleon spin is surprisingly smaller than expected, leading to the so called "spin puzzle". The results from polarized Deep-Inelastic-Scattering (DIS) experiments in the past 30 years have shown that the spins of the quarks and antiquarks account for only about 30% of proton's spin in the measured x-range.

Since 2001, RHIC has been providing pp collisions at center of mass energies of 200 GeV and 500 GeV with beams longitudinally or transversely polarized. There have been two main detectors, PHENIX and STAR, at RHIC to perform spin experiments as well as heavy ion ones. PHENIX completed its data taking in 2016. STAR is currently the only experiment running at RHIC.

In the following, we will highlight recent results on nucleon spin structure from the STAR and PHENIX experiments. RHIC spin experiments have been providing information on how much gluon spin contributes to the spin of the proton, which is observed to be sizable. Significant constraints on the sea quark polarization are also obtained beyond the semi-inclusive DIS measurements, in particular the SU(2) flavor asymmetry of the light sea quark polarization is confirmed through W boson production. Information on how quarks move around inside the proton can be gained by studying

⋆e-mail: xuqh@sdu.edu.cn

transverse spin effects at RHIC, which offer new implications on parton distribution functions in the collinear and transverse momentum dependent (TMD) frameworks. Finally we will discuss future plans for the STAR forward detector upgrade and prospects for pp and proton-ion (pA) collisions in the years beyond 2021 at RHIC.

2 Recent highlights from RHIC spin

2.1 Gluon polarization determination with jet/hadron production

The gluon polarization was once expected to account for most of the "missing" part of the proton spin. In lepton-nucleon DIS process, it can be determined through scaling violations and gluon splitting process, but was limited by kinematic range and statistics. In pp collisions at RHIC, the gluon helicity distribution $\Delta g(x)$ can be accessed with strongly interacting probes, in jet or hadron production, by measuring the longitudinal double-spin asymmetries (A_{LL}),

$$A_{LL} = \frac{(\sigma^{++} + \sigma^{--}) - (\sigma^{+-} + \sigma^{-+})}{(\sigma^{++} + \sigma^{--}) + (\sigma^{+-} + \sigma^{-+})} = \frac{1}{P_1 P_2} \frac{(N^{++} + N^{--}) - R(N^{+-} + N^{-+})}{(N^{++} + N^{--}) + R(N^{+-} + N^{-+})}, \quad (1)$$

where σ denotes the cross section for jet/hadron production and the superscripts "+" or "-" denote the helicities of the two proton beams. P_1 and P_2 are the beam polarizations, $R = (\mathcal{L}^{++}+\mathcal{L}^{--})/(\mathcal{L}^{+-}+\mathcal{L}^{-+})$ is the relative luminosity ratio, and N's are the corresponding jet/hadron yields. In QCD, assuming factorization, A_{LL} is proportional to the convolution of parton helicity distributions, the double spin asymmetry for the hard partonic scattering, and the fragmentation function for hadron production. A significant fraction of the partonic scatterings is gluon-involved such as gluon-gluon or quark-gluon scattering for jet and hadron production at RHIC, which thus provide sensitivity to the gluon helicity distribution function [1].

The first evidence of non-zero gluon polarization has been provided by A_{LL} measurements for inclusive jets at STAR [4] and for π^0 at PHENIX [5] in pp collisions at 200 GeV. The global analyses from the DSSV and NNPDF groups indicated that the integral of the gluon helicity distribution within the kinematic range of $0.05 < x < 0.2$ could be as large as about 0.2 [2, 3]. To probe the gluon polarization in the lower x region, STAR recently published A_{LL} results for inclusive jets at 510

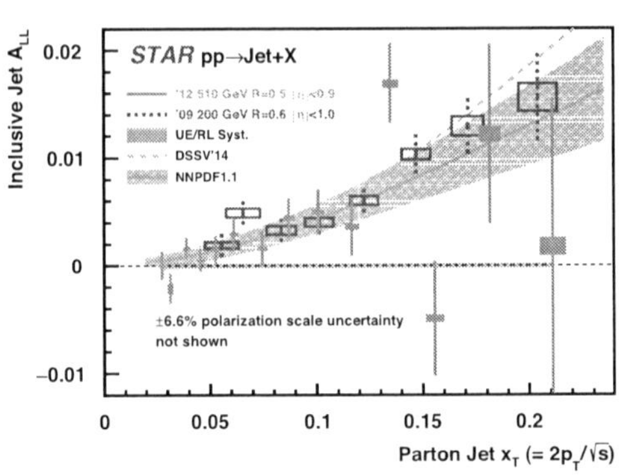

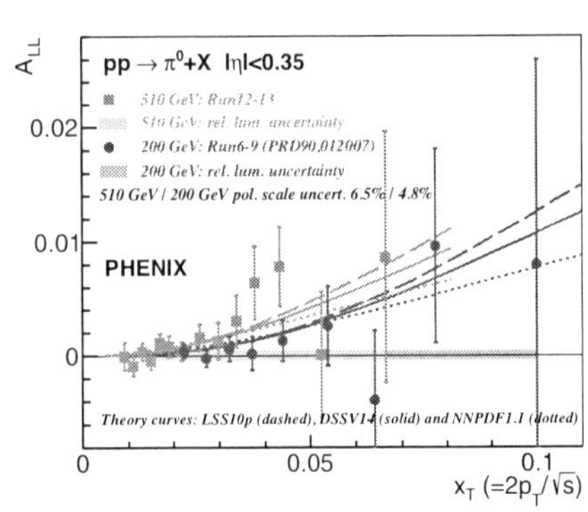

Figure 1. (left) Results of longitudinal double spin asymmetry A_{LL} for inclusive jet production in pp collisions at STAR at 200 GeV and 510 GeV [6]. (right) Results of longitudinal double spin asymmetry A_{LL} for π^0 production in pp collisions at PHENIX at 200 GeV and 510 GeV [7].

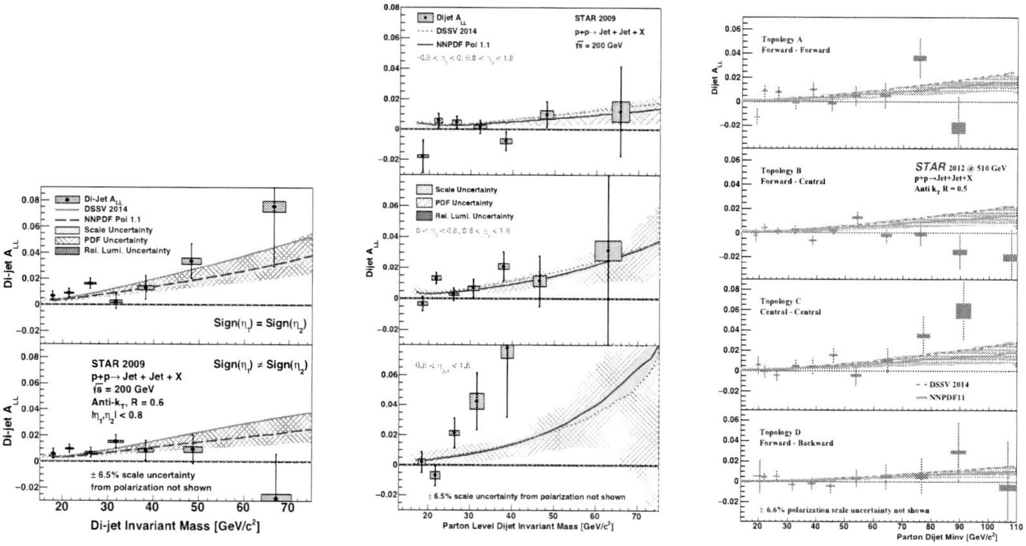

Figure 2. (left) A_{LL} results for di-jet production in mid-rapidity in pp collisions at STAR at 200 GeV [8]. (middle) A_{LL} results for di-jet production in mid & near forward rapidity in pp collisions at STAR at 200 GeV [9]. (right) A_{LL} results for di-jet production in mid-rapidity in pp collisions at STAR at 510 GeV [6].

GeV [6], and PHENIX also published their A_{LL} results for inclusive π^0 at 510 GeV [7]. Both results are shown in Fig. 1, which can provide constraints on gluon polarization with x down to 0.015. In addition, di-jet correlation measurements can provide information on the x-dependence of the gluon helicity distribution. Figure 2 shows the A_{LL} results for di-jet measurements with STAR at 200 GeV and 510 GeV with different detector coverages [6, 8, 9]. The recent STAR di-jet results have been included in a global analysis and indicated their significance in constraining the shape of $\Delta g(x)$ [10].

2.2 Probing sea quark polarization via W boson production

The spin contribution of the sea quarks is also an important piece to the complete understanding of the nucleon spin structure. The production of $W^{\pm}$ bosons in pp collisions with one beam longitudinally polarized provides an unique clean probe of the sea quark helicity distributions without the complication of hadron fragmentation, as in semi-inclusive DIS [1]. At RHIC, the longitudinal single-spin asymmetry A_L for W boson production in pp collisions is defined as: $A_L \equiv (\sigma^+ - \sigma^-)/(\sigma^+ + \sigma^-)$, where $\sigma^{+/-}$ is the W cross section with a positive/negative longitudinal spin orientation of the proton beam. At RHIC, the W's are detected via leptonic decay, which are characterized by an isolated $e^{\pm}$ or $\mu^{\pm}$ with a sizable transverse energy, which peaks near half the W mass.

The final A_L results for $W^{\pm}$ boson versus lepton pseudorapidity from the STAR 2013 data sample [13] are shown in the left panel of Fig. 3, in comparison with STAR results from the 2011+2012 data [12] and the final PHENIX results of A_L for leptons from W/Z decay [14, 15]. The 2013 data sample corresponds to an integrated luminosity of about 250 pb^{-1} with an average beam polarization of about 56%. This is about 3 times the previous data sample taken in 2011 and 2012 [12], and thus the most precise measurement of W A_L at RHIC. The combined STAR data from years 2011-2013 are shown in the right panel of Fig. 3, in comparison with theoretical expectations based on the different inputs of polarized parton densities.

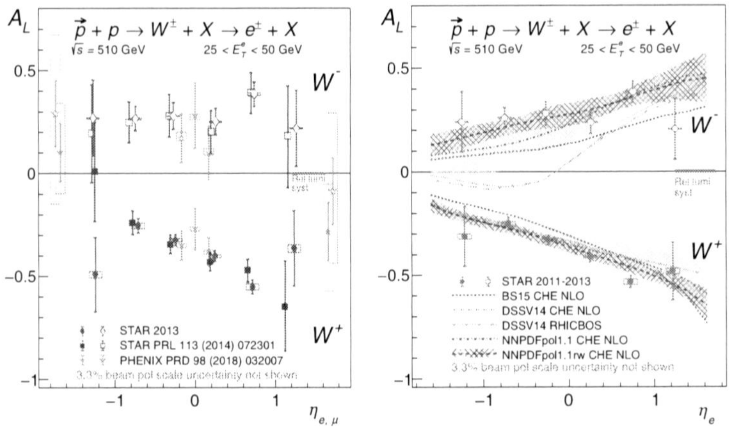

Figure 3. (left)Results of longitudinal single spin asymmetry A_L for W boson from STAR [13] and PHENIX [14, 15] from pp data at 510 GeV. (right) Combined results on A_L for W boson from STAR data 2011-2013 [13], and compared with theoretical calculations.

To assess their impact in constraining the sea quark helicity distribution, the STAR 2013 A_L data were used in the reweighting procedure of NNPDF global fit based on NNPDFpol1.1 parton densities [3]. The results from the reweighting are shown in Fig. 3 (right plot, blue hatched bands), and the uncertainties are significantly reduced compared to before the reweighting. The helicity distributions of $\bar{u}$ and $\bar{d}$ quarks in the proton and their difference $\Delta\bar{u}(x) - \Delta\bar{d}(x)$ from the reweighting are shown in Fig. 4 [13]. The new results confirm the existence of a flavor asymmetry in the polarized quark sea, $\Delta\bar{u}(x) > 0 > \Delta\bar{d}(x)$, in the range of $0.05 < x < 0.25$ at a scale of $Q^2 = 10\,(\mathrm{GeV}/c)^2$. This is opposite to the flavor asymmetry observed in the unpolarized quark distributions, where $\bar{d}(x) > \bar{u}(x)$ over a wide x range has been observed in the Drell-Yan process [16].

2.3 Transverse spin physics results

Significant transverse single spin asymmetries (A_N) have been observed for different hadron production in hadron-hadron collisions over a wide range of colliding energies since the 1970's. STAR measurements have demonstrated the persistence of sizeable A_N for forward π^0 production at RHIC energies [17], where different mechanisms including the higher twist effect, and TMD effects like the Sivers or Collins fragmentation effect could all contribute. Therefore, it is important to study different effects separately for a full understanding of the underlying mechanism.

The hadron production within a jet in pp collisions provides direct access to the Collins fragmentation function, and the corresponding A_N is also connected to proton transversity. Figure 5 shows the results of the mid-rapidity Collins asymmetries as a function of z for charged pions in pp collisions at 500 GeV at STAR [18], where z is the momentum fraction of the jet carried by the pion.

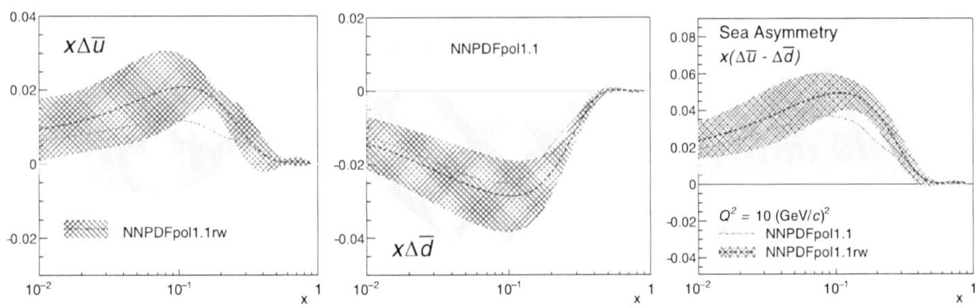

Figure 4. The helicity distributions of $\bar{u}$ and $\bar{d}$ quarks and their difference $\Delta\bar{u} - \Delta\bar{d}$ as a function of x at a scale of $Q^2 = 10\,(\mathrm{GeV}/c)^2$. The green curve/band shows the NNPDFpol1.1 result [3] and the blue curve/band shows the corresponding distribution after the 2013 $W^{\pm}$ data are included by reweighting [13].

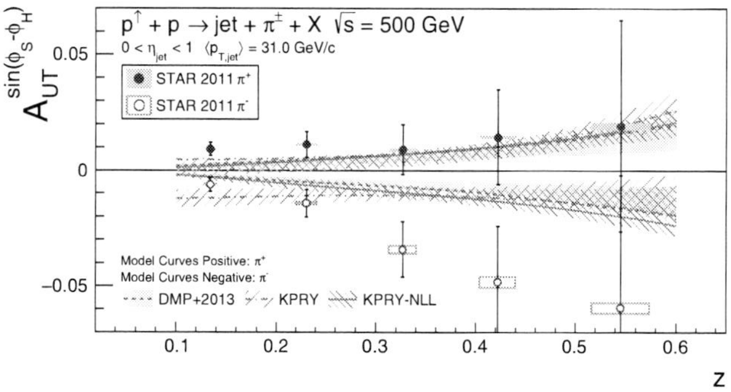

Figure 5. Collins asymmetries as a function of pion z for jets reconstructed with $< p_T >$=31 GeV and $0 < \eta < 1$ in pp collisions at 500 GeV at STAR [18]. The asymmetries are shown in comparison with model calculations.

STAR also made measurements of transverse single spin asymmetries in di-hadron production at mid-rapidity [19], which provide another channel to access proton transversity through the analyzing power of the interference fragmentation functions (IFF). Figure 6 shows the results for the IFF asymmetry $A_{UT}^{sin\phi}$ at mid-rapidity as a function of di-hadron invariant mass at 200 GeV and 500 GeV.

The transverse single spin asymmetry for W and Z production in pp collisions provides an excellent opportunity to test the sign-change of the Sivers function, in comparison to semi-inclusive DIS. STAR published the first results on W boson A_N [20], which indicated a preference for the sign-change. The analysis of a 14 times larger data sample taken in 2017 at STAR is underway, and will provide an important test of the sign-change of the Sivers function.

The possible nucleus dependence of single spin asymmetry has been predicted and also measured at RHIC. PHENIX measurements of the single spin asymmetry of forward neutron in proton-nucleus collisions indicated a possible A-dependence [21], while the STAR measurements of forward π^0 do not show this trend [22]. In addition, STAR made the first measurement of the transverse spin transfer for Λ and $\bar{\Lambda}$ hyperons in transversely polarized pp collisions at 200 GeV [23], which may provide insights into the transversity distribution of the strange quark.

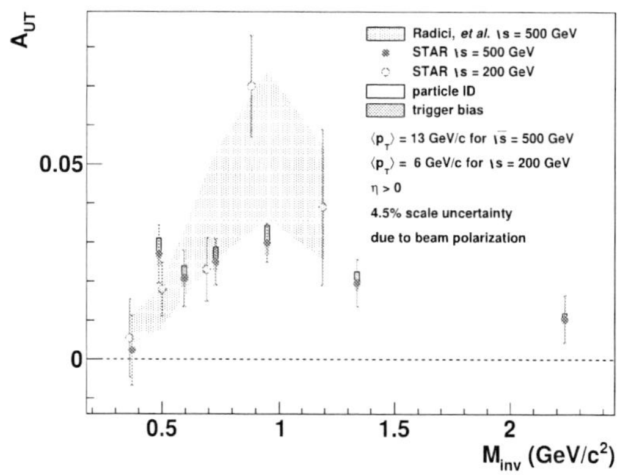

Figure 6. The azimuthal asymmetry A_{UT} for $\pi^+\pi^-$ pairs as a function of invariant mass at 200 GeV and 500 GeV [19], and compared with predictions.

3 Summary and future plan for cold-QCD physics at RHIC

RHIC continues its efforts to deepen our understanding of the nucleon spin structure. Sizable gluon polarization in the proton has been determined from RHIC measurements with polarized proton beams. New results on double spin asymmetries A_{LL} for inclusive jets, di-jets and π^0 production from the STAR and PHENIX experiments in pp collisions at 200 and 510 GeV, are providing further constraints on the gluon helicity distribution. Both PHENIX and STAR experiments also published their final results on the single-spin asymmetry A_L for W boson production from the largest data sample taken in 2013. The new A_L results from STAR further confirmed the SU(2) flavor asymmetry in sea quark polarization, i.e., $\Delta\bar{u}(x) > 0 > \Delta\bar{d}(x)$ in the range $0.05 < x < 0.25$ at $Q^2 \sim 10\,(\mathrm{GeV}/c)^2$. On the transverse spin program, new results on single spin asymmetries of the Collins effect for pion production within a jet at mid-rapidity are reported from STAR. STAR also made measurements of di-hadron spin asymmetries (IFF), which provide insights into the proton transversity distribution. The transverse single spin asymmetry for W and Z production with a large data sample taken in 2017 at STAR will provide a great opportunity to test the sign-change of the Sivers function.

STAR is currently performing a detector upgrade in the forward rapidity region $2.5<\eta<4$, which includes a Forward Tracking System (FTS) and a Forward Calorimeter System (FCS). The FTS consists of 3 layers of silicon mini-strip disks and 4 layers of small-strip Thin Gap Chamber, and the FCS includes both electro-magnetic and hadron calorimeters. The upgrade will be completed in late 2021 and will provide improved calorimetry, tracking, and charge identification and photon hadron separation in the forward region, which will enable the measurements of full jets, Drell-Yan, and prompt photons in pp and pA collisions. Dedicated polarized pp running is expected at RHIC-STAR in the year of 2021/2022, and then STAR will continue to run pp/pA/AA in parallel with sPHENIX beyond 2021. In particular, the proposed pp/pA measurements will be essential to fully realize the scientific promise of the EIC collider [24].

The author is supported partially by the National Natural Science Foundation of China (No. 11520101004).

References

[1] G. Bunce, N. Saito, J. Soffer and W. Vogelsang, Ann. Rev. Nucl. Part. Sci. **50**, 525 (2000).
[2] D. de Florian, R. Sassot, M. Stratmann and W. Vogelsang, Phys. Rev. Lett. **113**, 012001 (2014).
[3] E. R. Nocera *et al.* [NNPDF Collaboration], Nucl. Phys. B **887**, 276 (2014).
[4] L. Adamczyk *et al.* [STAR Collaboration], Phys. Rev. Lett. **115**, 092002 (2015).
[5] A. Adare *et al.* [PHENIX Collaboration], Phys. Rev. D **90**, 012007 (2014).
[6] J. Adam *et al.* [STAR Collaboration], Phys. Rev. D **100**, 052005 (2019).
[7] A. Adare *et al.* [PHENIX Collaboration], Phys. Rev. D **93**, 011501 (2016).
[8] L. Adamczyk *et al.* [STAR Collaboration], Phys. Rev. D **95**, 071103 (2017).
[9] J. Adam *et al.* [STAR Collaboration], Phys. Rev. D **98**, 032011 (2018).
[10] D. De Florian, G. A. Lucero, R. Sassot, M. Stratmann and W. Vogelsang, arXiv:1902.10548.
[11] J. Adam *et al.* [STAR Collaboration], Phys. Rev. D **98**, 032013 (2018).
[12] L. Adamczyk *et al.* [STAR Collaboration], Phys. Rev. Lett. **113**, 072301 (2014).
[13] J. Adam *et al.* [STAR Collaboration], Phys. Rev. D **99**, 051102R (2019).
[14] A. Adare *et al.* [PHENIX Collaboration], Phys. Rev. D **93**, 051103 (2016).
[15] A. Adare *et al.* [PHENIX Collaboration], Phys. Rev. D **98**, 032007 (2018).
[16] R. S. Towell *et al.* [NuSea Collaboration], Phys. Rev. D **64**, 052002 (2001).
[17] B. I. Abelev *et al.* [STAR Collaboration], Phys. Rev. Lett. **101**, 222001 (2008).
[18] L. Adamczyk *et al.* [STAR Collaboration], Phys. Rev. D **97**, 032004 (2018).
[19] L. Adamczyk *et al.* [STAR Collaboration], Phys. Lett. B **780**, 332 (2018).
[20] L. Adamczyk *et al.* [STAR Collaboration], Phys. Rev. Lett. **116**, 132301 (2016).
[21] C. Aidala *et al.* [PHENIX Collaboration], Phys. Rev. Lett. **123**, 122001 (2019).
[22] S. Heppelmann, talk at the 22nd International Symposium for Spin Physics (SPIN2016).
[23] J. Adam *et al.* [STAR Collaboration], Phys. Rev. D **98**, 091103 (2018).
[24] E.C. Aschenauer et al., arXiv:1602.03922.

Study of soft photon yield in pp and AA interactions at JINR

Elena Kokoulina[1,2,★], *Nurlan* Barlykov[1], *Vladimir* Dudin[1], *Vladimir* Dunin[1], *Andrey* Kutov[3], *Vladimir* Nikitin[1], *Vasilii* Riadovikov[4], and *Roman* Shulyakovsky[5]

[1]*Joint Institute for Nuclear Research 6 Joliot-Curie St 141980 Dubna, Moscow region, Russia*
[2]*Sukhoi State Technical University of Gomel Prospect Octiabria, 48, 246746 Gomel, Belarus*
[3]*Institute of Physics and Mathematics Komi SC UrD RAS Kommunisticheskaja st., 24, 167000 Syktyvkar, Russia*
[4]*Institute of High Energy Physics Science sq. 1, 142281 Protvino, Russia*
[5]*Institute of Applied Physics Akademicheskaya str. 16, 220072 Minsk, Belarus*

Abstract. Over 30 years there has been no comprehensive understanding of the mechanism of soft photons (energy smaller than 50 MeV) formation. Experimental data indicate an excess of their yield in hadron and nuclear interactions in comparison with calculations performed in QED. For a more thorough study of this phenomenon at the Nuclotron (a superconducting accelerator in JINR), preliminary measurements have been carried out with using an electromagnetic calorimeter based on BGO crystals. These results are consistent with the world data. In JINR, in connection with the building of a future accelerator complex NICA, it has become possible to carry out such studies in *pp*, *p*A and AA interactions at energies up to 25 A GeV. Our group develops the conception of an heterogeneous electromagnetic calorimeter as "spaghetti" and "shashlik" types based on gadolinium gallium garnet (GaGG) crystals with a low threshold for registration of photons. The first tests of prototypes of them manufactured at JINR on the basis of the GaGG and a mixture of tungstate and copper as an absorber are reported.

1 Introduction

We call soft photons (SPh) such photons that are the direct products of high energy interactions. They are not decay products of secondary particles and their energy is smaller than 50 MeV. More than thirty years ago the excess yield of SPh has been observed [1]. The following studies of this phenomenon have confirmed the excess yield in hadron and nuclear interactions from low up to high energies [2–6]. The excess gets 4–8 fold surplus for reactions with secondary charged particles. The unexpected result has been obtained in e^+e^- annihilation at DELPHI setup [6]. The excess of SPh yield depending on neutral particle multiplicity exceeds the theoretical estimation seventeen times over at 6 neutral pions. Such excess was observed only in the hadron channels. It was absent in the lepton channel: $e^+e^- \to \mu^+\mu^-$. The existing theoretical calculations based on the quantum electrodynamics can not predict and explain this excess. Physics only offer various phenomenological models taking into account quantum chromodynamics and experimental data.

★e-mail: kokoulina@jinr.ru

Why are we interested in SPh? First of all, up to now, the nature of SPh remains obscure. Apparently, they are formed in the region of non perturbative quantum chromodynamics. So physicists build phenomenological models. Among them, the model of P. Lichard and L. Van Hove [7] is the most successful. It is based on hypothesis of the cold quark-gluon plasma (QGP) formation. This model supposes formation of a quark-gluon system which consists of a few quarks, antiquarks and gluons (about 40 partons). These partons are encountering with each other and reradiate soft photons because there is not enough of their energy to produce hadrons. At that, the main reactions are Compton scattering, $q+g \rightarrow q+\gamma$, and annihilation, $q+\overline{q} \rightarrow \gamma$. This model describes well the experimental yield of SPh. There are also other phenomenological schemes explaining this yield [8, 9].

Our SVD-2 Collaboration develops the gluon dominance model (GDM) [10] in order to describe the region of high multiplicity (events with multiplicity way more than an average one). It takes into account two stages of multiparticle production (MP). In accordance to this model, at first stage, called quark-gluon cascade, a plentiful gluon splitting leads to abundance of gluons. We distinguish active and soft gluons. The active gluons have sufficiently high energy and can decay to quark-antiquark pairs. Further, these pairs form secondary hadrons by recombination mechanism. This model is evidence the main sources of secondary hadrons are active gluons. But soft gluons can not split to quark pairs. So they are caught up by new formed hadrons and scattering on their valence quarks with formation of SPh. We have estimated the fraction of such gluons. It is about 50 % of all gluons [11]. We could estimate the SPh emission region in the case of an almost equilibrium state using the black body emission spectrum in proton-proton interactions at the U-70 accelerator with 70 GeV/c proton beams. It represents a cube with side about 4–6 fm. This value exceeds the conventional region of hadronisation and raises some questions. Therefore, study of the soft gluon role will help understand deeper the mechanism of hadronisation.

Our experiment at U-70 was dedicated to study of the collective phenomena of secondary particles in events with high multiplicity. We managed to get the evidence of the Bose-Einstein condensate formation in this region [12]. Our Collaboration would like to answer question: Is there any connection between high multiplicity events and the excess yield of SPh.

And of course, the study of the soft gluon component will be very important for more detail description and understanding of the nucleon structure where an application of pQCD is hampered. We can not help but repeat the important words about them [13] : "Gluons are carriers of the strong force, bind quarks together inside nucleons and nuclei and generate nearly all of the visible mass in the universe. Despite their importance, fundamental questions remain about the role of gluons in nucleon and nuclei. These questions can only be answered with a powerful new electron ion collider (EIC) providing unprecedented precision and versatility." Xiangdong Ji recalled them at the "Confinement and hadron spectrum" conference in Ireland in 2018.

2 First results of SPh study at Nuclotron

Among with theoretical estimations of the SPh yield we carry out active experimental study of them in proton and nuclear collisions at Nuclotron (the JINR's setup) and plan extending them [14] at future accelerator center NICA [15]. This task was one of the main goals of Thermalization project that planned to solve at the U-70 accelerator. We could get down to this study at the ending of our project. Toward this end our Collaboration manufactured the electromagnetic calorimeter (ECal) on the base of BGO (bismuth ortogermanate) scintillators [16]. BGO crystals have a small radiation length $X_0 = 1.12$ cm that permits reducing considerably the volume of our device. Moreover, this scintillator has small sensibility to neutrons. It was important at the measurement of a gamma-radiation. At manufacturing of such calorimeter, the problems of uniform distributions of activator in the crystal volume did not appear.

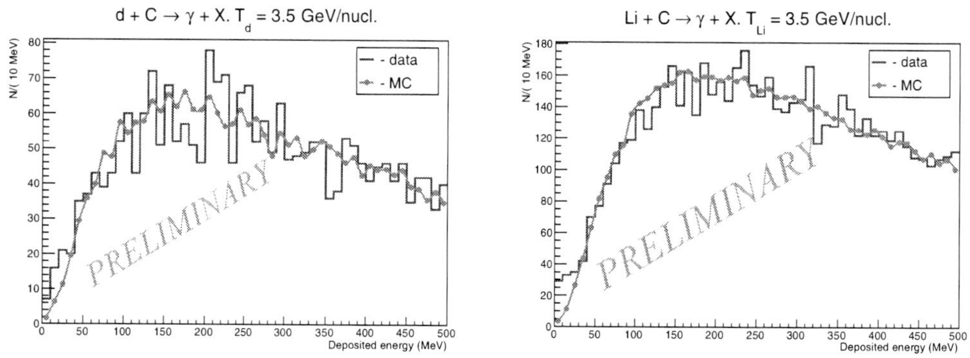

Figure 1. Energy release in BGO calorimeter with a pre-shower in d+C (the left panel) and in Li+C (the right panel) interactions at Nuclotron. Monte Carlo simulation (the red line) and data (the blue line).

This calorimeter presents a square matrix composed of 49 (7×7) counters. The counter with a demultiplier and a preliminary amplifier are placed directly on the panel of PMT, every crystal has the parallelepiped form, 30×30×180 mm^3. 180 mm correspond to 16 radiation lengths. The lateral facets of crystals are wrapped up in Tyvic (the thickness 120 μm) for increasing of light gathering. The PMT 9106SB are used (ET Enterprises). They have 8 dynodes and high quantum efficiency in the green part of spectrum. PMT is glued to the crystal by the optic EPO-TEK 301 glue.

This calorimeter with counters is surrounded by the scintillator detectors of a guard veto-system. It was placed inside of the thermostat. The thermo stabilisation was realised by a cooling system Huber 006B. The temperature of liquid in the thermostat can vary from -20° up to +40° C. After a few trials the temperature has been chosen +18° C.

The plastic veto-detector of charged particles (23×23×1 cm^3) was placed before the crystals and behind it an assembly of 4 plastics of a pre-shower (18×4.5×1 cm^3) was installed. A lead 2 mm-convertor was put between the front-veto and plastics. The trigger was produced by the signal from any 2 of 4 pre-shower counters. There were two large veto-counters in front of the target to forbid a response from the beam halo. This BGO calorimeter with veto-detectors has been laid at the Nuclotron hall at the NIS-GIBS setup.

During the Nuclotron run (2015 year) this BGO calorimeter was set at an angle of 16° relative to the beam direction. The front plane of the crystals was away from the target at the distance 203 cm. The digitisation of plastic scintillators was realised with a CAMAC ADCs (Lecroy 2249A) and TDCs (LeCroy 2228A), the digitisation of the analog signals of calorimeter – by ADC CC-008.

Monte Carlo simulation of this calorimeter has been carried out at the conditions of the last assembly and the beam energy – 3.5 AGeV. Monte-Carlo simulation (uRQMD+Geant-3.21) is used. Geant-4 has shown the same result. After data processing we have obtained SPh spectra of energy release in deuterium-carbon (Fig. 1, the left panel) and lithium-carbon (Fig. 1, the right panel) interactions. In the region of energy below 50 MeV, a noticeable excess over Monte-Carlo simulation has been observed. It agrees well to other SPh experiments [17].

3 Simulation and testing of "spaghetti" and "shashlik" calorimeters

To continue SPh study at JINR's setups we started designing a heterogeneous ECals with a low (about 10 MeV) threshold of registration [18]. At present, physicists use homogeneous, pure crystal ECals

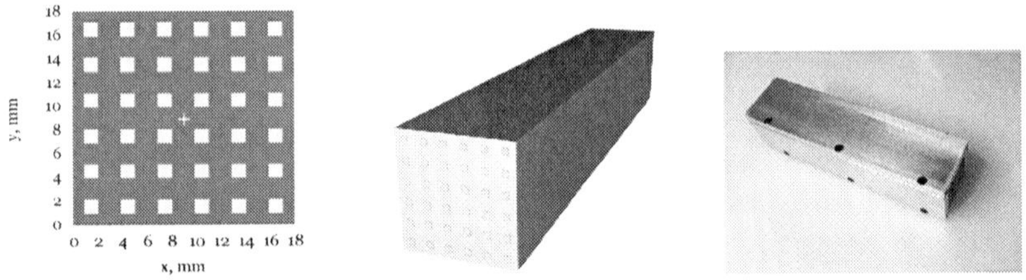

Figure 2. Scheme of the cell for "spaghetti" ECal and its assembly.

and heterogeneous, such types as "spaghetti" [19] and "shashlik" [18]. Physicists use the abridgment SpaCal saying about "spaghetti" ECal. Crystal ECals are very expensive. Their cost gets 50 $/cm^3. In the second case, we lose a little with the energy resolution but we gain with a cost which gets 5÷30 $/cm^3. We would like to occupy a niche between heterogeneous structures ("spaghetti" or "shashlik" type) with light yield ~ 3-6 ph/MeV and crystal detectors with light yield ~10,000 - 40,000 ph/MeV and tune them for the region 10–50 MeV (SPh). It is known that the threshold for shashlik types currently used starts from 100 MeV which is not suitable for our task. The main cause making an impact on such choice was using of heavy crystal in the capacity of scintillator.

We are aimed at the creation of "heavy" ECals (by analogy with heavy ions). We are going to substitute light scintillator material [20] for very dense (heavy) crystal with high specific light yield. It will allow us to build the device that is capable to keep the considerable compactness (space-saving about 30%). To produce the "spaghetti" type we have chosen a mono-crystal of gadolinium-gallium garnet, $Gd_3Al_2Ga_3O_{12}$:Ce (GaGG), as a scintillator and tungsten+copper composite by way of the absorber. We wait the decay time ~ 90 ns; the light yield ~ 45,000-55,000 ph/MeV; price about 25-35 $/cm^3 of volume, and good radiation resistance.

Comparison of properties of several crystals are presented in table 1. We choose crystal GaGG because it is a fast-acting scintillator, has a high light yield. The light yield in GaGG is 4 times more than in BGO. What is more, irradiation tests demonstrate good radiation resistance of GaGG. And we can acquire these crystals from the well-known domestic firm "Fomos-Materials" [21]. It has a long history, rich in scientific researches and technological developments that is competitive with the foreign analogues (and even surpass them) on an equal basis. Thanks to knowledge and professionalism of founding developers it was possible to obtain a material of high quality.

The first prototype of the detector cell SpaCal was an assembly of W+Cu composite plates and GaGG: Ce rods, with the shape of a rectangular parallelepiped: 18×18×100 mm^3. It had 6×6 (1×1×100 mm^3) scintillator rods surrounded by absorber. The surfaces of plates and absorber rods were coated with a 10 μm polymer dim white reflector. We tested two such assemblies. "Spaghetti" technology is a type of the sampling calorimeter with scintillation fibres running along a shower direction [19]. Granularity of whole module is defined by the granularity of read-out system. The scheme of a single cell and its view in an assembly is shown in Fig. 2.

In spring of 2019 year the experimental tests have been carried out by ourselves in Mainz (Mainz Microtron, Germany). Two manufactured cells were irradiated by 41, 51 and 69 MeV-photons. In Fig. 3, both Monte Carlo simulation of such a cell and our experimental result demonstrate the same behaviour in the case of 51 MeV-photon beam. It is seen that in the case of an even number of rods a narrow photon beam hitting in the centre of the cell turns out in the absorber plate. We are only observing the tail of the Gauss distribution.

Table 1. Comparison of scintillator properties

parameters	GaGG	$Bi_4Ge_3O_{12}$	NaI:Tl
light yield, 10^3ph/MeV	57	8	4.5
energy resolution at 662 keV	5.2	12	7.1
decay time, ns	88	300	250
density, g/cm3	6.63	7.13	3.67

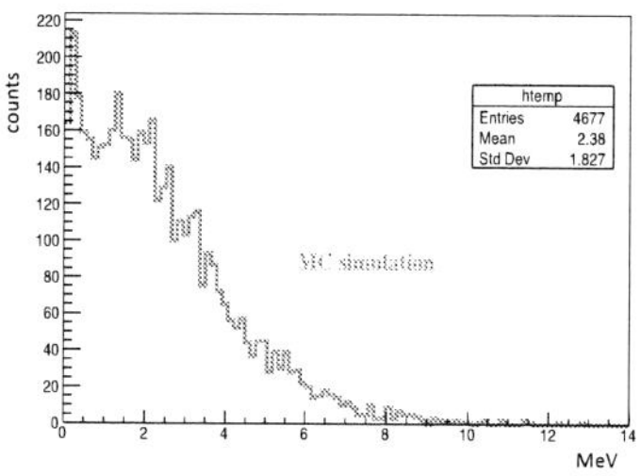

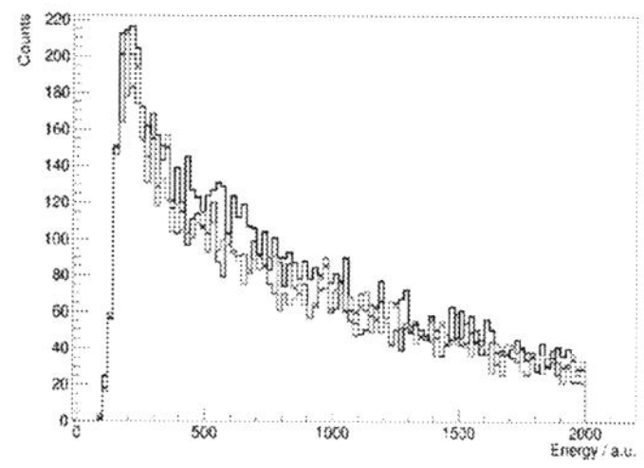

Figure 3. Energy release: Monte Carlo simulation and experimental data for 51 MeV-photon irradiation of SpaCal. The different colours of line correspond to a few small falling angles of a photon beam (the right panel)

To optimize the properties of ECals, modeling of both types has been carried out. We have opted for "spaghetti" simulation the following parameters: the matrix 25×25 GaGG crystal rods with 3×3 mm^2 cross section, the ratio between the tungsten and copper components for the absorber was chosen 5:95. The total size of such a cell was equal to 101×101×150 mm^3. The gap between rods consisted of 1 mm. For "shashlik" we constrained of sixteen GaGG plates (100×100×3 mm^3) and fifteen 2-mm absorber plates with the same ratio of W/ Cu components and the same cross section of every plate like in the SpaCal case. The total thickness of that prototype consists of 78 mm. In Fig. 4 the energy resolution of both types of ECals is presented. It is seen that at that choice of ECal parameters the "shashlik" type gives the energy resolution much better and can be used in our future experimental study.

4 Conclusions

The unique physical program of SPh study [14] can be carried out at accelerator setups of JINR or elsewhere. Simulation both "spaghetti" and "shashlik" types of heterogeneous ECals gives evidence that our program are quite reasonable and feasible. Their characteristics (the energy resolution and others) can be improved by changing of the cross-section of crystal rods, adjusting the absorber density, and optimising the length of ECals for the energy field of SPh. The energy resolution will be 10 and better for photons with energy above 50 MeV with SiPM and the correct scheme for light transporting to a photo detector.

Simulation of "shashlik" type with GaGG as a scintillator and W/Cu as an absorber demonstrate the energy resolution at low energy much better than with "spaghetti" type. Our experimental tests at photon beams are in progress. We hope to carry out these investigations in this year. We are also learning possibilities of using of Glass and Glass Ceramic Stoichiometric and Gd3+ heavy loaded BaO*2SiO2:Ce(DSB:Ce) scintillation material for ECal application. We express our sincere gratitude

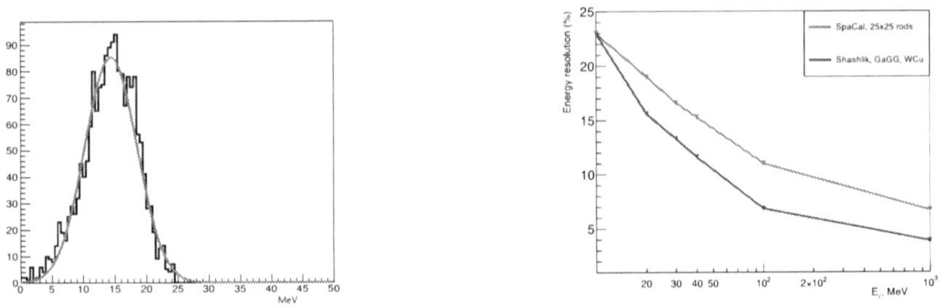

Figure 4. The left panel: spectrum of the energy release in "shashlik" (E_γ=30 MeV). The right panel: the energy resolution in the"spaghetti" (the red line) versus the "shashlik" (the blue line) calorimeters.

to the SVD-2 Collaboration for the fruitful work in the manufacture of SpaCal, getting and discussing the main results.

References

[1] P.V. Chliapnikov et al., Physics Letters **B141**, 276-280 (1984)
[2] J. Schukraft. HELIOS Collaboration, Nuclear Physics **A498**, 79-92 (1989)
[3] S. Banerjee et al. SOPHIE/WA83, Physics Letters **B305**, 82-186 (1993)
[4] A. Belogianni et al. WA91, Physics Letters **B408**, 487-492 (1997)
[5] A. Belogianni et al. WA102, Physics Letters **B548**, 129-139 (2002)
[6] J. Abdallah et al., DELPHI Collaboration, Eur. Phys. J **C47**, 273-294 (2006)
[7] P. Lichard, L. Van Hove, Physics Letters **B245**, 605-608 (1990)
[8] S. Barshay, Physics Letters **B227**, 279-284 (1989)
[9] Wong Cheuk-Yin, Phys. Rev. Lett. **C81**, 064903 (2010)
[10] E. Kokoulina, Acta Phys. Polon. **B35**, 295-302 (2004)
[11] E.S. Kokoulina, Phys. Part. Nucl. Lett. **13**, 74-78 (2016)
[12] Afonin, A.G. et al. SVD Collaboration EPJ Web of Conferences **37**, 06002 (2013)
[13] Xiangdong Ji and Tsung-Dao Lee. Report at the XIIIth Quark Confinement and Hadron Spectrum Aug. 3, 2018, Maynooth University, Ireland
[14] V.V. Avdeichikov et al. *NICA White Paper* (JINR, Dubna, 2013) p.135 http://theor.jinr.ru/twiki-cgi/view/NICA/NICAWhitePaper
[15] V. Golovatyuk, et al. Eur.Phys.J. **A52**, 212 (2016)
[16] E.N. Ardashev et al. Instr. Exp. Tech. **58**, 18 (2015)
[17] J. Schukraft. HELIOS Collaboration. Nucl.Phys. **A498**, 79 (1989)
[18] C. Grupen and B. Shwartz. *Particles detectors* (Cambridge University Press, New York, 2011) 652 p.
[19] D. Acosta et al. Nucl.Instrum.Meth. **A294**, 193 (1990)
[20] Paul Lecoq, Alexander Gektin, and Mikhail Korzhik. *Inorganic Scintillators for Detector Systems* (Second Edition. Springer, Berlin, 2017) 408 p.
[21] Nina S. Kozlova et al. Modern Electronic Materials **4**, 7 (2018)

EPJ Web of Conferences **235**, 03004 (2020) https://doi.org/10.1051/epjconf/202023503004
ISMD 2019

Theoretical Progress at the Frontiers of Small-x Physics

Matthew Sievert[1,*]

[1]University of Illinois at Urbana-Champaign, Urbana, IL 61801, USA

Abstract. In recent years, the theoretical foundations of small-x physics have made significant advances in two frontiers: higher-order (NLO) corrections and power-suppressed (sub-eikonal) corrections. Among the former are the NLO calculations of the linear (BFKL) and nonlinear (BK-JIMWLK) evolution equations, as well as cross sections for various processes. Among the latter are corrections to the whole framework of high-energy QCD, including new contributions from quarks and spin asymmetries. One common element to both of these frontiers is the appearance of collinear logarithms beyond the leading-order framework. The proper treatment of these logarithms is a major challenge in obtaining physical cross sections at NLO, and they lead to a new double-logarithmic resummation parameter which governs spin at small x. In this paper, I will focus on the role of these collinear logarithms in both frontiers of small-x physics, as well as give a brief sample of other recent advances in its theoretical foundations.

1 Introduction: Small-x Physics at Leading Order

In the limit of high energies and fixed transverse momenta, QCD enters what is known as the "small-x regime," with $x \sim p_T^2/s$ where p_T is a particle's transverse momentum and s is the center-of-mass energy squared of the collision (see [1] and references therein for a review). When $x \ll \alpha_s \ll 1$, one has the onset of "small-x kinematics" or "Regge kinematics." In this regime, processes which can be higher order in α_s but leading in x dominate. One example is the transition of Deep Inelastic Scattering from a "knockout" process, in which a virtual photon ejects a quark from the proton, to a dipole process in which the virtual photon fluctuates into a $q\bar{q}$ pair and scatters hadronically on the proton [2–4]. The features of small-x kinematics, like boost invariance, are ubiquitous at top RHIC and LHC energies.

Features of small-x kinematics, however, must be distinguished from signals of small-x *evolution*. Quantum evolution is triggered when systematic large logarithms of $1/x$ begin to spoil the pQCD perturbation series; this occurs when $\alpha_s \ln \frac{1}{x} \gtrsim 1$, or equivalently $x \lesssim \exp[-1/\alpha_s]$. The deviation from pQCD is reflected in the tension of pure DGLAP fits to the proton structure functions at HERA; these can be considered experimental signs of the onset of *linear* small-x evolution [5]. The linear small-x evolution leads to a power-law divergence of the F_2 structure function as $x \to 0$ which is irreconcilable with the unitarity of QCD. This contradiction is indicative of a third regime at even smaller x, when the saturation momentum $Q_s(x)$ induced by the small-x evolution has grown to become a semi-hard scale, $Q_s(x) \gg \Lambda_{QCD}$. This high-energy, high-density regime of QCD is characterized by gluon

*e-mail: msievert@illinois.edu

saturation: the softening of the power-law divergence as $x \to 0$ due to a maximum gluon number density. As a consequence of gluon saturation, this regime admits a powerful semi-classical field theoretic description, known as the color-glass condensate (CGC).

Various formulations of the CGC framework exist, employing different languages to describe the same physics. These include the "background field" or "rapidity factorization" methods [6]; light-front Hamiltonian methods [7]; and functional integral methods [8]. While the languages used to describe small-x physics are very different, these methods are carefully calibrated to reproduce the known results at leading order. Important unanswered questions remain, however, about the these descriptions beyond leading order.

2 Active Frontiers of Small-x Theory

Much of the recent progress in small-x theory focuses on corrections to this LO picture. The calculation of small-x evolution and specific processes at next-to-leading order (NLO) in the strong coupling α_s is picking up speed, and it is of vital importance for the quantitative comparison between theory and experiment at a future electron-ion collider (EIC). More challenging is extending the small-x framework beyond leading *power* in x as $x \to 0$. Since $x \sim p_T^2/s$, this is equivalent to keeping terms which are suppressed by powers of $1/s$. These power-suppressed corrections are often referred to as "sub-eikonal corrections" in this context, with the "eikonal approximation" referring to keeping only leading powers $O\left(x^0\right) \sim O\left(s^0\right)$. Another important frontier of progress consists not of improving the underlying small-x framework, but using it to compute many-body correlations.

2.1 Small-x Theory at NLO

NLO corrections are nominally suppressed by a factor of α_s compared to the leading order. But when combined with small-x evolution equations which resum powers of $\alpha_s \ln \frac{1}{x} \sim O(1)$, a suppression by α_s is equivalent to a suppression by one logarithm: $O\left(\frac{1}{\ln 1/x}\right)$. These corrections are therefore much more numerically important at high energy / low x than fully $O\left(x^1\right)$ power-suppressed corrections. NLO corrections generally preserve the structure of the small-x framework, but can introduce new physical effects. Perhaps the most well-known of these is the running coupling, as implemented in the rcBK code [9]. The running coupling is of course just one of many NLO corrections to the small-x evolution equations, which have been fully computed to NLO now for both the linear and non-linear evolution equations. These NLO evolution equations must then be coupled to specific physical processes computed to NLO accuracy, which has been performed for a handful of observables (e.g., [10]).

Another new physical channel which opens up at NLO involves the QCD odderon [11]. Many small-x observables are expressed in terms of the dipole amplitude

$$D_{xy} = \frac{1}{N_c} \mathrm{tr}\left[V_x V_y^\dagger\right], \tag{1}$$

where N_c is the number of quark colors and V_x is a quark Wilson line at transverse position $\vec{x}_\perp$. In general, one can decompose the dipole amplitude into a piece which is symmetric and antisymmetric under exchange of the quark/antiquark coordinates:

$$D_{xy} = \frac{1}{2}\left[D_{xy} + D_{yx}\right] + \frac{1}{2}\left[D_{xy} - D_{yx}\right] \equiv S_{xy} + iO_{xy}. \tag{2}$$

The symmetric part S_{xy} of the dipole amplitude is known as the Pomeron, while the antisymmetric part O_{xy} is known as the odderon. The Pomeron possesses all the quantum numbers of the vacuum (even under C, P, and T), while the odderon changes sign under C and P. At lowest order in pQCD, the Pomeron consists of two gluons in a color-singlet configuration, while the odderon consists of *three* gluons. The odderon is, therefore, a type of $O(\alpha_s)$ NLO contribution opening up new antisymmetric scattering channels. All of this describes the low-p_T regime $p_T \lesssim Q_s(x)$ where the semihard scale $Q_s(x)$ underlies a perturbative description. At higher $p_T \gg Q_s(x)$ the odderon enters instead as a higher-twist (twist-3) correction instead.

2.2 Sub-Eikonal Corrections to Small-x Theory

The incorporation of $O(1/s)$ sub-eikonal corrections begins to fundamentally change the structure of the small-x formalism. In eikonal (Regge) kinematics $s \to \infty$, a hadronic state is infinitely Lorentz-contracted (i.e., a delta function). But for finite s, the collision changes from being an instantaneous shockwave to occurring over a finite time. In the case of heavy-ion collisions, one can no longer simply compute the "pre-hydrodynamic" phase of the collision using the classical Yang-Mills equations of small-x theory and then switch suddenly to a hydrodynamic description. Instead, one needs a dynamical picture in which hydrodynamics is initiated continuously as the colliding ions pass through each other [12].

Another consequence of a finite-duration collision is that in-medium radiation can occur. In the eikonal approximation, all partonic branchings occur in vacuum before or after the collision. But at sub-eikonal accuracy, branchings which occur *during* the propagation of a high-energy particle through the medium begin to contribute. The resulting medium-induced radiation is given the well-studied non-Abelian Landau-Pomeranchuk-Migdal (LPM) effect for the radiative energy loss of jets [13–17]. An extension of small-x theory to include $O(1/s)$ sub-eikonal corrections will thus make contact with the physics of jet energy loss.

Aside from these *kinematic* changes to particle production at small x, the interactions themselves can change as well. At eikonal accuracy, a high-energy particle propagates through a background field of gluons, with one component of the gluon field A^μ leading to a total color rotation of the traversing particle. For a particle moving along the $+z$ axis through a background field of gluons, this leads to the standard eikonal Wilson line $V_x \equiv \mathcal{P}\left[ig \int dz^+ A^-(z^+, 0^-, \vec{x}_\perp)\right]$, where $v^\pm \equiv \frac{1}{\sqrt{2}}(v^0 \pm v^3)$ are the standard light-front coordinates. This expression is a consequence of the dominance of the background field A^- in, say, Feynman gauge. Beyond eikonal accuracy, other components of the gluon field A^μ begin to enter, which change the structure of the Wilson line. The new field components (notably $\vec{A}_\perp$) transfer different information to the target, including the exchange of spin [18, 19].

Even more challenging is the complete breakdown of the "background field" concept beyond eikonal accuracy. Gluons dominate in the regime of small x, but at $O(1/s)$, quark-mediated interactions play an important role as well. The operators for quark exchange differ significantly from standard Wilson lines, allowing a high-energy parton to change its identity (e.g. between a quark and a gluon) by exchanging quarks with the medium [20, 21]. These complications also directly reflect the fact that baryon stopping, the transfer of valence quarks from the forward direction to mid-rapidity, arises at $O(1/s)$. As such, sub-eikonal corrections also capture the substantial qualitative changes in the theory of heavy-ion collisions from top collider energies down to the RHIC Beam Energy Scan, where all of these difficult sub-eikonal effects are large and non-negligible [22].

2.3 Multiparticle Correlations Within Small-x Theory

Many-body correlations such as the multiparticle anisotropic flow cumulants $v_n\{m\}$ in heavy-ion collisions are quite challenging to compute within the CGC framework. One approach is to use a toy "CGC parton model" [23]. This approach resembles the "hybrid factorization" formalism for particle production at forward rapidities [24], although the CGC parton model is generally applied to mid-rapidity. Valence partons from a dilute projectile become correlated due to scattering in the same correlated color fields. While far from realistic, the CGC parton model does have the virtue of singling out the impact of color correlations alone.

In a complete CGC calculation, particle production at mid-rapidity is dominated by gluons. Full dilute/dense calculations with many particles in the final state are challenging; at present two- and three-gluon correlations exist in the literature [25, 26]. In the simplifying dilute/dilute limit, this has been extended up to four-gluon correlations [27]. Even the structure of two-gluon correlations has yielded new insights recently. While the leading two-gluon correlation function can describe features of the elliptic flow cumulant $v_2\{2\}$, it possesses a mirror symmetry such that it cannot generate triangular flow $v_3\{2\}$ [25]. This feature of the dilute/dense calculation was puzzling, since this mirror symmetry was observed to be broken numerically in "dense/dense" classical Yang-Mills simulations [28]. Recently, it was shown analytically that the first high-density correction to the projectile ("next-to-dilute / dense") breaks the symmetry which of the dilute/dense case and does generate $v_3\{2\}$ [29]. Moreover, the mechanism by which this symmetry is broken – through a relative complex phase arising from rescattering – is analogous to the breaking of PT symmetry in the Sivers function [30].

Finally, while gluons dominate the particle production cross section at mid-rapidity, quark-antiquark pairs can be produced with only a relative penalty of α_s. These quark correlations differ from the gluon correlations in three significant ways. First, because the quarks exist in the fundamental representation of the gauge group $SU(3)$, their dipole amplitude (1) can be complex and lead to a nontrivial odderon contribution (2). Second, because the quarks are fermions, quark correlations will exhibit "Pauli blocking" due to their quantum statistics [31] in contrast to the well-known Bose enhancement of two-gluon correlations [32, 33]. Finally, because quarks carry a variety of conserved charges, correlations among (anti)quarks are responsible for the spatial distribution of conserved charge fluctuations. This feature has recently been implemented in a new Monte Carlo event generator for the initial conditions of conserved charges in ultrarelativistic heavy-ion collisions [34, 35].

3 A Common Thread: New Collinear Logarithms

Many CGC cross sections at NLO suffer from a negativity problem. This problem was first encountered in the context of "hybrid factorization," which describes forward particle production in $p + A$ dilute/dense collisions [24]. While the LO hybrid factorization cross section is well-behaved, the NLO cross section inevitably becomes negative (and unphysical) for large transverse momentum [36]. A closely-related problem arises in the small-x evolution equations themselves: the rate of growth of the dipole amplitude (1) with energy starts out initially positive, but at NLO turns over and goes negative [37]. The source of the negativity was identified to be terms containing an additional large collinear logarithm [38]:

$$\frac{dS_{xy}}{dY} \supset -\frac{\bar{\alpha}_s^2}{4\pi} \int d^2z \, \frac{(x-y)_\perp^2}{(x-z)_\perp^2 \, (z-y)_\perp^2} \left(S_{xz} S_{zy} - S_{xy} \right) \ln \frac{(x-z)_\perp^2}{(x-y)_\perp^2} \ln \frac{(y-z)_\perp^2}{(x-y)_\perp^2} . \tag{3}$$

These numerically large logarithms spoil the stability of the perturbation series and lead to unphysical results. Clearly, to stabilize the perturbation series, these problematic logarithms need to be resummed, but the scheme for how to do so is not unique. One proposal involves completing the first two terms of the perturbation series to a resummed function which matches at NLO, akin to replacing $1 + x \approx e^x$ at NLO accuracy. An alternative scheme modifies the evolution equations to be non-local instead. All of these proposals are equivalent at NLO and differ only by higher-order terms. The result of these schemes are observables which remain physical, but are subject to a scheme dependence on the final results [38].

Physically, the new large collinear logarithms at NLO arise from a corner of phase space associated with lifetime ordering (or angular ordering) of the gluon cascade. While the LO evolution equations do not need to explicitly enforce lifetime ordering at that order, this "kinematic constraint" cannot be ignored at NLO. This same problem also arises for longitudinal spin asymmetries at small x [20]. Because helicity dependence vanishes in the strict eikonal limit, longitudinal spin asymmetries are directly sensitive to sub-eikonal physics. In the small-x evolution of polarized observables, the lifetime ordering condition leads to a second parametrically large logarithm. While the unpolarized small-x evolution is leading-logarithmic (resumming $\alpha_s \ln \frac{1}{x} \sim 1$), polarized small-$x$ evolution is double-logarithmic (resumming $\alpha_s \ln^2 \frac{1}{x} \sim 1$) [39]. Thus lifetime ordering, which enters only at NLO in the unpolarized (eikonal) sector, is a leading-order effect in the polarized (sub-eikonal) sector.

The important difference between these cases is that polarized evolution is governed by a resummation parameter. In the polarized case, the logarithm that results is $\ln \frac{1}{x}$, and the evolution equations must systematically resum all double logarithms of $1/x$. But in the NLO unpolarized case, the logarithms do not correspond to a large external parameter like $1/x$; they are simply numerically unstable large logarithms. As a result, the NLO unpolarized evolution suffers from scheme dependence associated with the non-unique ways of resumming the problematic logarithms. The LO polarized evolution, on the other hand, is uniquely specified by the double-logarithmic parameter $\alpha_s \ln^2 \frac{1}{x} \sim 1$.

The similarities between the NLO unpolarized evolution and the LO polarized evolution go further. In both cases, the ordering of the cascade in longitudinal momentum k^+ makes the lifetime-ordering condition unwieldy. For polarized evolution, this leads to a messy dependence on an auxiliary "neighbor dipole amplitude" even in the large-N_c limit [20]. It is interesting that, in the NLO unpolarized case, simpler equations arise from using an inverse ordering variable k^-; perhaps this could simplify the polarized evolution equations as well. In any case, the common origins of the large double logarithms suggests possible deep connections between the NLO frontier of small-x theory and the sub-eikonal frontier. Perhaps exploring these connections further will lead to better control of both processes.

4 Conclusions

The theory of particle production at small x, long established at leading order and leading power, is now making significant progress on three active fronts: the generalization to NLO, the incorporation of "sub-eikonal" power-suppressed corrections, and multiparticle correlations. While there are a range of very distinct formalisms expressing the theory of small-x physics that all agree at LO, it remains an open question whether these formalisms will continue to agree at NLO. More generally, the program of computing corrections to the LO picture of small-x physics is an essential exercise to establish the validity and convergence of the small-x picture. If this regime of QCD can really be understood as an effective theory at high energies and high densities, then it is crucial to establish that its approximations can be systematically improved and compared to data.

References

[1] Y.V. Kovchegov, E. Levin, *Quantum chromodynamics at high energy*, Vol. 33 (Cambridge University Press, 2012), ISBN 9780521112574
[2] V.N. Gribov, Sov. Phys. JETP **30**, 709 (1970), [Zh. Eksp. Teor. Fiz.57,1306(1969)]

[3] J.D. Bjorken, J.B. Kogut, Phys. Rev. **D8**, 1341 (1973)
[4] L.L. Frankfurt, M.I. Strikman, Phys. Rept. **160**, 235 (1988)
[5] I. Abt, A.M. Cooper-Sarkar, B. Foster, V. Myronenko, K. Wichmann, M. Wing, Phys. Rev. **D94**, 034032 (2016), 1604.02299
[6] I. Balitsky, A. Tarasov, JHEP **10**, 017 (2015), 1505.02151
[7] A. Kovner, M. Lublinsky, Y. Mulian, JHEP **08**, 114 (2014), 1405.0418
[8] L.D. McLerran, R. Venugopalan, Phys. Rev. **D49**, 2233 (1994), hep-ph/9309289
[9] Y.V. Kovchegov, H. Weigert, Nucl. Phys. **A784**, 188 (2007), hep-ph/0609090
[10] K. Roy, R. Venugopalan (2019), 1911.04530
[11] Y. Hatta, E. Iancu, K. Itakura, L. McLerran, Nucl. Phys. **A760**, 172 (2005), hep-ph/0501171
[12] C. Shen, B. Schenke, Nucl. Phys. **A982**, 411 (2019), 1807.05141
[13] L.D. Landau, I. Pomeranchuk, Dokl. Akad. Nauk Ser. Fiz. **92**, 535 (1953)
[14] A.B. Migdal, Phys. Rev. **103**, 1811 (1956)
[15] R. Baier, Y.L. Dokshitzer, A.H. Mueller, S. Peigne, D. Schiff, Nucl. Phys. **B484**, 265 (1997), hep-ph/9608322
[16] M. Gyulassy, P. Levai, I. Vitev, Nucl. Phys. **B594**, 371 (2001), nucl-th/0006010
[17] M.D. Sievert, I. Vitev, Phys. Rev. **D98**, 094010 (2018), 1807.03799
[18] T. Altinoluk, N. Armesto, G. Beuf, M. Martínez, C.A. Salgado, JHEP **07**, 068 (2014), 1404.2219
[19] Y.V. Kovchegov, D. Pitonyak, M.D. Sievert, JHEP **10**, 198 (2017), 1706.04236
[20] Y.V. Kovchegov, D. Pitonyak, M.D. Sievert, JHEP **01**, 072 (2016), [Erratum: JHEP10,148(2016)], 1511.06737
[21] Y.V. Kovchegov, M.D. Sievert, Phys. Rev. **D99**, 054032 (2019), 1808.09010
[22] S. Rao, M. Sievert, J. Noronha-Hostler (2019), 1910.03677
[23] K. Dusling, M. Mace, R. Venugopalan (2017), 1705.00745
[24] G.A. Chirilli, B.W. Xiao, F. Yuan, Phys. Rev. **D86**, 054005 (2012), 1203.6139
[25] Y.V. Kovchegov, D.E. Wertepny, Nucl. Phys. **A906**, 50 (2013), 1212.1195
[26] T. Altinoluk, N. Armesto, A. Kovner, M. Lublinsky, Eur. Phys. J. **C78**, 702 (2018), 1805.07739
[27] Ş. Özönder, Phys. Rev. **D91**, 034005 (2015), 1409.6347
[28] T. Lappi, S. Srednyak, R. Venugopalan, JHEP **01**, 066 (2010), 0911.2068
[29] Y.V. Kovchegov, V.V. Skokov, Phys. Rev. **D97**, 094021 (2018), 1802.08166
[30] S.J. Brodsky, D.S. Hwang, Y.V. Kovchegov, I. Schmidt, M.D. Sievert, Phys. Rev. **D88**, 014032 (2013), 1304.5237
[31] T. Altinoluk, N. Armesto, G. Beuf, A. Kovner, M. Lublinsky, Phys. Rev. **D95**, 034025 (2017), 1610.03020
[32] T. Altinoluk, N. Armesto, G. Beuf, A. Kovner, M. Lublinsky, Phys. Lett. **B751**, 448 (2015), 1503.07126
[33] T. Altinoluk, N. Armesto, D.E. Wertepny, JHEP **05**, 207 (2018), 1804.02910
[34] M. Martinez, M.D. Sievert, D.E. Wertepny, J. Noronha-Hostler (2019), 1911.10272
[35] M. Martinez, M.D. Sievert, D.E. Wertepny, J. Noronha-Hostler (2019), 1911.12454
[36] A.M. Stasto, B.W. Xiao, D. Zaslavsky, Phys. Rev. Lett. **112**, 012302 (2014), 1307.4057
[37] T. Lappi, H. Mäntysaari, Phys. Rev. **D91**, 074016 (2015), 1502.02400
[38] B. Ducloué, E. Iancu, A.H. Mueller, G. Soyez, D.N. Triantafyllopoulos, JHEP **04**, 081 (2019), 1902.06637
[39] R. Kirschner, L.n. Lipatov, Nucl. Phys. **B213**, 122 (1983)

EPJ Web of Conferences **235**, 04001 (2020) https://doi.org/10.1051/epjconf/202023504001
ISMD 2019

Recent results from BABAR

J. William Gary[1,*]

[1]Department of Physics and Astronomy, University of California, Riverside, CA 92521, USA

Abstract. The BABAR Collaboration at SLAC continues to produce a wide variety of publications on interesting and unique topics despite having ceased data-collection operations more than 10 years ago. Presented here are the results of three recent studies. The first is on the direct measurement of the $B^{\pm} \to K^{\pm}X(3872)$ branching fraction, allowing the first determination of the branching fractions of the exotic $X(3872)$ charmonium state. The second is a publication on the first observation of the $D^0 \to K^-\pi^+e^+e^-$ branching fraction. The third is on searches for lepton flavor and lepton number violating decays in D^0 meson decays, resulting in improvements over existing limits by factors that vary from 20 to almost 800 depending on the D^0 decay channel.

1 Introduction

The BABAR detector collected data at the SLAC PEP-II asymmetric e^+e^- collider from 1999 to 2008. Despite having ceased operation more than 10 years ago, the data analysis is still active, with 7-8 papers submitted to journals each of the past several years. The PEP-II collider acted as a "B factory", built to study CP violation in B meson decays. Nonetheless, the general-purpose design of BABAR makes its data sample suitable for a wide variety of other studies, such as the ones discussed here.

Most of the BABAR data were collected at the energy of the $\Upsilon(4S)$ resonance, which decays to a $B\overline{B}$ pair essentially 100% of the time, thus yielding a large sample of B mesons. The final BABAR data sample contains around 471 million $B\overline{B}$ pairs, corresponding to an integrated luminosity of $424\,\text{fb}^{-1}$. Beside B mesons, the PEP-II collider also produced a large number of charmed mesons, because of the strong coupling of charm quarks to the intermediate photons from the e^+e^- annihilation.

The main topic described here is the first measurement of the absolute $B^{\pm} \to K^{\pm}X(3872)$ branching fraction. Recent results on the first observation of $D^0 \to K^-\pi^+e^+e^-$ decays, and a search for rare and forbidden decays of the D^0 meson, are also described.

2 Absolute measurement of the $B^{\pm} \to K^{\pm}X(3872)$ branching fraction

This study [1] was recently submitted to Physical Review Letters.

A number of studies of exclusive 2-body B meson decays of the form $B^{\pm} \to K^{\pm}X_{c\bar{c}}$ have been presented, where $X_{c\bar{c}}$ is a charmonium state. For example, $X_{c\bar{c}}$ could be a J/ψ, $\chi_{cJ}(1P)$, η_c, or $\psi(2S)$ meson, or an exotic state such as the $X(3872)$. A Feynman diagram is shown in Fig. 1. These studies are focused on tests of B hadron dynamics and CP violation. Examples of such studies are given in Refs. [2, 4, 5].

*e-mail: bill.gary@ucr.edu

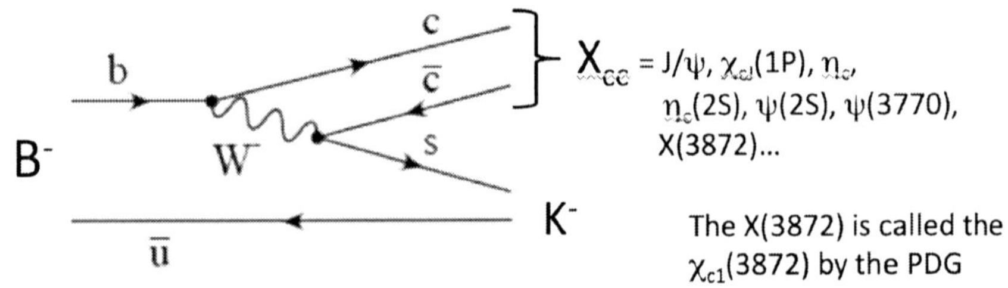

Figure 1. Feynman diagram for $B^{\pm} \to K^{\pm} X_{c\bar{c}}$, where $X_{c\bar{c}}$ is a charmonium state.

In these studies, the $X_{c\bar{c}}$ is reconstructed in its decay to a specific final state "f". Thus these studies do not measure the absolute branching fractions $\mathcal{B}(B^{\pm} \to K^{\pm} X_{c\bar{c}})$ or $\mathcal{B}(X_{c\bar{c}} \to f)$ but rather their product. To obtain the *absolute* branching fraction $\mathcal{B}(B^{\pm} \to K^{\pm} X_{c\bar{c}})$ or $\mathcal{B}(X_{c\bar{c}} \to f)$, the other of the two must already be known. For well established particles like the J/ψ, η_c, etc., this is not a problem, because the $X_{c\bar{c}} \to f$ branching fractions of these particles are well known, and the focus of these studies is on the B hadron branching fraction $\mathcal{B}(B^{\pm} \to K^{\pm} X_{c\bar{c}})$. However, for the exotic charmonium state $X(3872)$, the absolute branching fractions $\mathcal{B}(X(3872) \to f)$ are *not* known. The $X(3872)$ has only been observed in the decay $B^{\pm} \to K^{\pm} X(3872)$ followed by the decay to some specific reconstructed final state $X(3872) \to f$, and we therefore only know the product $\mathcal{B}(B^{\pm} \to K^{\pm} X(3872))\,\mathcal{B}(X(3872) \to f)$. However, knowledge of the absolute values of $\mathcal{B}(X(3872) \to f)$ is of significant importance because it would greatly aid in understanding the nature of the $X(3872)$, as discussed in the following paragraph. The new BABAR analysis aims to measure the absolute $B^{\pm} \to K^{\pm} X(3872)$ branching fraction and thus, in conjunction with the already known product $\mathcal{B}(B^{\pm} \to K^{\pm} X(3872))\,\mathcal{B}(X(3872) \to f)$, allow the absolute $X(3872) \to f$ branching fractions to be determined.

A multitude of exotic charmonium states have been discovered over the past two decades. Their masses and/or decay properties do not correspond to those expected for yet-undiscovered non-exotic states and thus they do not fit into the framework of normal $c\bar{c}$ mesons. These exotic charmonium states have been dubbed the "X", "Y", and "Z" particles. The Y particles are those with $J^{PC} = 1^{-}$, the Z particles are those with isospin 1, and the X particles are all the rest [3]. The $X(3872)$ was the first XYZ particle discovered [4] and is the most studied. There are two main interpretations of the $X(3872)$: that it is a $D^0\overline{D}^{*0}$ molecule, namely two color singlet systems loosely bound together [6], or that it is a point-like tetraquark, namely two tightly bound color triplet systems [7]. Other interpretations include a $\chi_{c1}(2P)$–DD^* hybrid or a $c\bar{c}$–gluon hybrid. (See Refs. [3, 8] for reviews.) For the decay $X(3872) \to J/\psi\pi^+\pi^-$, which is the $X(3872)$ discovery channel [4], the simple tetraquark model predicts a large branching fraction, $\mathcal{B}(X(3872) \to J/\psi\pi^+\pi^-) \gtrsim 50\%$ [6], while the $D^0\overline{D}^{*0}$ molecular model predicts a small one, $\mathcal{B}(X(3872) \to J/\psi\pi^+\pi^-) \approx 10\%$ [9]. Thus, knowledge of the absolute branching fractions of the $X(3872)$ allows the different interpretations to be distinguished and can greatly aid in understanding the nature of the $X(3872)$.

The $X(3872)$ has a narrow width, less than 1.2 MeV [10], making it a good candidate for a search in a missing-mass analysis. This is the strategy of the new BABAR study, implemented as follows. It employs the well-established and very powerful hadronic tag method [11], unique to e^+e^- experiments and illustrated in Fig. 2. Events are selected in which one of the B mesons in the $e^+e^- \to \Upsilon(4S) \to B\overline{B}$ events can be fully reconstructed in one of its hadronic decay modes (it therefore has no undetected momentum from a neutrino). This fully

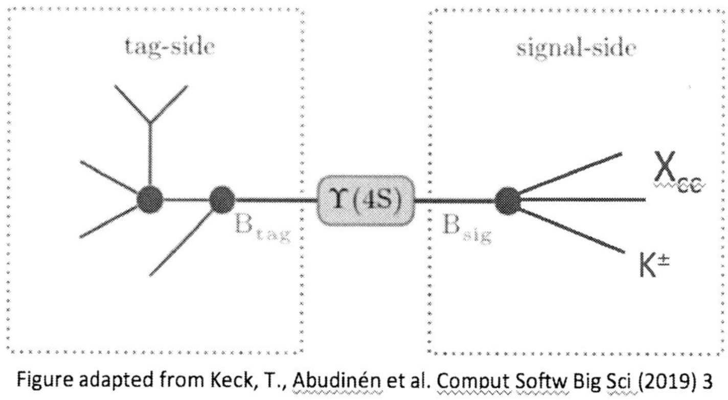

Figure 2. Illustration of the hadronic tag method to determine $\mathcal{B}(B^\pm \to K^\pm X_{c\bar{c}})$.

reconstructed B meson is denoted B_{tag}. One then retains those events with an identified $K^\pm$ in the *other* B meson. This other B meson is denoted B_{sig}, i.e., it is the signal B meson. Since the energies and momenta of the initial e^+ and e^- are completely known, as are those of B_{tag}, so are the energy and momentum of the B_{sig} meson, and one can thus boost to the center-of-mass (c.m.) of B_{sig} and plot the momentum $p_K^{\text{c.m.}}$ of the $K^\pm$ in this frame. The momentum $p_K^{\text{c.m.}}$ in the two-body $B^\pm \to K^\pm X_{c\bar{c}}$ events is related to the missing mass m_X recoiling against the $K^\pm$ through the relation $m_X = \sqrt{m_B^2 + m_K^2 - 2E_K^{\text{c.m.}} m_B}$, with $E_K^{\text{c.m.}}$ the c.m. energy of the $K^\pm$. We choose to examine $p_K^{\text{c.m.}}$ rather than m_X because the experimental resolution is better. If a peak is observed at the $p_K^{\text{c.m.}}$ value corresponding to the $X(3872)$ mass, one can thereby determine the absolute $B^\pm \to K^\pm X(3872)$ branching fraction directly, with no knowledge required of the $X(3872) \to f$ branching fractions, since the $X(3872)$ is not reconstructed but rather detected only through its recoil. Then the $X(3872) \to f$ branching fractions can be determined as explained above, from the already known product $\mathcal{B}(B^\pm \to K^\pm X(3872))\mathcal{B}(X(3872) \to f)$.

In fact, this missing mass method to search for the $X(3872)$ has already been used twice previously, but without success. The technique was pioneered by BABAR in 2006 [12] using a data sample of 211 fb^{-1}. No signal was seen and the limit $\mathcal{B}(B^\pm \to K^\pm X(3872)) < 3.2\times10^{-4}$ was obtained. The Belle experiment repeated the study in 2018 [13], similarly not observing a signal, and improving the limit to $\mathcal{B}(B^\pm \to K^\pm X(3872)) < 2.6 \times 10^{-4}$.

The new BABAR study makes use of the final BABAR data set, corresponding to 424 fb^{-1}, thereby doubling the data sample compared to the 2006 analysis. More importantly, the $X(3872)$ signal efficiency has been improved by a factor of 3 compared to 2006, primarily because, in the new analysis, *all* B_{tag} candidates in an event are accepted, in contrast to the previous BABAR and Belle studies, for which only the *best* B_{tag} candidate (one per event) was accepted. In the new study, the mean number of B_{tag} candidates, for events that have at least one such candidate, is 1.9. Thus by accepting all B_{tag} candidates, and not merely the best one, the $X(3872)$ signal efficiency is improved by about a factor of 2. In the previous analysis, the signal B meson (i.e., the reconstructed B meson containing the $K^\pm$) was often the best B_{tag} candidate as well, and was retained exclusively as such, thus removing it from the signal candidate pool. The remainder of the factor of 3 improvement in the $X(3872)$ signal efficiency relative to the 2006 study is achieved through improvements in the hadronic tag and background-subtraction procedures. Thus, the new analysis is optimized for the $X(3872)$ signal efficiency rather than the B_{tag} purity.

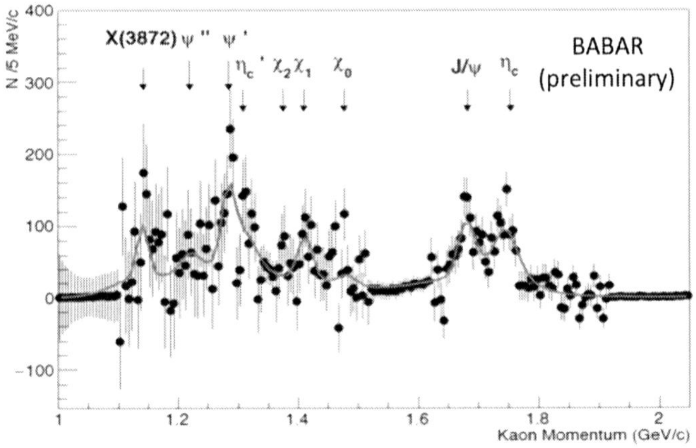

Particle	Yield
J/ψ	1463±133
η_c	1334±129
χ_{c0}	287±181
χ_{c1}	1035±193
χ_{c2}	200±164
$\eta_c(2S)$	527±271
ψ'	1278±285
$\psi(3770)$	497±308
$X(3872)$	992±285

Figure 3. The background-subtracted distribution of the $K^{\pm}$ momentum in B_{sig} signal B mesons, plotted in the B_{sig} center-of-mass. The curve shows the result of a maximum likelihood fit to the data, accounting for the resonances shown in the table to the right. The table shows the fitted signal yield of each resonance, with its statistical uncertainty.

The results for the $p_K^{\mathrm{c.m.}}$ spectrum in the selected events are shown in Fig. 3. A clear peak is observed for the $X(3872)$, corresponding to $p_K^{\mathrm{c.m.}} \approx 1.13\,\mathrm{GeV}/c^2$. We thus obtain the first signal for the $X(3872)$ using a missing-mass technique. A maximum likelihood fit of the background-subtracted data is performed to determine the yields of the observed resonances. We find 992 ± 285 $X(3872)$ candidates, as indicated in the table on the right side of Fig. 3. By repeating the fit with the $X(3872)$ signal yield fixed to zero, the significance of the $X(3872)$ signal, after accounting for the effects of systematic uncertainties, is found to be 3.0 standard deviations.

From the observed yield, we determine the absolute $B^{\pm} \to K^{\pm}X_{c\bar{c}}$ branching fraction to be $\mathcal{B}(B^{\pm} \to K^{\pm}X(3872)) = (2.1 \pm 0.7) \times 10^{-4}$. In conjunction with the result for the product $\mathcal{B}(B^{\pm} \to K^{\pm}X(3872))\mathcal{B}(X(3872) \to J/\psi\pi^+\pi^-) = (8.6 \pm 0.8) \times 10^{-6}$ [10], we thus obtain $\mathcal{B}(X(3872) \to J/\psi\pi^+\pi^-) = (4.1 \pm 1.3)\%$ (preliminary). This rules out the simple tetraquark model, which predicts $\mathcal{B}(X(3872) \to J/\psi\pi^+\pi^-) \gtrsim 50\%$ as mentioned above. Molecular models, with $\mathcal{B}(X(3872) \to J/\psi\pi^+\pi^-) \approx 10\%$ or less, are much more consistent with our measurement. We should note, however, that the pure molecular model has difficulty describing other results, such as the branching fractions in $X(3872)$ radiative decays.

3 First observation of the rare decay $D^0 \to K^-\pi^+e^+e^-$

Searches for rare decay modes of particles are sensitive to new physics, which can modify the decay rates and angular distributions (thus the detection efficiencies). Since the standard model (SM) decay rate is small or negligible for rare processes, the new physics contributions are not necessarily overwhelmed and might be visible. The decay $D^0 \to K^-\pi^+e^+e^-$ is rare in the SM. It cannot occur at tree level. Short distance contributions occur through box and penguin diagrams, with SM branching fractions too small ($O(10^{-9})$) [14] to be currently detectable. Long distance contributions from photon pole or vector meson dominance processes can increase the SM branching fraction up to $O(10^{-6})$ [14]. New physics such as supersymmetry might substantially increase the rate [14, 15]. The current limit on the branching fraction is $\mathcal{B}(D^0 \to K^-\pi^+e^+e^-) < 4.1 \times 10^{-5}$ [16].

https://doi.org/10.1051/epjconf/202023504001

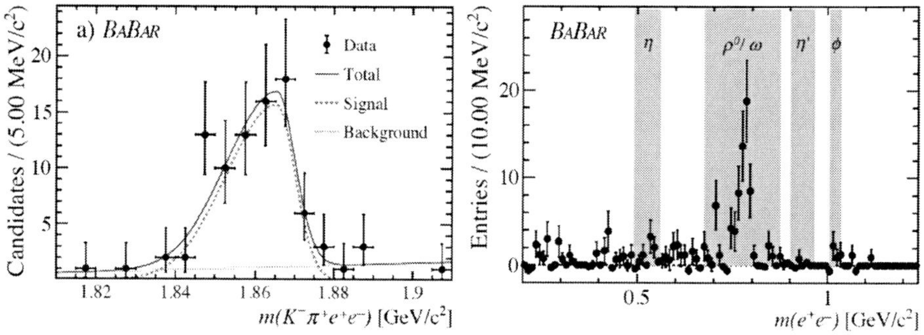

Figure 4. (Left) Reconstructed $D^0 \to K^-\pi^+e^+e^-$ invariant mass peak. (Right) Invariant mass distribution $m(e^+e^-)$ of the final state e^+e^- pair in $D^0 \to K^-\pi^+e^+e^-$. The shaded regions are excluded in the search for $D^0 \to K^-\pi^+e^+e^-$ from short-distance interactions.

We perform a search [17] for $D^0 \to K^-\pi^+e^+e^-$ based on a data sample of 468 fb^{-1} and find 68 ± 9 signal events. The signal peak is shown in Fig. 4 (left). The background is seen to be very small. The significance of the signal is >9 standard deviations. This is the first observation of this decay mode. We find $\mathcal{B}(D^0 \to K^-\pi^+e^+e^-) = (4.0 \pm 0.5) \times 10^{-6}$, consistent with the expectation for long-distance SM interactions.

To search away from regions dominated by long distance interactions, where the e^+e^- in the final state arises from vector meson decay (η, ρ^0, η', and ϕ), we repeat the search excluding the invariant mass $m(e^+e^-)$ regions indicated by the shaded bands in Fig. 4 (right). This allows us to probe regions where short-distance interactions might dominate. We do not observe a significant signal and set the limit $\mathcal{B}(D^0 \to K^-\pi^+e^+e^-)_{\text{non-resonant}} < 3.1 \times 10^{-6}$.

4 Search for rare or forbidden decays of the D^0 meson

Decay mode $D^0 \to$	N_{sig} (candidates)	ϵ_{sig} (%)	$\mathcal{B}$ ($\times 10^{-7}$)	$\mathcal{B}^{U.L.}_{90\%}$ ($\times 10^{-7}$)	$\mathcal{B}^{U.L.}_{90\%}$ [PDG] ($\times 10^{-7}$)
$\pi^-\pi^-e^+e^+$	$0.22 \pm 3.15 \pm 0.54$	4.38	$0.27 \pm 3.90 \pm 0.67$	9.1	1120
$\pi^-\pi^-\mu^+\mu^+$	$6.69 \pm 4.88 \pm 0.80$	4.91	$7.40 \pm 5.40 \pm 0.91$	15.2	290
$\pi^-\pi^-e^+\mu^+$	$12.42 \pm 5.30 \pm 1.45$	4.38	$15.41 \pm 6.59 \pm 1.85$	30.6	790
$\pi^-\pi^+e^\pm\mu^\mp$	$1.37 \pm 6.15 \pm 1.28$	4.79	$1.55 \pm 6.97 \pm 1.45$	17.1	150
$K^-\pi^-e^+e^+$	$-0.23 \pm 0.97 \pm 1.28$	3.19	$-0.38 \pm 1.60 \pm 2.11$	5.0	28†
$K^-\pi^-\mu^+\mu^+$	$-0.03 \pm 2.10 \pm 0.40$	3.30	$-0.05 \pm 3.34 \pm 0.64$	5.3	3900
$K^-\pi^-e^+\mu^+$	$3.87 \pm 3.96 \pm 2.36$	3.48	$5.84 \pm 5.97 \pm 3.56$	21.0	2180
$K^-\pi^+e^\pm\mu^\mp$	$2.52 \pm 4.60 \pm 1.35$	3.65	$3.62 \pm 6.61 \pm 1.95$	19.0	5530
$K^-K^-e^+e^+$	$0.30 \pm 1.08 \pm 0.41$	3.25	$0.43 \pm 1.54 \pm 0.58$	3.4	1520
$K^-K^-\mu^+\mu^+$	$-1.09 \pm 1.29 \pm 0.42$	6.21	$-0.81 \pm 0.96 \pm 0.32$	1.0	940
$K^-K^-e^+\mu^+$	$1.93 \pm 1.92 \pm 0.83$	4.63	$1.93 \pm 1.93 \pm 0.84$	5.8	570
$K^-K^+e^\pm\mu^\mp$	$4.09 \pm 3.00 \pm 1.59$	4.83	$3.93 \pm 2.89 \pm 1.45$	10.0	1800

Figure 5. Results for the branching fractions of lepton flavor violating and lepton number violating decays of the D^0 meson.

Lepton flavor violation (LFV) and lepton number violation (LNV) are essentially forbidden in the standard model but arise in many models of new physics. The BABAR data sample, with its large number of charmed mesons and well reconstructed events, is ideally suited for a search for LFV and LNV in D^0 meson decays. Figure 5 shows the results obtained in our recent search [18] for such decays. No significant signal for LFV or LNV is found. The table in this figure presents the decay mode, the number of observed candidates, the detection efficiency, the branching fraction, the 90% confidence level upper limit on the branching fraction, and the current such limit from the Particle Data Group [10]. Where uncertainties are given, the first is statistical and the second systematic. The BABAR results improve on the current limits by very large factors: from around 20 to almost 800 depending on the channel.

5 Summary

The BABAR Collaboration continues to produce a wide variety of publications on interesting and unique topics despite having ceased data-collection operations more than 10 years ago. Presented here are the results of a study that provides the first direct measurement of the $B^{\pm} \to K^{\pm}X(3872)$ branching fraction, allowing the first determination of the branching fractions of the exotic $X(3872)$ charmonium state. Also presented are the results of a publication on the first observation of the $D^0 \to K^-\pi^+e^+e^-$ branching fraction, and the results of a search for lepton flavor and lepton number violating decays in D^0 meson decays that improves existing limits by factors from 20 to almost 800, depending on the D^0 decay channel.

References

[1] BABAR Collaboration, arXiv:1911.11740.
[2] Belle Collaboration, Phys. Rev. Lett **89**, 102001 (2002); Phys. Rev. D **67**, 032003 (2003); Phys. Rev. Lett. **90**, 071801 (2003).
[3] S. Olsen et al., Rev. Mod. Phys. **90**, 015003 (2018).
[4] Belle Collaboration, Phys. Rev. Lett. **91**, 262001 (2003).
[5] BABAR Collaboration, Phys. Rev. D **65**, 032001 (2002); Phys. Rev. Lett. **94**, 141901 (2005); Phys. Rev. D **93**, 071103 (2004).
[6] N. Toernqvist, Phys. Lett. B **590**, 209 (2004).
[7] L. Maiani et al., Phys. Rev. D **71**, 014028 (2005); Phys. Lett. B **778**, 247 (2018).
[8] H. X. Chen et al., Phys. Rep. **639**, 1 (2016).
[9] E. Braaten et al., Phys. Rev. D **72**, 054002 (2005).
[10] Particle Data Group, Phys. Rev. D **98**, 030001 (2018).
[11] BABAR Collaboration, Phys. Rev. Lett. **92**, 071802 (2004); Phys. Rev. D **88**, 072012 (2013).
[12] BABAR Collaboration, Phys. Rev. Lett. **96**, 052002 (2006).
[13] Belle Collaboration, Phys. Rev. D **97**, 012005 (2018).
[14] S. Fajfer, S. Prelovsek, and P. Singer, Phys. Rev. D. **64**, 114009 (2001).
[15] G. Burdman et al., Phys. Rev. D **66**, 014009 (2002).
[16] BESIII Collaboration, Phys. Rev. D **97**, 072015 (2018).
[17] BABAR Collaboration, Phys. Rev. Lett. **122**, 081802 (2019).
[18] BABAR Collaboration, arXiv:1905.00608.

A New Heavy Flavor Program for the Future Electron-Ion Collider

Xuan Li[1,⋆], *Ivan* Vitev[1], *Melynda* Brooks[1], *Lukasz* Cincio[1], *J. Matthew* Durham[1], *Michael* Graesser[1], *Ming X.* Liu[1], *Astrid* Morreale[1], *Duff* Neill[1], *Cesar* da Silva[1], *Walter E.* Sondheim[1], *Boram* Yoon[1].

[1] *Los Alamos National Laboratory*

Abstract. The proposed high-energy and high-luminosity Electron-Ion Collider (EIC) will provide one of the cleanest environments to precisely determine the nuclear parton distribution functions (nPDFs) in a wide x-Q^2 range. Heavy flavor production at the EIC provides access to nPDFs in the poorly constrained high Bjorken-x region, allows us to study the quark and gluon fragmentation processes, and constrains parton energy loss in cold nuclear matter. Scientists at the Los Alamos National Laboratory are developing a new physics program to study heavy flavor production, flavor tagged jets, and heavy flavor hadron-jet correlations in the nucleon/nucleus going direction at the future EIC. The proposed measurements will provide a unique way to explore the flavor dependent fragmentation functions and energy loss in a heavy nucleus. They will constrain the initial-state effects that are critical for the interpretation of previous and ongoing heavy ion measurements at the Relativistic Heavy Ion Collider and the Large Hadron Collider. We show an initial conceptual design of the proposed Forward Silicon Tracking (FST) detector at the EIC, which is essential to carry out the heavy flavor measurements. We further present initial feasibility studies/simulations of heavy flavor hadron reconstruction using the proposed FST.

1 Introduction

Heavy flavor production is an ideal probe to study the full evolution of the nuclear medium created in heavy ion collisions at the Relativistic Heavy Ion Collider (RHIC) and the Large Hadron Collider (LHC). Heavy quarks are produced early on in the scattering process due to their high masses ($m_{c/b} \gg \Lambda_{QCD}$). The precise mechanisms by which heavy flavor interacts with the nuclear medium, however, are still not well understood in heavy ion collisions. The production process includes both cold nuclear matter effects such as modification of the nuclear Parton Distribution Functions (nPDF), and hot nuclear matter effects such as parton energy loss within the Quark Gluon Plasma (QGP). The future Electron Ion Collider (EIC), recommended as top priority for new construction by the 2015 Long Range Plan for nuclear science and strongly endorsed by the 2018 National Academy of Sciences committee, will utilize high-energy high-luminosity electron-nucleon (electron-nucleus) collisions to study several fundamental questions in the nuclear physics field. The deep inelastic scattering (DIS) processes of electrons on light/heavy nuclei at the future EIC will provide a clean

⋆e-mail: xuanli@lanl.gov

environment to directly explore the initial-state cold nuclear matter effects and the final-state fragmentation/hadronization in the nuclear environment. These studies will improve our understanding of the current and future RHIC and LHC measurements and shed light on the non-perturbative aspects of quantum chromodynamics (QCD).

2 A new heavy flavor and jet program for the future EIC

The future EIC will provide electron+nucleus collisions with multiple nuclear species ranging from ^{3}He to ^{208}Pb at a series of center-of-mass energies from 20 GeV to 141 GeV [2]. Unlike heavy ion experiments at RHIC and LHC, these asymmetric collisions will provide a clean environment to precisely study the initial-state nPDFs in the large Bjorken-x region [4, 5] and the fragmentation/hadronization processes for heavy flavor production. Heavy flavor measurements in polarized electron+nucleon/nuclei collisions at the future EIC will provide opportunities to study the nucleon/nucleus spin structure such as the gluon Sivers effect.

2.1 Exploration of the flavor-dependent parton energy energy loss and in medium hadronization at the EIC

Quarks and gluons lose energy when traversing a nuclear medium and their energy loss is mass/flavor dependent ($\Delta E_g > \Delta E_{u,d,s} > \Delta E_c > \Delta E_b$) [1]. The quark/gluon fragmentation and hadronization processes can be modified in the nuclear medium as shown in figure 1. The cross sections of final-state hadrons and jets to be measured in electron+proton and electron+nuclei collisions at the future EIC are proportional to the initial quark/gluon PDFs, the quark/gluon hard scattering part which can be calculated in perturbative QCD, and the quark/gluon fragmentation and hadronization processes. Comparison of different final-state cross sections can help separate the information about the initial-state nuclear PDFs and the final-state fragmentation and hadronization processes. Comparison of cross sections measured in electron+proton and electron+nucleus collisions for the same final hadron states can precisely determine the nuclear medium effects such as the parton energy loss contributions.

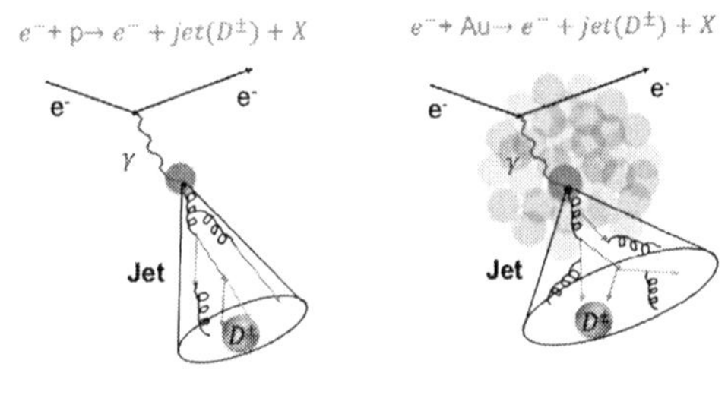

Figure 1. Heavy flavor hadron and jet production in electron+proton (left) and electron+nuclei (right) collisions. The heavy flavor fragmentation and hadronization process in electron+nuclei collisions is modified when compared to the process in electron+proton collisions.

There are competing theoretical models of nuclear modification in DIS reactions within nuclei. One is the absorption model, which describes hadronization that happens inside nuclear matter. The other is quark/gluon energy loss, which is associated with hadronization outside the nuclear medium. Differentiation between these two models is challenging with only light flavor hadron measurements. Figure 2 shows the predicted nuclear modification factor R_{eA} for heavy flavor hadron measurements

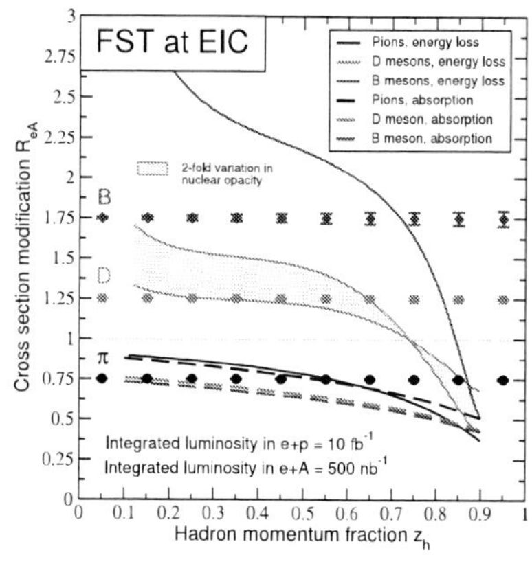

Figure 2. Projected statistical uncertainties of the nuclear modification factor R_{eA} for pions, D-mesons and B-mesons in electron+gold collisions at $\sqrt{s} = 69$ GeV. The projected R_{eA} mean value is based on assumptions. Theory calculations include parton energy loss (solid lines) and absorption within nuclei (dashed lines) [3]. The integrated luminosity in electron+proton (electron+gold) collisions is 10 fb^{-1} (500 nb^{-1}).

versus the hadron momentum fraction relative to their parent parton in $e + Au$ collisions at $\sqrt{s} = 69$ GeV. We show nuclear absorption estimates by dashed lines and parton energy loss calculations by solid lines. The statistical uncertainties are evaluated at the particle generation level with estimated sampling and reconstruction efficiencies in the forward pseudorapidity region of $1.0 < \eta < 4.5$, which is the proposed Forward Silicon Tracking coverage. The integrated luminosity in electron+proton (electron+gold) collisions is 10 fb^{-1} (500 nb^{-1}). Clear separation between these two models can only be achieved through heavy flavor hadron measurements, as they have very different fragmentation functions and formation time when compared to light flavor hadrons [3]. The future EIC with the proposed FST detector can provide sufficient statistical precision for these heavy flavor measurements with enhanced sensitivity to nuclear transport properties.

2.2 Initial design of the proposed Forward Silicon Tracking detector and its performance

The initial design of the proposed Forward Silicon Tracking (FST) detector in fast simulation is shown in figure 3. The proposed FST consists of 2 barrel layers and 5 forward planes to cover the pseudorapidity range of $1.0 < \eta < 4.5$ in the nucleon/nucleus going direction at the future EIC. This detector can precisely determine the transverse decay length of particles (see the schematics in figure 3), which is critical for heavy flavor decay product reconstruction and tagging. Single track performance based on a 3 layer mid-rapidity Monolithic Active Pixel Sensor (MAPS) [8] central vertex detector and the proposed FST with a hybrid design of MAPS plus other silicon technology is shown in figure 4. A uniform 3T magnetic field is applied in the fast simulation and the magnetic field provided by the EIC interaction point design will be included in the full simulation later. With the help of the FST, better than 70 μm spatial resolution can be obtained for forward tracks. Additionally, charged tracks with low transverse momentum can get better than 2% momentum resolution.

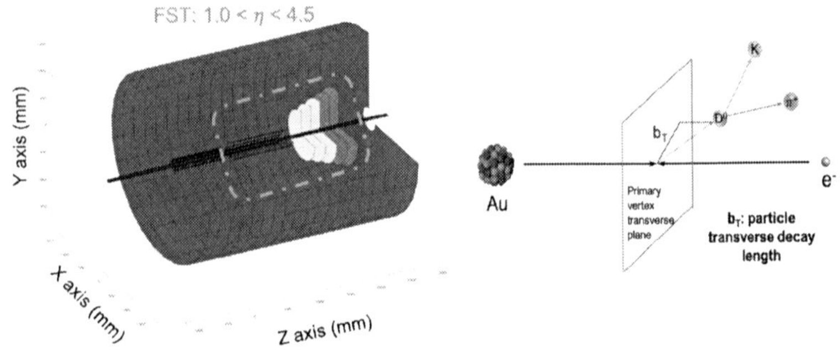

Figure 3. Initial design of the proposed Forward Silicon Tracker, which covers the pseudorapidity of $1.5 < \eta < 4.5$ in the nucleon/nuclei going direction at the future EIC, is highlighted inside the green box on the left. This detector can precisely determine the transverse decay length of particles from the primary vertex (b_T) as shown on the right.

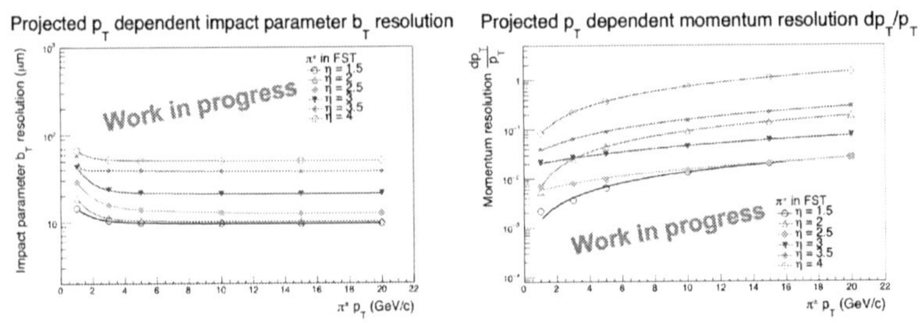

Figure 4. Single track performance for the proposed Forward Silicon Tracker in different pseudorapidity regions from $\eta = 1.5$ to $\eta = 4.0$. A uniform 3T magnetic filed is applied for the track performance evaluation. The resolution of the transverse decay length b_T versus the track p_T is shown on the left and the track p_T dependent momentum resolution $\Delta p_T / p_T$ is shown on the right.

2.3 Heavy flavor hadron reconstruction with the proposed Forward Silicon Tracking detector

To demonstrate the proposed FST capability, the track performance discussed in Sect. 2.2 is embedded into the PYTHIA8 [6] simulation for hadron reconstruction studies. One $e + p$ collision is embedded with on average of 0.02 $p + p$ collisions, which account for the beam interaction background. As the EIC background evaluation is underway, we will include other backgrounds such as the synchrotron radiation in future studies. Charm mesons including D^0, $D^{\pm}$ and $D_s^{\pm}$ can decay into $K^{\pm}$, $\pi^{\pm}$ or lighter D-mesons with a relatively large branching ratio. The reconstruction algorithm in simulation is developed to cluster charged tracks by matching their transverse decay length (b_T) ($\Delta b_T < 50\mu m$) measured by the FST (see the geometry in figure 3). To enhance the probability of finding charm mesons, the charged track clusters are required to contain at least one $K^{\pm}$ track with 100% PID.

Figure 5 shows the invariant mass distributions of charged track clusters which includes at least one $K^{\pm}$ track in 18 GeV electron and 100 GeV proton collisions in different mass regions. Two different FST pixel pitch values are used to compare the reconstructed D-meson mass distributions. With 10 fb^{-1} integrated luminosity, clear signals of D^0 ($\bar{D^0}$), $D^{\pm}$ and $D_s^{\pm}$ are found with the help of

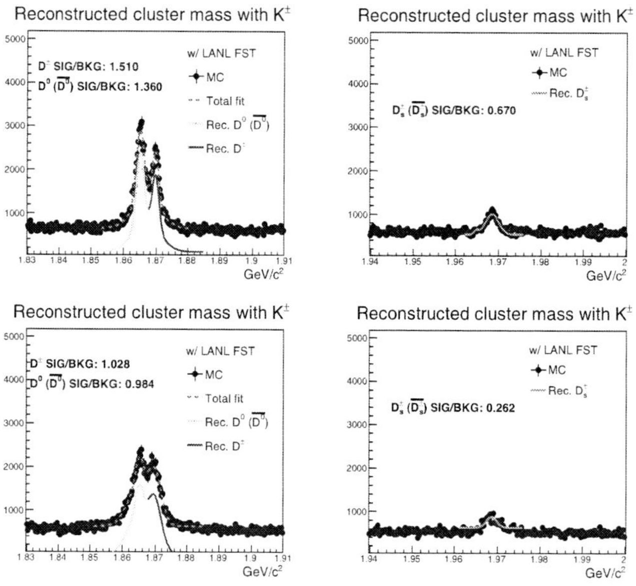

Figure 5. Invariant mass of charged track clusters with at least one $K^{\pm}$ track in 18 GeV electron on 100 GeV proton collisions. No charge separation is applied. The integrated luminosity is 10 fb^{-1}. Clear D^0 ($\bar{D^0}$), $D^{\pm}$ and $D_s^{\pm}$ signals are found with the help of the proposed FST. The performance in the top panel is based on the proposed FST has 30 μm pixel pitch, 500 kHZ trigger rate and 0.4% X_0 radiation length per detector layer. The distributions in the bottom panel are based on the proposed FST has 200 μm pixel pitch, 500 kHZ trigger rate and 0.4% X_0 radiation length per detector layer.

the proposed FST, which has 30 μm pixel pitch, 500 kHZ trigger rate and 0.4% X_0 radiation length per detector layer. The signal to background ratio for D-meson reconstruction is determined from a fit which includes the signal and combinatorial background contributions. When the pixel pitch increases to 200 μm, the mass resolution for the reconstructed D-mesons gets broader and the signal to background ratio is reduced.

To study the physics results, such as the D-meson reconstruction, dependence on the detector performance, we scan the 3D parameter space which includes the pixel pitch, the trigger integration time, and the average material budgets per layer. Table 1 summarizes the parameter values in the simulation studies.

Table 1. The proposed FST parameter table

Parameter name		values		
Pixel pitch	30 μm	50 μm	100 μm	200 μm
Material budgets per layer	0.3% X_0	0.4% X_0	0.5% X_0	1.0% X_0
Trigger integration time	2 μs	4 μs	8 μs	

By varying the pixel pitch, the material budget per layer and the trigger integration time, the extracted signal over background ratios for reconstructed D^0 ($\bar{D^0}$), $D^{\pm}$ and $D_s^{\pm}$ in 18 GeV electron on 100 GeV proton collisions are shown in figure 6. The reconstruction statistical uncertainties, which are determined with 10 fb^{-1} integrated luminosity and include 95% tracking efficiency, are better than 0.1%.

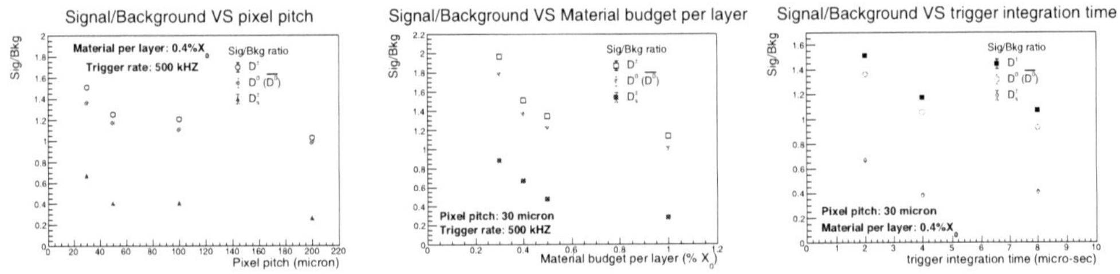

Figure 6. Signal to background ratio for reconstructed D^0 ($\bar{D}^0$), $D^{\pm}$ and $D_s^{\pm}$ versus pixel pitch (left), materials budgets per detector layer (middle) and the trigger integration time (right) for different detector performance combinations. These reconstruction values are achieved with 10 fb^{-1} integrated luminosity.

The D-meson reconstruction signal to background ratio decreases as pixel pitch, material budgets per detector layer and trigger integration time increases. It has a strong dependence on the material budgets per detector layer with fixed pixel pitch at 30 μm and fixed trigger rate at 500 kHZ. Further studies include a wider detector parameter phase space scan, higher mass hadron (e.g. B-mesons) and jet reconstruction, more realistic interaction point design and backgrounds in full simulation are underway.

3 Summary and Outlook

The future EIC will provide unique opportunities to utilize heavy flavor and jet probes to precisely determine the nuclear PDFs, explore the gluon Sivers function and study parton energy loss mechanism in the nuclear medium within a wide Bjorken-x and Q^2 kinematic region. A new heavy flavor and jet program has started at Los Alamos National Laboratory (LANL) to develop new observables and initialize a conceptual detector design for a proposed Forward Silicon Tracking detector for the future EIC. Heavy flavor production in $e + A$ collisions at the future EIC will provide strong discriminating power between competing theoretical models for the parton transport coefficients in nuclear medium. A fast simulation version of the FST design is completed and its tracking performance is consistent with or better than the EIC detector R&D handbook requirements [7]. Initial studies in the fast simulation demonstrate the capability for D-meson reconstruction by the proposed FST. A silicon detector R&D lab is being set up to verify the performance of the proposed silicon techniques, such as MAPS. Ongoing theoretical and experimental studies at LANL, which include new heavy flavor and jet observables determination, physics projections in full simulation, detector design optimization and system integration, will bring about key advances in theory and experimentation and provide important contributions to the EIC physics and detector developments.

References

[1] Azfar Adil, Ivan Vitev, Phys. Lett. B **649**: 139-146 (2007).
[2] A. Accardi et al., Eur. Phy. J. A., **52** 9 (2016).
[3] Ivan Vitev, *Radiative process and jet modification at the EIC* (2019), arXiv: 1912.10965.
[4] E. Chudakov et al., JLAB-THY-16-2354, arXiv: 1610.08536.

[5] E. C. Aschenauer et al., Phys. Rev. D **96**, 114005 (2017).

[6] T. Sjöstrand, S. Mrenna and P. Skands, Comput. Phys. Comm. **178**, 852 (2008), arXiv:0710.3820.

[7] E. C. Aschenauer et al., "Electron-Ion Collider Detector Requirements and R&D Handbook".

[8] M. Mager, the ALICE collaboration, Nuclear Instruments and Methods in Physics Research Section A: Accelerators, Spectrometers, Detectors and Associated Equipment, **824**, 434-438 (2016).

EPJ Web of Conferences **235**, 04003 (2020) https://doi.org/10.1051/epjconf/202023504003
ISMD 2019

Connection between soft and hard probes of small collision systems at RHIC and LHC

Dennis V. Perepelitsa[1,⋆]

[1] *University of Colorado Boulder, 2000 Colorado Ave, Boulder, CO 80309 USA*

Abstract.
The expected connections between signatures in the soft and hard sectors of small collision systems, and the status of experimental attempts to identify them, are discussed. These proceedings summarize the talk as given at the International Symposium on Multi-Particle Dynamics in September 2019 in Santa Fe, NM (ISMD19). As such, the choice of content and focus are selective and not intended to be comprehensive.

1 Introduction

A significant open challenge in the study of QCD systems and their interactions is to understand the nature of the system formed in high-multiplicity proton-proton (pp) or proton-nucleus (p+A) collisions [1, 2] at the Relativistic Heavy Ion Collider (RHIC) and the Large Hadron Collider (LHC).

Analyses of the correlations between final-state particles in these systems reveal that they behave collectively, with their yields at fixed transverse momentum (p_T) following common, preferential orientations in the transverse plane. These patterns are popularly represented as a series of n-th order azimuthal anisotropy coefficients, v_n. In nucleus-nucleus (A+A) collisions, this bulk particle behavior can be well described using theoretical frameworks based on a viscous hydrodynamical description of final-state interactions in the Quark-Gluon Plasma (QGP) medium. Remarkably, these models are also able to describe the v_n values as a function of p_T across pp and p+A collisions, in a common paradigm with A+A collisions [3].

This picture is strengthened by a large body of corroborating data, including but not limited to: the "projectile scan" program at RHIC [4], in which p+Au, d+Au and ^{3}He+Au collisions are used to produce systems with differing initial-state geometric anisotropies $\varepsilon_{2,3}$ which are then observed to be translated into corresponding differences in the measured $v_{2,3}$ (shown in Fig. 1); the observation of hadron mass ordering in the v_n coefficients [6], as expected from motion in a common velocity field; and measurements of multi-particle correlations which verify that the observed anisotropies are truly event-wide phenomena affecting all particles [7].

2 Expectations for the hard sector

For measurements of particles with $p_T \lesssim 3$ GeV (the so-called "soft sector"), the paradigm described above with final-state interactions between strongly-interacting particles is dominant and widely supported by data. The purpose of these proceedings is to discuss what, given this paradigm, may be

⋆e-mail: dvp@colorado.edu

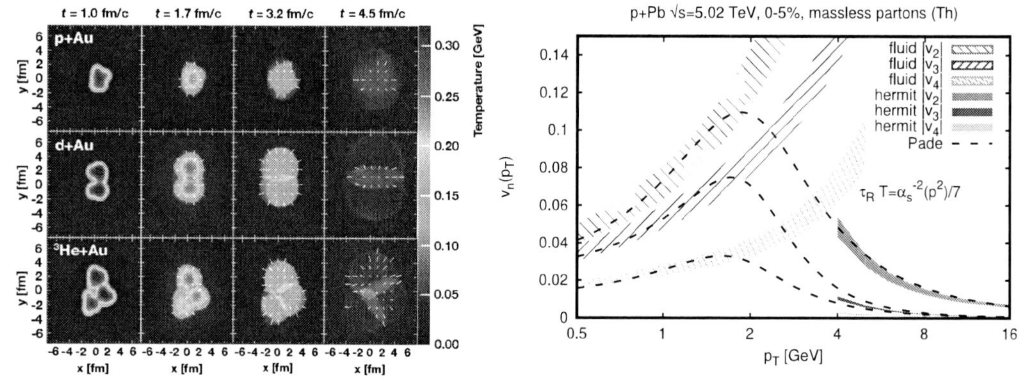

Figure 1. *Left:* Demonstration of hydrodynamic expansion of QGP medium in small collision systems, as part of the RHIC "projectile scan," from Ref. [4]. *Right:* Calculated $v_{2,3,4}$ values in central p+Pb collisions at low and high p_{T}, from Ref. [5].

expected in the so-called "hard sector", where particle production is dominated by the fragmentation of high-p_{T} partons that emerge as jets. Rather than being treated as two unrelated phenomena, one should strive to understand the two behaviors together, since ultimately they are taking place concurrently in the same physical system.

In A+A collisions, the dominant behaviors in soft and hard sectors are indirectly connected through the common relationship of each to the underlying geometry of the collision: low-p_{T} particles acquire an azimuthal anisotropy from the larger pressure gradient along the minor axis of the elliptic overlap region during the hydrodynamic expansion, while high-p_{T} particles acquire it from azimuthally-differential energy loss. Thus both low-p_{T} and high-p_{T} particle pair yields follow a $\cos(2\Delta\phi)$ modulation. In fact, attempting to simultaneously understand differential energy loss and hydrodynamic flow in A+A collisions has spurred important theoretical developments [8].

One direct calculation of what may be expected at high-p_{T} in p+A collisions is performed in Ref. [5], and is summarized in Fig. 1. The calculation uses a common hydrodynamically evolving background and determines the $v_{2,3,4}$ for particles in many-scattering and few-scattering limits at low and high p_{T}, respectively. This calculation indicates that appreciable v_2 values on the order of 2% for $p_{\mathrm{T}} > 10$ GeV particles should inescapably accompany a non-trivial v_2 at low p_{T}. However, in most models [9] this necessarily results in a significant effect in the nuclear modification factor, $R_{p\mathrm{A}}$, which is not borne out by data: a large set of jet and charged-particle measurements have indicated that in minimum-bias p+A collisions, there is no detectable modification.

In the first p+A data taken in 2013 at the LHC, the total event statistics prohibited a detailed study of this high p_{T} region, even in ultra high-multiplicity events enhanced by special triggers. Nevertheless, the early results in Ref. [12] suggested that v_2 could have an appreciable value even at $p_{\mathrm{T}} \approx 10$ GeV in 0-1% centrality 5.02 TeV p+Pb events. Shortly after ISMD19, new results from ATLAS were presented in which jet-triggered events were used in the significantly higher-luminosity 8.16 TeV p+Pb data taken in 2016 [10]. In these results, shown in Fig. 2, a non-zero azimuthal anisotropy was observed for particles up to $p_{\mathrm{T}} \approx 50$ GeV, far from the hydrodynamic regime, and over a broad centrality range. The overall shape of the v_2 as a function of p_{T} is remarkably similar to that in A+A collisions, again suggesting that A+A-like effects are present in p+A collisions. These results exacerbate the already significant tension with a minimum-bias $R_{p\mathrm{A}}$ factor of unity.

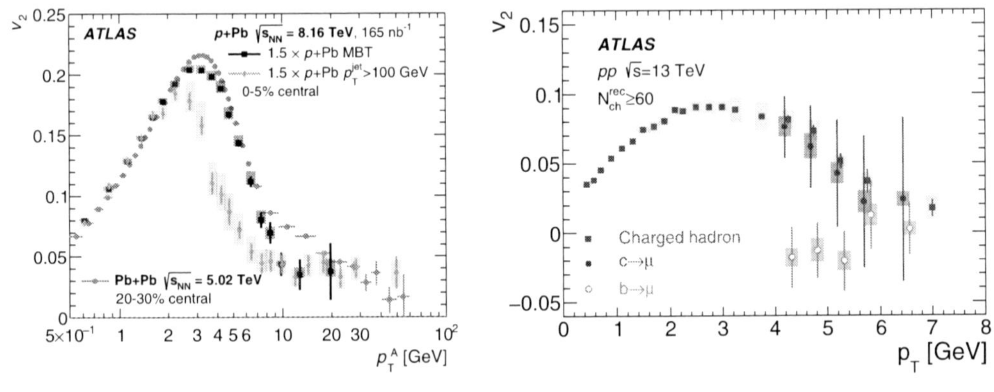

Figure 2. *Left:* Measured azimuthal anisotropy coefficients v_2 in p+Pb collisions using 8.16 TeV data, from Ref. [10]. *Right:* Measured v_2 for muons from charm and bottom hadron decays, compared to that for light hadrons, in high-multiplicity 13 TeV pp collisions, from Ref. [11].

2.1 Jet rates and modification

Of course, azimuthal anisotropies at high p_T are not the only possible manifestation of A+A-style partonic energy loss. Following the analogy with A+A collisions, it is useful to have measurements of the nuclear modification factor for various strongly-interacting probes, particularly for system sizes in the transition region from peripheral A+A to central p+A, over which energy loss effects are rapidly turning off. Unfortunately, this region is bracketed by the opposite extrema of each type of system, and each is potentially affected by large biases. Central p+A collisions feature auto-correlations between hard processes and soft particle production [13, 14], while selections on very peripheral A+A collisions may suffer from opposite (i.e. jet veto) effects [15]. Model-dependent corrections can be applied to account for these, but it is difficult to obtain the needed level of precision to observe or rule out energy loss effects.

Coincidence measurements where jets are paired with an electroweak boson, another jet, or a high-p_T trigger particle can avoid this specific limitation. These measurements are sensitive to energy loss effects which distort the expected kinematic balance between final-state objects, and can be reported as a function of event activity. For example, a measurement by ALICE of semi-inclusive recoil jet distributions in 5.02 TeV data [16] provides constraints on the out-of-cone energy loss. However, this strategy also comes with potential pitfalls: they may have a reduced sensitivity to energy loss (e.g. a di-jet measurement where both jets lose energy, leaving them balanced on average) and the multiplicity selection itself may incur significant biases (e.g. in pp collisions, the resulting enhancement of multi-jet event topologies may significantly modify particular observables). Nevertheless, additional coincidence measurements enabled by the high-statistics 8.16 TeV p+Pb data would be very valuable.

Finally, the jet-medium interaction in p+A collisions may result only in modest in-cone modification of the parton shower, and thus give a null result in the above approaches. In this case, precision measurements of jet structure in p+A collisions are needed. The techniques learned from successful measurements A+A collisions [17, 18], along with comparisons to state-of-the-art pQCD calculations such as SCET$_G$ [19], can be applied here. Some initial measurements in this direction of fragmentation functions [20] and jet masses [17] have been performed but only in minimum-bias p+A collisions where jet modification effects are likely to be very modest.

2.2 Heavy flavor probes

Since high-p_T light hadrons are produced as part of the jet fragmentation process, it is possible they are not sensitive to the QGP medium in p+A collisions. This may occur if there is a finite formation time, or a virtuality separation between the developing shower of the hard-scattered parton and the short-lived medium. This possibility can be avoided with heavy flavor quarks, which due to their large mass must be created early in the collision and thus are present as a conserved flavor charge throughout the full evolution of the medium. Historically, the first measurements of simultaneous flow and suppression of heavy quarks in A+A collisions [21] were important in indicating that a strongly-coupled medium was indeed being formed at RHIC. Studies at the LHC indicate that charm quarks approach equilibrium with the QGP and are thus sensitive to the shape of the produced region [22].

Current data on heavy flavor quark production in p+A collisions indicates no modification at the 10% level in production spectra compared to pp collisions [23], except perhaps at very low p_T. However, identified D-mesons show a significant azimuthal anisotropy in high-multiplicity events over a wide p_T range [6]. The combination of no modification in the p_T distribution with an appreciable modulation results in tension: in a final-state-effect picture, how can the trajectories of the heavy quarks be redirected to such a large degree without a corresponding modification to the p_T distribution?

New ATLAS data on the azimuthal anisotropies of heavy flavor muons bend this contradiction to its breaking point [11], and are summarized in Fig. 2. Whereas charm hadrons are observed to flow with a similar magnitude as light hadrons for a broad range of pp multiplicities, the anisotropy for bottom hadrons is compatible with zero! This is one of the rare strongly interacting probes which is observed to *not* participate in collective motion [1]. An alternative possibility is that this signal arises from initial-state momentum correlations [25], however this picture has challenges of its own in describing the multiplicity dependence of the data and will not be discussed further here.

These data indicate the special role of heavy flavor quarks in delineating boundaries for final-state collectivity, such as the degree of thermalization or the system size and lifetime. To determine the limiting conditions for b-quark flow to appear, a precise measurement in p+A collisions would be very valuable. Shortly after ISMD19, a measurement of non-prompt D^0 flow was made by CMS [26], but within the reported uncertainties it is compatible with the level of charm flow, or no flow at all.

3 Novel collision systems

If a deconfined QGP medium is indeed created in p+A and pp collisions, one may expect several possible experimental signatures. For the case of azimuthal anisotropies, the signature primarily focused on here, a sufficiently high multiplicity is not by itself determinative. In the final state interaction picture, it is the underlying transverse spatial geometry of the system which is translated to a momentum anisotropy in the transverse plane. To better test this picture, it is interesting to study exotic collision systems where the transverse geometry is very different and any possible initial-state effects [29] have a different relative contribution. For example, consider the AMPT study described in Ref. [27], which is shown in Fig. 3. If one models a $e^+e^- \to q\bar{q}$ event as a single string stretched between the outgoing quarks, even with final-state interactions, no v_2 is generated since there is no common event-wide geometry. On other hand, the fictitious scenario in which the total energy or string tension is split between two nearby, parallel strings does generate a v_2.

These kinds of pictures can be tested in data: there have been tremendous efforts to recover old collider data and re-analyze them in the modern context of two- and multi-particle correlation analyses which characterize azimuthally anisotropic particle production. The publically available results so far

[1] A similarly interesting example is the recent indication that the v_2 for Upsilon particles - a $b\bar{b}$ bound state - is compatible with zero in Pb+Pb collisions [24]. Perhaps the common heavy quark content is not coincidental.

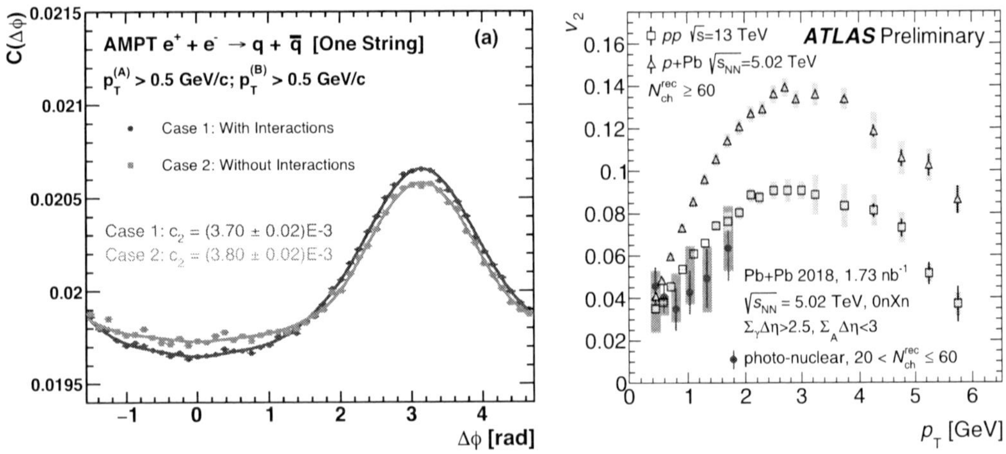

Figure 3. *Left:* Two-particle $\Delta\phi$ distributed in an AMPT simulation of $e^+e^- \rightarrow q\bar{q}$ events with a single string between the outgoing quarks, from Ref. [27]. *Right:* Measured v_2 in photo-nuclear events, compared to that in pp and p+A events, from Ref. [28].

are a re-analysis of 91 GeV e^+e^- collisions with ALEPH at LEP [30] and 316 GeV ep collisions with ZEUS at HERA [31], with others underway. In the first measurement, upper statistical limits are set on the possible magnitude of v_2, while the latter does not observe compatible patterns in the data. Since these experiments and facilities are no longer operating, it is difficult to imagine significant improvements in the archived data, and insight may come only from future data on novel collision systems at RHIC, the LHC, and eventually the Electron Ion Collider (EIC).

Another interesting possibility is to explore the so-called ultra-peripheral A+A processes, electromagnetic interactions in which the nuclear beams remain separated. Among these, photo-nuclear (γ+A) interactions are those in which one nucleus emits a quasi-real photon which strikes the other nucleus. These events feature a significantly cleaner environment for exploring nuclear effects than p+A collisions and allow for the possibility of accessing "nuclear-DIS-like" physics (at least, at the photoproduction limit of DIS) before the EIC is operating.

These events can also be used to search for collective phenomena, such as the analysis of high-multiplicity photo-nuclear events in Ref. [28], shown in Fig. 3. Using standard two-particle correlation methods, a symmetric ridge can be observed in these events. When interpreted as a single-particle anisotropy under the assumptions of factorization and the non-flow subtraction procedure, it results in v_2 values compatible with that in pp collisions but lower than that in p+Pb. However, this may not be surprising in a final-state picture: in the vector-meson dominance (VMD) paradigm, the photon may fluctuate to a meson, especially under a high multiplicity selection. Thus many of these collisions may proceed as ρ+A or ω+A collisions with a non-trivial transverse geometric structure which may induce momentum-space anisotropies. A more quantitative comparison to theory which matches the particular γ+A kinematics and experimental acceptance would be very interesting.

4 Conclusion

To better understand the connection between soft and hard probes of small collision systems, several specific approaches can be proposed. Precision jet-coincidence and jet-structure measurements should continue to be performed in p+A collisions, ideally with guidance from theoretical approaches that correctly describe A+A collisions. Measurements of azimuthal anisotropy should be extended to higher p_T and should particularly be performed for charm and bottom quarks as key discriminants. New collision systems at RHIC and the LHC, such as possible O+O or Ar+Ar running [32], are important for filling in the p+A / A+A "gap" without biases arising from extreme multiplicity selections. Finally, novel systems such as those accessible in ultra-peripheral collisions and at the future EIC will be important for testing the final-state interaction picture in situations with starkly different geometries.

References

[1] J.L. Nagle, W.A. Zajc, Ann. Rev. Nucl. Part. Sci. **68**, 211 (2018), 1801.03477
[2] J. Adolfsson et al. (2020), 2003.10997
[3] R.D. Weller, P. Romatschke, Phys. Lett. B **774**, 351 (2017), 1701.07145
[4] C. Aidala et al. (PHENIX), Nature Phys. **15**, 214 (2019), 1805.02973
[5] P. Romatschke, Eur. Phys. J. C **78**, 636 (2018), 1802.06804
[6] A.M. Sirunyan et al. (CMS), Phys. Rev. Lett. **121**, 082301 (2018), 1804.09767
[7] C. Aidala et al. (PHENIX), Phys. Rev. Lett. **120**, 062302 (2018), 1707.06108
[8] J. Noronha-Hostler et al., Phys. Rev. Lett. **116**, 252301 (2016), 1602.03788
[9] X. Zhang, J. Liao (2013), 1311.5463
[10] G. Aad et al. (ATLAS), Eur. Phys. J. C **80**, 73 (2020), 1910.13978
[11] G. Aad et al. (ATLAS), Phys. Rev. Lett. **124**, 082301 (2020), 1909.01650
[12] G. Aad et al. (ATLAS), Phys. Rev. C **90**, 044906 (2014), 1409.1792
[13] J. Adam et al. (ALICE), Phys. Rev. C **91**, 064905 (2015), 1412.6828
[14] D.V. Perepelitsa, P.A. Steinberg (2014), 1412.0976
[15] C. Loizides, A. Morsch, Phys. Lett. B **773**, 408 (2017), 1705.08856
[16] S. Acharya et al. (ALICE), Phys. Lett. B **783**, 95 (2018), 1712.05603
[17] S. Acharya et al. (ALICE), Phys. Lett. B **776**, 249 (2018), 1702.00804
[18] M. Aaboud et al. (ATLAS), Phys. Rev. Lett. **123**, 042001 (2019), 1902.10007
[19] Z.B. Kang et al., Phys. Rev. Lett. **114**, 092002 (2015), 1405.2612
[20] M. Aaboud et al. (ATLAS), Nucl. Phys. A **978**, 65 (2018), 1706.02859
[21] A. Adare et al. (PHENIX), Phys. Rev. Lett. **98**, 172301 (2007), nucl-ex/0611018
[22] S. Acharya et al. (ALICE), JHEP **02**, 150 (2019), 1809.09371
[23] J. Adam et al. (ALICE), Phys. Rev. C **94**, 054908 (2016), 1605.07569
[24] S. Acharya et al. (ALICE), Phys. Rev. Lett. **123**, 192301 (2019), 1907.03169
[25] C. Zhang et al., Phys. Rev. Lett. **122**, 172302 (2019), 1901.10320
[26] CMS Collaboration, CMS-PAS-HIN-19-009 (2019)
[27] J. Nagle et al., Phys. Rev. C **97**, 024909 (2018), 1707.02307
[28] ATLAS Collaboration, ATLAS-CONF-2019-022 (2019)
[29] B. Schenke, C. Shen, P. Tribedy, Phys. Lett. B **803**, 135322 (2020), 1908.06212
[30] A. Badea et al., Phys. Rev. Lett. **123**, 212002 (2019), 1906.00489
[31] I. Abt et al. (ZEUS), JHEP **04**, 070 (2020), 1912.07431
[32] Z. Citron et al., CERN Yellow Rep. Monogr. **7**, 1159 (2019), 1812.06772

Hot-medium effects on Υ yields in p-Pb and Pb-Pb collisions

Georg Wolschin[1,a]

[1] *Institut für Theoretische Physik der Universität Heidelberg, Philosophenweg 16, D-69120 Heidelberg, Germany, EU*

Abstract. The modification of bottomonia yields in Pb-Pb and p-Pb collisions at LHC energies with respect to the expectation from p-p is investigated in a theoretical approach. Dissociation of the $\Upsilon(nS)$ and $\Upsilon(nP)$ states in the hot quark-gluon plasma (QGP) occurs due to screening of the real quark-antiquark potential, collisional damping through the imaginary part of the potential, and gluon-induced dissociation. Reduced feed-down plays a decisive role. Transverse-momentum and centrality-dependent data are well reproduced. In the asymmetric p-Pb system, alterations of the parton density functions in the lead nucleus account for the leading fraction of the modifications in cold nuclear matter (CNM), but the hot-medium effects turn out to be relevant in spite of the small initial spatial extent of the fireball.

1 Introduction

Heavy mesons such as J/ψ or Υ provide sensitive probes for the properties of the quark-gluon plasma (QGP) that is generated in relativistic heavy-ion collisions. They are produced in hard collisions at short formation times, typically at $\tau_F = 0.3 - 0.6$ fm/c. Since the spin-triplet Υ(1S)-state is particularly stable, it has a sizeable probability to survive as a color-neutral state in the colored hot quark-gluon medium of light quarks and gluons that is generated in a central heavy-ion collision at LHC energies, even at initial medium temperatures of the order of 400 MeV or above.

There exists a considerable literature on the dissociation of quarkonia, in particular of the Υ meson [1–4], in the hot quark-gluon medium; see [5] and references therein for a review. In minimum-bias PbPb-collisions at LHC energies of $\sqrt{s_{\rm NN}} = 5.02$ TeV in the midrapidity range, the strongly bound Υ(1S)-state is found to be suppressed down to about 38% as compared to the expectation from scaled pp collisions at the same energy. The Υ(2S)-state has a smaller binding energy and is even more suppressed, down to 12% [6]. Regarding bottomonia physics at RHIC energies of 200 GeV, the STAR collaboration has results for AuAu and UU collisions [3, 4]. The AuAu data [4] show more suppression of the Υ(1S) state compared to our hot-medium model, indicating that cold nuclear matter (CNM) effects are relatively more pronounced as compared to PbPb at LHC energies.

The asymmetric pPb system at $\sqrt{s_{NN}} = 8.16$ TeV has been investigated theoretically in great detail with respect to CNM effects in Ref. [7]. In addition, we have recently considered the hot-medium contribution to Υ suppression as functions of transverse momentum, centrality, and rapidity. I report

[a] e-mail: wolschin@uni-hd.de
Work done in collaboration with V.H. Dinh, now PhD student at Orsay, and J. Hölck, PhD student at Heidelberg.

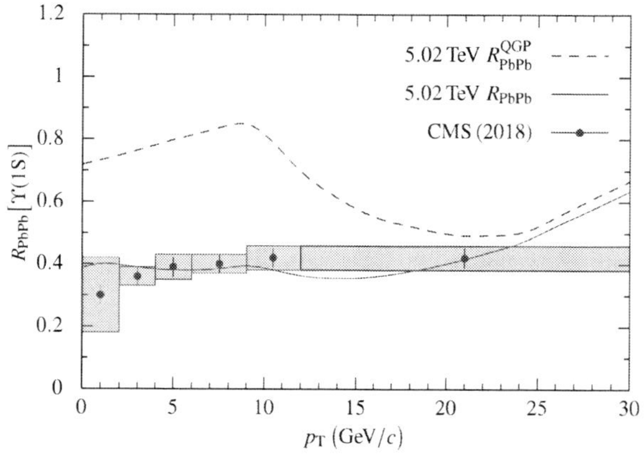

Figure 1: Transverse-momentum dependence of the modification factor $R^{QGP}_{PbPb}[\Upsilon(1S)]$ in the hot medium (dashed curve), and of the total suppression including reduced feed-down (solid curve) for the $\Upsilon(1S)$ state in minimum-bias PbPb collisions at $\sqrt{s_{NN}} = 5.02$ TeV ($T_0 = 513$ MeV). Recent CMS data from Ref. [6] are compared with our theoretical prediction from Ref. [11].

some of the results of our multiparticle dynamics group from Ref. [8], and compare with recent LHCb and ALICE data in this conference contribution. It turns out that in pPb, the bottomonia dissociation in the hot medium is relevant in spite of the small spatial extent of the fireball.

2 The model

In Refs. [9–11] we have devised a model that accounts for the screening of the real part of the potential, the gluon-induced dissociation of the various bottomonium states in the hot medium (gluodissociation), and the damping of the quark-antiquark binding due to the presence of the medium which generates an imaginary part of the temperature-dependent potential. Screening is less important for the strongly bound Υ(1S) ground state, but it is relevant for the $b\bar{b}$ excited states, and also for all $c\bar{c}$ bound states.

Due to screening and depopulation of the excited states in the hot medium, the subsequent feed-down cascade towards the Υ(1S) ground state differs considerably from what is known based on pp collisions. The LHCb collaboration has measured a feed-down fraction of Υ(1S) originating from $\chi_b(1P)$ decays in pp collisions at $\sqrt{s} = 7$ TeV of 20.7% [12], and the total feed-down from excited states to the ground state is estimated to be around 40% [13] at LHC energies. If feed-down was completely absent because of screening and depopulation of excited states in the hot medium, a modification factor $R_{AA}(\Upsilon(1S)) \simeq 0.6$ would thus result, whereas the measured modification factor of the Υ(1S) state in minimum-bias PbPb collisions at 2.76 TeV is 0.453 ± 0.014 (stat) ±0.046 (syst) [14], and 0.378 ± 0.013 (stat)±0.035 (syst) at 5.02 TeV [6]. Hence, there clearly exist in-medium suppression mechanisms for the strongly bound Υ(1S) state which we aim to account for in detail, together with the suppression of the excited states, and the reduced feed-down.

In our model calculation [11], we thus determine the respective contributions from in-medium suppression, and from reduced feed-down for the Υ(1S) ground state, and the Υ(2S) first excited state in PbPb collisions at both LHC energies, 2.76 TeV and 5.02 TeV. The p_T-dependence and the role of the relativistic Doppler effect on the measured transverse-momentum spectra is considered. For the

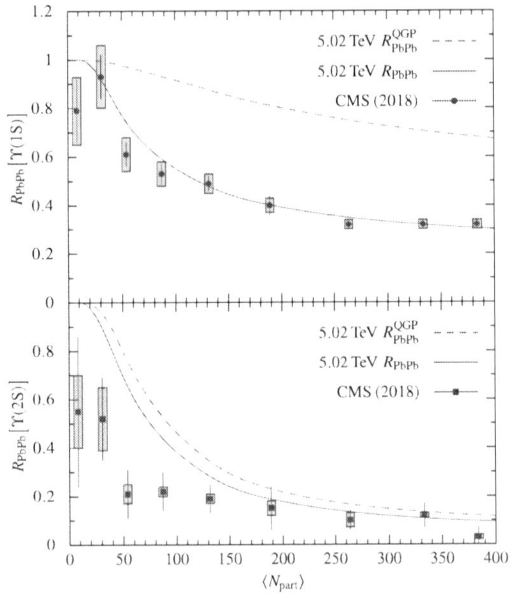

Figure 2: Top: Predicted modification factor $R_{PbPb}[\Upsilon(1S)]$ in PbPb collisions at $\sqrt{s_{NN}} = 5.02$ TeV (solid curve) with centrality-dependent data from CMS ($|y| < 2.4$, [6]) as function of the number of participants $\langle N_{part}\rangle$ (averaged over centrality bins). The modification factor in the QGP-phase without the effect of reduced feed-down is shown as dashed (upper) curve. The formation time is $\tau_F = 0.4$ fm/c, the initial central temperature $T_0 = 513$ MeV. Bottom: Predicted modification factor for the first excited state $R_{PbPb}[\Upsilon(2S)]$ in PbPb collisions at $\sqrt{s_{NN}} = 5.02$ TeV (solid curve) with data from CMS [6]. The modification factor in the QGP-phase (dashed) accounts for most of the calculated total suppression (solid) of the Υ(2S) state. Predictions are from Ref. [11].

Υ(2S) state, the QGP effects are expected to be much more important than reduced feed-down. We compare in Ref. [11] with centrality-dependent CMS data [1, 14] for the Υ(1S) and Υ(2S) states in 2.76 TeV PbPb collisions. The $p_\perp$- and centrality-dependent suppression at the higher LHC energy of $\sqrt{s_{NN}} = 5.02$ TeV has also been predicted in Ref. [11].

For symmetric systems such as AuAu at RHIC or PbPb at LHC, we do not include an explicit treatment of CNM effects such as shadowing in the present study. These are, however, important in asymmetric collisions such as pPb where most of the system remains cold during the interaction time, and we have considered them in our corresponding calculations [8]. Statistical recombination of the heavy quarks following bottomonia dissociation is disregarded: Although this is certainly a relevant process in the J/ψ case, the cross section for Υ production is significantly smaller.

The anisotropic expansion of the hot fireball is accounted for using hydrodynamics for a perfect fluid that includes transverse expansion. Such a simplified nonviscous treatment [10, 11] of the bulk evolution appears to be tolerable because conclusions on the relative importance of the in-medium suppression versus reduced feed-down are not expected to depend much on the details of the background model. When calculating the in-medium dissociation, we consider the relativistic Doppler effect that arises due to the relative velocity of the bottomia with respect to the expanding medium. It leads to more suppression at high $p_\perp$, and to an overall flat dependence of R_{AA} on $p_\perp$.

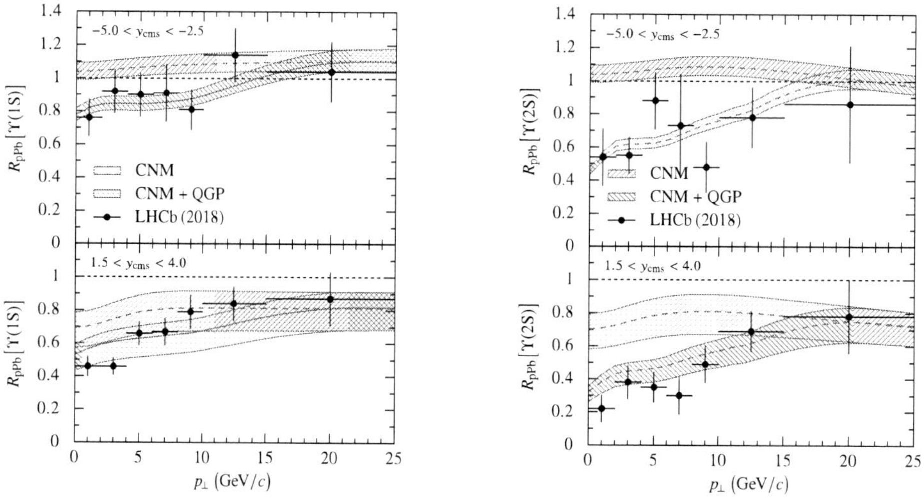

Figure 3: Calculated $p_\perp$-dependent nuclear modification factors R_{pPb} for the $\Upsilon(1S)$ spin-triplet ground state (left) and the excited $\Upsilon(2S)$ state (right) in pPb collisions at = 8.16 TeV are compared with LHCb data [15] in the backward (Pb-going, top) and forward (p-going, bottom) region, for minimum-bias centrality. Results for CNM effects that include shadowing, energy loss, and reduced feed-down (dashed curves, blue) are shown together with calculations that incorporate also QGP effects (solid curves, red). The error bands result from the uncertainties of the parton distribution functions that enter the calculations. From Dinh, Hoelck and Wolschin [8].

3 Results and comparison with data

In Ref. [11] we have calculated predictions for the $p_\perp$-dependent Υ-suppression in 5.02 TeV PbPb collisions, which are shown to be in agreement with recent CMS data [6] in Fig. 1; see the caption for details. For the $\Upsilon(1S)$ state, a substantial fraction of the suppression, in particular at low $p_\perp$, is due to reduced feed-down. The corresponding centrality-dependent suppression (integrated over $p_\perp$) is shown in Fig. 2, in agreement with the data [6] for the $\Upsilon(1S)$ state. Related ALICE data at more forward rapidities $2.5 < y < 4$ are roughly consistent within the error bars [16].

The suppression of the $\Upsilon(2S)$ state is mostly in-medium, with only a small contribution due to reduced feed-down. The prediction shows less suppression than the data in peripheral collisions.

For bottomonia in asymmetric collisions, pPb at $\sqrt{s_{NN}} = 8.16$ TeV has been investigated experimentally by the LHCb [15] and ALICE [16] collaborations, and cold nuclear matter predictions had been published by a group of theorists [7]. Clearly, CNM effects are much more relevant than in symmetric systems, because the bulk of the hadronic matter remains cold during the interaction. There is, however, a spatially small hot zone (fireball) with an initial central temperature that is comparable to the one in a symmetric system, and during its expansion and cooling, it contributes to bottomonia dissociation in regions where the temperature remains above the critical value. We have investigated the respective cold-matter and hot-medium effects on Υ-dissociation in 8.16 pPb collisions [8].

Representative results from this work are shown in Fig. 3 for the transverse-momentum dependence, and Fig. 4 for the rapidity dependence. The plots show CNM (blue, upper bands) and CNM plus QGP (red, lower bands) effects on the $\Upsilon(1S)$ and $\Upsilon(2S)$ yields in 8.16 TeV pPb collisions at the LHC. The transverse-momentum dependence in the backward direction (top) shows enhancement due

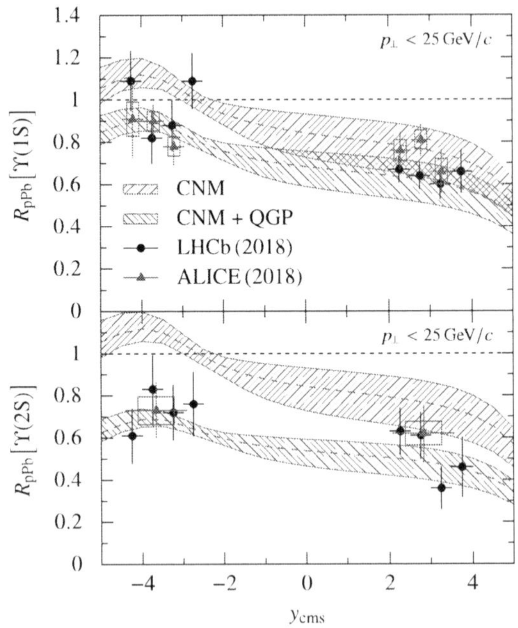

Figure 4: Calculated rapidity-dependent nuclear modification factors R_{pPb} for the $\Upsilon(1S)$ (top) and $\Upsilon(2S)$ state (bottom) in pPb collisions at $\sqrt{s_{\mathrm{NN}}} = 8.16$ TeV with preliminary ALICE data, triangles [16], and with LHCb data, circles [15]. Results for cold nuclear matter (CNM) effects that include shadowing, energy loss, and reduced feed-down (dashed curves, blue) are shown together with calculations that incorporate also QGP effects (solid curves, red). The error bands result from the uncertainties of the parton distribution functions that enter the calculations. The initial central temperature in the fireball region is $T_0 \simeq 460$ MeV. From Dinh, Hoelck and Wolschin [8], with added $\Upsilon(2S)$ LHCb data.

to antishadowing when only the CNM effects are considered, whereas the data for Υ(1S) are clearly suppressed at $p_\perp < 10$ GeV/c, and for Υ(2S) at all measured $p_\perp$ values. This discrepancy is cured through the consideration of dissociation in the QGP, as shown in our cold-matter plus hot-medium calculation (red).

The forward/backward asymmetric shape of the nuclear modification factors as functions of rapidity (Fig. 4) arises from the different cold-matter effects in the forward and backward regions (in particular, shadowing/ antishadowing of the parton distribution functions, but also energy loss in the relatively cold medium). The additional suppression due to the dissociation in the hot fireball is again shown in the lower (red) curves, which are in better agreement with the data for the $\Upsilon(1S)$ ground state not only in the backward, but also in the forward direction. The substantial role of the hot-medium effects is even more pronounced for the $\Upsilon(2S)$ first excited state, where the CNM-calculation shows enhancement in the backward region, whereas the full calculation with in-medium dissociation displays a suppression down to almost 70% – in agreement with the LHCb data [15], and the ALICE data point from Refs. [16, 17].

4 Conclusions

Our phenomenological model for Upsilon suppression in relativistic heavy-ion collisions incorporates gluodissociation, damping, and reduced feed-down. It has been shown to predict [11] the $\Upsilon(1S)$-suppression in PbPb collisions at $\sqrt{s_{\rm NN}} = 5.02$ TeV accurately when compared to recent CMS data [6]. Screening is unimportant for the $\Upsilon(1S)$ state, whereas reduced feed-down is responsible for a considerable fraction of the suppression.

In contrast, for the excited $\Upsilon(2S)$ state the model reveals substantial screening effects and – together with the other dissociation processes that we consider – more suppression than for $\Upsilon(1S)$, with only a small contribution from reduced feed-down. In very peripheral collisions, however, the current CMS data for $\Upsilon(2S)$ [6] show more suppression than the model, leaving room for future improvement. Electromagnetic field effects [18] are unlikely to be the origin of the discrepancy.

Regarding bottomonia in asymmetric systems, we have investigated pPb collisions at LHC energies of 8.16 TeV [8], with the result that for a proper understanding of the data not only the well-established CNM effects need to be considered, but also the hot-medium suppression in the spatially small, rapidly expanding fireball. This is consistent with the presence of a quark-gluon droplet in asymmetric systems at LHC energies.

Acknowledgments

This work was performed in collaboration with Johannes Hölck (ITP Heidelberg), and Viet Hung Dinh (now IJC Paris-Saclay) who contributed in his MSc thesis CNM results for modifications of Υ-yields in pPb collisions at LHC energies. I am grateful to my colleagues from New Mexico and Texas – in particular, Ivan Vitev – for setting up the ISMD at Santa Fe.

References

[1] S. Chatrchyan et al. (CMS Collaboration), Phys. Rev. Lett. **109**, 222301 (2012)
[2] B. Abelev et al. (ALICE Collaboration), Phys. Lett. B **738**, 361 (2014)
[3] L. Adamczyk et al. (STAR Collaboration), Phys. Lett. B **735**, 127 (2014)
[4] Z. Liu et al. (STAR Collaboration), PoS (HardProbes 2018) p. 161 (2018)
[5] A. Andronic, F. Arleo, R. Arnaldi, A. Beraudo et al., Eur. Phys. J. C **76**, 107 (2016)
[6] A.M. Sirunyan et al. (CMS Collaboration), Phys. Lett. B **790**, 270 (2019)
[7] J.L. Albacete et al., Nucl. Phys. A **972**, 18 (2018)
[8] V.H. Dinh, J. Hoelck, G. Wolschin, Phys. Rev. C **100**, 024096 (2019)
[9] F. Nendzig, G. Wolschin, Phys. Rev. C **87**, 024911 (2013)
[10] F. Nendzig, G. Wolschin, J. Phys. G: Nuclear and Particle Physics **41**, 095003 (2014)
[11] J. Hoelck, F. Nendzig, G. Wolschin, Phys. Rev. C **95**, 024905 (2017)
[12] R. Aaij et al. (LHCb Collaboration), JHEP **11**, 031 (2012)
[13] G. Manca (LHCb Collaboration), private communication (2018)
[14] V. Khachatryan et al. (CMS Collaboration), Phys. Lett. B **770**, 357 (2017)
[15] R. Aaij et al. (LHCb Collaboration), JHEP **11**, 194 (2018)
[16] D. Adamová et al. (ALICE Collaboration), ALICE-PUBLIC-2018-008 (2018)
[17] S. Acharya et al. (ALICE Collaboration), Phys. Lett. B, submitted; 1910.14405 (2019)
[18] J. Hoelck, G. Wolschin, Eur. Phys. J. A **53**, 241 (2017)

Recent progress on jet substructure theory

Zhong-Bo Kang[1,2,3,⋆]

[1] *Department of Physics and Astronomy, University of California, Los Angeles, CA 90095, USA*
[2] *Mani L. Bhaumik Institute for Theoretical Physics, University of California, Los Angeles, CA 90095, USA*
[3] *Center for Frontiers in Nuclear Science, Stony Brook University, Stony Brook, NY 11794, USA*

Abstract. In these proceedings, we review recent theoretical progress on jet substructure at the colliders. Focusing on two observables – jet mass and groomed jet radius, we perform theoretical computations for jets measured in the single inclusive jet production in proton-proton collisions, $pp \to \text{jet} + X$. We consider both standard ungroomed jets as well as soft-drop groomed jets. Within the Soft Collinear Effective Theory (SCET), we establish QCD factorization theorems which allow for the joint resummation of several classes of logarithmic corrections to all orders in the strong coupling constant. We present numerical results and compare with the available data from the LHC.

1 Introduction

At present day collider experiments such as the Large Hadron Collider (LHC) and the Relativistic Heavy Ion Collider (RHIC), highly energetic jets play an important role as precision probes of the Standard Model and beyond [1]. In the past years, the capability of jet probes at a future Electron Ion Collider (EIC) and their usage of probing different aspects of nuclear dynamics has been realized [2–4]. In addition, jet substructure techniques have become important tools in high energy particle and nuclear physics, where people have realized that the internal structure of the identified jets contains additional valuable information. For example, jet substructure techniques are used to improve our understanding of the QCD hadronization mechanism, to discriminate between quark and gluon jets, or to identify jets originating from the decay of boosted objects. Jet substructure observables are also increasingly being studied in heavy-ion collisions where they provide important tests and insights of the hot and dense QCD medium [5]. More recently it has also been realized that jet substructure can probe both collinear and transverse momentum dependent parton-to-hadron fragmentation functions in ways that traditional inclusive observables do not offer [6–10].

One of the important techniques that have been developed for better studying jet substructure is jet grooming, which is designed to remove soft wide-angle radiation from the identified jets. We focus on the so-called soft-drop grooming algorithm developed in [11]. Such a grooming algorithm removes the soft contamination of jets and thus leads to reduced sensitivity to nonperturbative effects. Because of this, it allows for a more direct comparison of perturbative QCD calculations and experimental data. Both on the experimental and the theoretical side, significant progress has been made recently in improving our understanding of soft drop groomed jet observables. In the heavy-ion community, soft

⋆e-mail: zkang@physics.ucla.edu

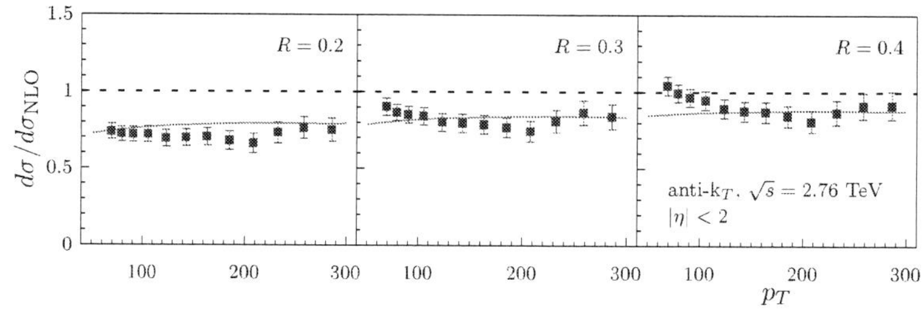

Figure 1. The ln R-resummed jet cross section is compared with the CMS experimental data [18].

drop groomed jet substructure observables have also received an increasing attention from both experiment and theory. Jet grooming techniques can be used to isolate different aspects of jet quenching and may help to discriminate between different model assumptions.

In these proceedings, we mainly review our recent studies on groomed and ungroomed jet mass distribution, and the groomed jet radius.

2 Jet cross section and jet mass

We consider inclusive jet production $pp \to$ jet + X, where jets are identified with the anti-k_T algorithm [12] with a given radius R. A factorized form has been shown in [13] within the framework of Soft Collinear Effective Theory (SCET) [14–17], where the double differential cross section for inclusive jets with a given transverse momentum p_T and rapidity η is given by

$$\frac{d\sigma^{pp\to \text{jet}X}}{dp_T d\eta} = \sum_{a,b,c} f_a \otimes f_b \otimes H_{ab}^c \otimes J_c(z, p_T R, \mu)\,. \tag{1}$$

Here, we suppressed the arguments of the various functions for better readability. See [13] for more details. Let us just mention in the above equation that $J_c(z, p_T R, \mu)$ is called semi-inclusive jet functions (siJFs), which describe the formation of a jet with the transverse momentum p_T and jet parameter R originating from a parent parton c at scale μ. They satisfy the same timelike DGLAP equations that govern the scale evolution of fragmentation functions. By solving the DGLAP equations and thus evolving the siJFs from the natural scale $\mu \sim p_T R$ to the hard scale $\mu \sim p_T$, the resummation of logarithms $\ln R$ can be achieved. As shown in Fig. 1, such a $\ln R$-resummed jet cross section leads to better agreement with the CMS experimental data [18]. A further improvement can be achieved with a joint threshold and $\ln R$ resummation [19].

One might generalize the above framework to study jet substructure. One of the most prominent and most often used jet substructure observables is the jet mass distribution, with $m_J^2 = (\sum_{i\in J} p_i)^2$ and p_i the four-momenta of all the particles inside the reconstructed jet. The factorized cross section can be written as

$$\frac{d\sigma}{d\eta dp_T d\tau} = \sum_{abc} f_a \otimes f_b \otimes H_{ab}^c \otimes \mathcal{G}_c(z, p_T R, \tau, \mu)\,, \tag{2}$$

where the dimensionless variable $\tau = m_J^2/p_T^2$, and the functions $\mathcal{G}_c(z, p_T R, \tau, \mu)$ are the semi-inclusive jet mass functions (siJMFs), which describe how a parton c initiates the signal jet that carries a momentum fraction z of that parton, and at the same time the jet mass τ is observed. In the kinematic

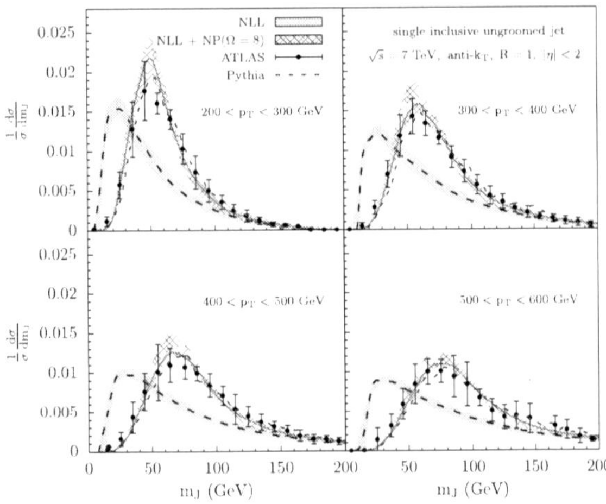

Figure 2. Comparison with the experimental data from the ATLAS collaboration [20] taken at $\sqrt{s} = 7$ TeV.

region where $\tau \ll R^2$, i.e. $m_J^2 \ll (p_T R)^2$, we obtain the following factorization for the siJMFs

$$\mathcal{G}_c(z, p_T R, \tau, \mu) = \sum_i \mathcal{H}_{c\to i}(z, p_T R, \mu) C_i\left(\tau^{C_i}, p_T, \mu\right) \otimes S_i\left(\tau^{S_i}, p_T R, \mu\right), \tag{3}$$

where $\mathcal{H}_{c\to i}$ are the hard matching functions, and are related to unconstrained radiation outside the jet, C_i are the collinear functions that take into account collinear radiation inside the observed jet, and S_i are the soft functions. In Fig. 2, we show comparison of our theoretical calculations and the experimental data from the ATLAS collaboration [20] taken at $\sqrt{s} = 7$ TeV. The dashed black lines and the yellow error bands show the purely perturbative results at next-to-leading logarithmic (NLL) accuracy. The red lines and the red hatched error bands show the theoretical results when the non-perturbative shape function is included. For details about the functional form of such a shape function, see Ref. [21]. The characteristic momentum scale of $\Omega = 8$ GeV signals the fact that, as expected, the ungroomed jet mass distribution is very sensitive to non-perturbative physics such as hadronization and the underlying event etc [22, 23].

3 Soft-drop groomed jet mass

The soft drop grooming procedure recursively removes soft wide-angle radiation from an identified jet [11]. The algorithm starts by re-clustering the constituents of an anti-k_T jet [12] with Cambridge-Aachen algorithm [24, 25]. The obtained angular ordered clustering tree is then declustered recursively where at each step the soft drop condition is checked

$$\frac{\min[p_{T1}, p_{T2}]}{p_{T1} + p_{T2}} > z_{\rm cut}\left(\frac{\Delta R_{12}}{R}\right)^{\beta}. \tag{4}$$

Here $p_{T1,2}$ are the transverse momenta of the two branches obtained at each declustering step and ΔR_{12} is their geometric distance. Soft branches that fail the criterion are removed from the jet. The algorithm terminates when the criterion is met and the particles in the remaining two branches constitute the groomed jet. The soft drop algorithm depends on two parameters: a soft threshold $z_{\rm cut}$ and an

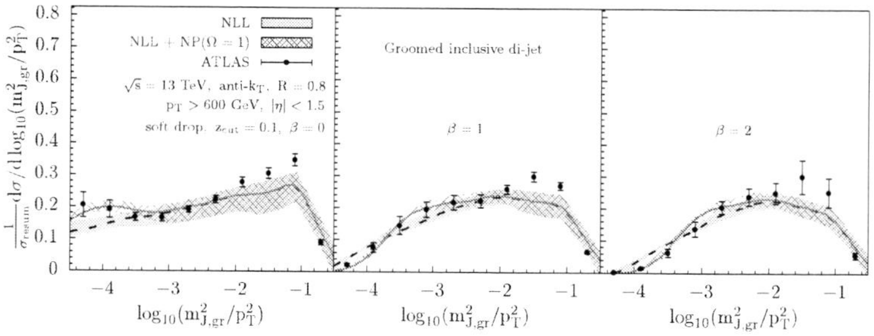

Figure 3. Comparison with the ATLAS data [26] for the soft drop groomed jet mass distribution at $\sqrt{s}$ = 13 TeV.

angular exponent β. While $z_{\rm cut}$ sets the energy scale below which soft branches are removed from the jet, β determines the sensitivity of the grooming algorithm to the wide-angle soft radiation. The mass of the resulting groomed jet is usually referred to as the soft drop groomed jet mass denoted as $m_{J,\rm gr}$.

One can derive a similar factorized form for the groomed jet mass distribution, where one replaces $\mathcal{G}_c(z, p_T R, \tau, \mu)$ in Eq. (2) by the groomed siJMFs $\mathcal{G}_c^{\rm gr}(z, p_T, R, \tau_{\rm gr}, \mu; z_{\rm cut}, \beta)$, where $\tau_{\rm gr} = m^2_{J,\rm gr}/p_T^2$. In practice, the LHC experiments usually choose $z_{\rm cut} \sim 0.1$ while $\tau_{\rm gr}$ can be as low as $O\left(10^{-4}\right)$, and a typical value for the jet radius parameter is $R = 0.8$. Therefore, we are particularly interested in the factorization of the cross section in the region where $\tau_{\rm gr}/R^2 \ll z_{\rm cut} \ll 1$. In this region, we have the following refactorized expression for the groomed siJMFs

$$\mathcal{G}_c^{\rm gr}(z, p_T, R, \tau_{\rm gr}, \mu; z_{\rm cut}, \beta) = \sum_i \mathcal{H}_{c\to i}(z, p_T R, \mu) S_i^{\notin \rm gr}(p_T, R, \mu; z_{\rm cut}, \beta) C_i \otimes S_i^{\rm gr}(\tau_{\rm gr}, p_T, R, \mu; z_{\rm cut}, \beta)\,.$$

Here the hard matching functions $\mathcal{H}_{c\to i}$ and the collinear functions C_i are the same as for the ungroomed case in Eq. (3). However, the soft functions are different where $S_i^{\notin \rm gr}$ characterizes the soft wide-angle radiation (failing the soft drop criterion), while $S_i^{\rm gr}$ is associated with soft particles that pass the soft drop criterion and thus remain inside the groomed jet.

The soft drop grooming procedure is designed to eliminate the sensitivity to the underlying event contribution. This is confirmed by our numerical results for the jet mass distributions. In Fig. 3, we compare our theoretical calculations with the experimental data from the ATLAS collaboration [26] at $\sqrt{s}$ = 13 TeV at the LHC. The dashed line and the corresponding yellow error band are the purely perturbative results at NLL accuracy, while the red line and the red hatched band are NLL results convolved with the non-perturbative shape function. By choosing a much smaller non-perturbative parameter $\Omega = 1$ GeV in the shape function, we find a very good description of the experimental data.

4 Groomed jet radius

Two variables that characterize important features of the soft drop groomed jet are the momentum sharing fraction z_g and the groomed jet radius R_g. Their values are obtained from the kinematics of the two remaining branches when the soft drop algorithm terminates: $z_g = \frac{\min[p_{T1}, p_{T2}]}{p_{T1}+p_{T2}}$, and $R_g = \Delta R_{12} = R\,\theta_g$. Often the variable θ_g is used which corresponds to the geometric distance of the two remaining branches normalized by the radius R of the ungroomed initial jet. The radius of a soft drop groomed jet is one of the key observables characterizing the impact of grooming on a jet and is related to the active area of the groomed jet. In [27], we establish a factorization formalism that allows for the resummation of all relevant large logarithms, which is based on demonstrating the all

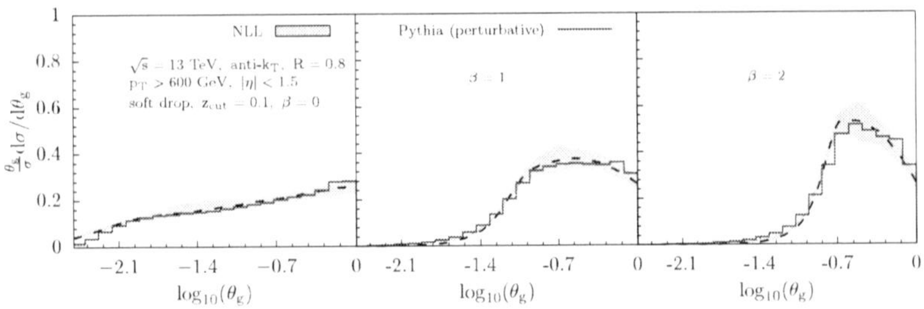

Figure 4. The soft drop groomed jet radius θ_g (dashed black, yellow band) in proton-proton collisions at $\sqrt{s}$ = 13 TeV, compared to Pythia 8 simulations [28] (purple) at the parton level, for $\beta = 0,\ 1,\ 2$, respectively.

order equivalence to a jet veto in the region between the boundaries of the groomed and ungroomed jet. Non-global logarithms including clustering effects due to the Cambridge/Aachen algorithm are resummed to all orders using a suitable Monte Carlo algorithm.

We start from the cross section $\Sigma(\theta_g)$ differential in the transverse momentum p_T and rapidity η of the observed jet, but cumulative in the groomed jet radius where any value below θ_g contributes. The distribution differential in θ_g can then be obtained as

$$\frac{\mathrm{d}\sigma}{\mathrm{d}\eta\,\mathrm{d}p_T\,\mathrm{d}\theta_g} = \frac{\mathrm{d}}{\mathrm{d}\theta_g}\frac{\mathrm{d}\Sigma(\theta_g)}{\mathrm{d}\eta\,\mathrm{d}p_T}\,. \tag{5}$$

The factorized formalism for the cumulative cross section $\mathrm{d}\Sigma(\theta_g)$ is similar to Eq. (2), but now with a new semi-inclusive jet function $\mathcal{G}_c(z, \theta_g, p_T R, \mu; z_{\rm cut}, \beta)$. In the kinematic region where $z_{\rm cut} \ll 1$ and $\theta_g \ll 1$, we develop a refactorization for $\mathcal{G}_c$. For details, see Ref. [27]. In Fig. 4, we present the numerical results (dashed black, yellow band) for the θ_g differential distribution $\theta_g/\sigma_{\rm incl}\,\mathrm{d}\sigma/\mathrm{d}\theta_g$ which is obtained by differentiating the cumulative cross section $\Sigma(\theta_g)$, see eq. (5). The three panels show the result for different values of $\beta = 0,\ 1,\ 2$ (left to right). For comparison, we also show Pythia 8.230 results using the default tune [28] (purple). Here we do not include the nonperturbative hadronization and the contribution from underlying event (UE). In general, we find very good agreement between the Pythia 8 simulation and our perturbative results concerning both the shape and the overall magnitude. More recently ATLAS released their measurements [29] on groomed jet radius, and our predictions agree well with the experimental data.

5 Conclusions

In these proceedings, we review recent theoretical progress on jet substructure at the colliders. We first reported the impact of the jet radius $\ln R$ resummation on the single inclusive jet cross section, and we then studied groomed and ungroomed jet mass, and the soft-drop groomed jet radius. Within the soft-collinear effective theory, we establish QCD factorization theorems for the observables that are measured for single inclusive jet production in proton-proton collisions, $pp \to$ jet + X. We find that the $\ln R$ resummation leads to better agreement with the experimental measurements on the jet cross section. For the groomed and ungroomed jet substructure observables, in general, we found that our theoretical calculations lead to a very good description of the experimental data in the regions where the factorization theorems hold. In addition, we found that the ungroomed/standard jet mass distribution is very sensitive to the non-perturbative contributions such as underlying events and hadronization. However, with soft-drop grooming applied, such strong sensitivity is much reduced for the groomed jet mass distribution.

6 Acknowledgements

This work is supported by the National Science Foundation under Grant No. PHY-1720486.

References

[1] A.J. Larkoski, I. Moult, B. Nachman, Phys. Rept. **841**, 1 (2020), 1709.04464

[2] E.C. Aschenauer, S. Fazio, J.H. Lee, H. Mantysaari, B.S. Page, B. Schenke, T. Ullrich, R. Venugopalan, P. Zurita, Rept. Prog. Phys. **82**, 024301 (2019), 1708.01527

[3] E.C. Aschenauer, K. Lee, B.S. Page, F. Ringer (2019), 1910.11460

[4] M. Arratia, Y. Song, F. Ringer, B. Jacak (2019), 1912.05931

[5] X.N. Wang, ed., *Quark-Gluon Plasma 5* (World Scientific, New Jersey, 2016), ISBN 9789814663700

[6] Z.B. Kang, F. Ringer, I. Vitev, JHEP **11**, 155 (2016), 1606.07063

[7] D.P. Anderle, T. Kaufmann, M. Stratmann, F. Ringer, I. Vitev, Phys. Rev. **D96**, 034028 (2017), 1706.09857

[8] Z.B. Kang, J.W. Qiu, F. Ringer, H. Xing, H. Zhang, Phys. Rev. Lett. **119**, 032001 (2017), 1702.03287

[9] R. Bain, L. Dai, A. Leibovich, Y. Makris, T. Mehen, Phys. Rev. Lett. **119**, 032002 (2017), 1702.05525

[10] Z.B. Kang, X. Liu, F. Ringer, H. Xing, JHEP **11**, 068 (2017), 1705.08443

[11] A.J. Larkoski, S. Marzani, G. Soyez, J. Thaler, JHEP **05**, 146 (2014), 1402.2657

[12] M. Cacciari, G.P. Salam, G. Soyez, JHEP **04**, 063 (2008), 0802.1189

[13] Z.B. Kang, F. Ringer, I. Vitev, JHEP **10**, 125 (2016), 1606.06732

[14] C.W. Bauer, S. Fleming, D. Pirjol, I.W. Stewart, Phys. Rev. **D63**, 114020 (2001), hep-ph/0011336

[15] C.W. Bauer, I.W. Stewart, Phys. Lett. **B516**, 134 (2001), hep-ph/0107001

[16] C.W. Bauer, D. Pirjol, I.W. Stewart, Phys. Rev. **D65**, 054022 (2002), hep-ph/0109045

[17] M. Beneke, A.P. Chapovsky, M. Diehl, T. Feldmann, Nucl. Phys. **B643**, 431 (2002), hep-ph/0206152

[18] V. Khachatryan et al. (CMS), Phys. Rev. **C96**, 015202 (2017), 1609.05383

[19] X. Liu, S.O. Moch, F. Ringer, Phys. Rev. Lett. **119**, 212001 (2017), 1708.04641

[20] G. Aad et al. (ATLAS), JHEP **05**, 128 (2012), 1203.4606

[21] Z.B. Kang, K. Lee, X. Liu, F. Ringer, JHEP **10**, 137 (2018), 1803.03645

[22] M. Dasgupta, K. Khelifa-Kerfa, S. Marzani, M. Spannowsky, JHEP **1210**, 126 (2012), 1207.1640

[23] I.W. Stewart, F.J. Tackmann, W.J. Waalewijn, Phys. Rev. Lett. **114**, 092001 (2015), 1405.6722

[24] Y.L. Dokshitzer, G.D. Leder, S. Moretti, B.R. Webber, JHEP **08**, 001 (1997), hep-ph/9707323

[25] M. Wobisch, T. Wengler, *Hadronization corrections to jet cross-sections in deep inelastic scattering*, in *Monte Carlo generators for HERA physics. Proceedings, Workshop, Hamburg, Germany, 1998-1999* (1998), pp. 270–279, hep-ph/9907280

[26] M. Aaboud et al. (ATLAS), Phys. Rev. Lett. **121**, 092001 (2018), 1711.08341

[27] Z.B. Kang, K. Lee, X. Liu, D. Neill, F. Ringer, JHEP **02**, 054 (2020), 1908.01783

[28] T. Sjöstrand, S. Ask, J.R. Christiansen, R. Corke, N. Desai, P. Ilten, S. Mrenna, S. Prestel, C.O. Rasmussen, P.Z. Skands, Comput. Phys. Commun. **191**, 159 (2015), 1410.3012

[29] G. Aad et al. (ATLAS) (2019), 1912.09837

EPJ Web of Conferences **235**, 05002 (2020) https://doi.org/10.1051/epjconf/202023505002
ISMD 2019

Mitigating Large Background in Jet Substructure Observables

Daekyoung Kang[1,⋆] *Yang-Ting* Chien[2,⋆⋆] *Kyle* Lee[3,4,⋆⋆⋆] *Yiannis* Makris[5,⋆⋆⋆⋆]

[1]*Key Laboratory of Nuclear Physics and Ion-beam Application (MOE) and Institute of Modern Physics, Fudan University, Shanghai, China 200433*
[2]*Center for Theoretical Physics, Massachusetts Institute of Technology, Cambridge, MA 02139, USA*
[3]*C.N. Yang Institute for Theoretical Physics, Stony Brook University, Stony Brook, NY 11794, USA*
[4]*Department of Physics and Astronomy, Stony Brook University, Stony Brook, NY 11794, USA*
[5]*Theoretical Division T-2, Los Alamos National Laboratory, Los Alamos, NM, 87545, USA*

Abstract. We discuss a new approach for jet physics analysis by using subtraction between cumulants of jet substructure observables. The subtracted cumulants are insensitive to soft-particle background uncorrelated with the hard process and allow comparisons between theoretical results and experimental measurements without the complication of soft background like underlying and pile-up events. We find our method using jet mass cumulants efficiently eliminates the background in Monte Carlo simulations and ATLAS jet mass measurements and they show a good agreement with our analytic calculations performed using soft-collinear effective theory.

1 Introduction

Jets produced everywhere in high-energy process are one of essential objects of study at high energy colliders. However, theoretical predictions are many times limited by soft radiations such as multiparton interactions (MPI) in hadron collisions and pile-up (PU) events in high-luminosity (HL) collisions. The predictions cannot be compared with experiments without plugging in model for the soft radiations. Or, one has to adopt techniques [1–10] developed to remove these contributions. Ideal goal of these techniques would be to correct jets back to their pure signal jet but it is impossible due to the intrinsic ambiguity in distinguishing signal from the background. In Ref. [11], we proposed an alternative approach using subtraction between cumulants which cancel the contributions from soft uncorrelated emissions (SUEs) and applied for the transverse energy observable of Drell-Yan processes. Then, we further applied this method to jet substructure observables in [12]. For the observables to which SUEs additively contribute, their contributions are cancelled in the subtraction between two cumulants. As a concrete example for this, we used the classic jet observable, jet mass m_J, and showed theoretical predictions for subtracted jet mass cumulant Δ agree with Monte Carlo simulations and with ATLAS measurements.

⋆e-mail: dkang@fudan.edu.cn
⋆⋆e-mail: ytchien@mit.edu
⋆⋆⋆e-mail: kunsu.lee@stonybrook.edu
⋆⋆⋆⋆e-mail: yiannis@lanl.gov

In this proceeding, we review and summarize of our result in [12]. In following sections we first define the subtracted cumulant Δ, then show that in Monte Carlo simulations the subtraction is robust against various SUE effect. We also show that our theoretical predictions are in a good agreement with ATLAS measurements in subtracted cumulant space.

2 Definition of the observable

Let us consider a jet substructure observable, e, which can be expressed as an addition of individual particle contribution. Then, SUEs contribution can be separated from signal contribution as

$$e = e_{\rm S} + e_{\rm B} \,, \tag{1}$$

where $e_{\rm B}$ is background contribution by SUEs and it is statistically independent of $e_{\rm S}$. In other words, their probability densities are uncorrelated each other by the definition of SUEs. The cumulants $\kappa_n(e)$ is defined by its generating function: $\sum_{\rm n=1}^{\infty} \kappa_n \frac{t^n}{n!} = \log\langle \exp(te)\rangle$. The cumulants inherit the additivity of observable and the background contribution remains separated.

$$\kappa_n(e) = \kappa_n(e_{\rm S}) + \kappa_n(e_{\rm B}) \tag{2}$$

We use this property to cancel uncorrelated contributions in the subtraction between cumulants.

Let us define an additive observable $\hat{\tau}$

$$\hat{\tau} = 2\cosh(\eta) \sum_{i\in \rm jet} p_i^+ = \frac{m_J^2}{p_T}\left[1 + O\Big(\frac{m_J^2}{p_T^2}\Big)\right] \,, \tag{3}$$

where η is pseudorapidity of jet and p_T is the jet transverse momentum. This can be written as an addition of signal and background as in Eq. (1) and is equivalent to the jet invariant mass m_J^2/p_T up to relative corrections of $(m_J/p_T)^2$.

For the dimensionless version $\tau = \hat{\tau}/p_T$, the first cumulant $\langle\tau\rangle$ is given by

$$\langle \tau \rangle = \left(\frac{d\sigma}{dp_T d\eta}\right)^{-1} \int d\tau\, \tau \frac{d\sigma}{d\tau dp_T d\eta} = \langle \tau_{\rm corr}\rangle + \frac{\Omega_f}{p_T} \,, \tag{4}$$

where $\langle \tau_{\rm corr}\rangle$ is the signal contribution correlated with hard scale, while Ω_f is the background and uncorrelated with hard scale.

For comparison with experimental results, we consider a binned cross section $\sigma^{[i,j]}$ where the first index represents the i-th bin in τ and the second represents j-th bin in jet p_T. The cumulant at j-th p_T bin is denoted by $\langle\tau\rangle^{[j]}$. Now let us define the subtracted cumulant of τ

$$\Delta_\tau^{jk} = \langle\tau\rangle^{[j]} - \langle\tau\rangle^{[k]} \frac{\langle p_T^{-1}\rangle^{[j]}}{\langle p_T^{-1}\rangle^{[k]}} = \langle \tau_{\rm corr}\rangle^{[j]} - \langle \tau_{\rm corr}\rangle^{[k]} \frac{\langle p_T^{-1}\rangle^{[j]}}{\langle p_T^{-1}\rangle^{[k]}} \,. \tag{5}$$

Here the background contribution Ω_f is cancelled in the subtraction and the subtraction is rewritten in terms of pure signal contribution $\tau_{\rm corr}$.

3 Robustness of subtracted cumulant

3.1 Pythia simulations

Here, we verify that the subtracted cumulants are robust again to the background such as MPI and PU by using Pythia simulations in pp collisions. We also compare that the subtracted cumulants obtained

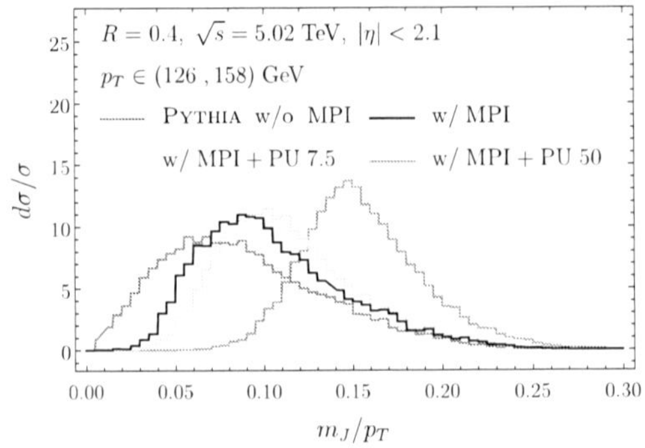

Figure 1. Jet mass distributions using PYTHIA at parton level (red), hadron level with MPI (black), PU with $\langle N_{\rm PU}\rangle = 7.5$ (green) and 50 (magenta).

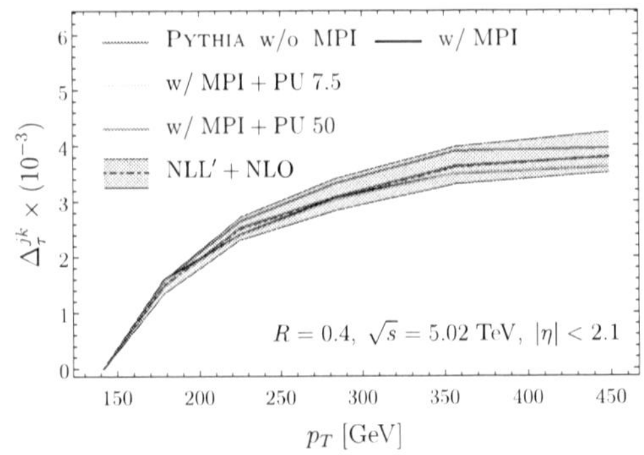

Figure 2. Subtracted cumulants Δ_τ^{jk} for jet mass for p_T between 126 and 500 GeV. perturbative calculation (blue band) and PYTHIA simulations.

by the simulation agree with theoretical predictions at next-to-leading logarithmic and next-to-leading order accuracy (NLL′+NLO) using SCET.

FIG. 1 shows the jet mass distributions with and without MPI and PU effects. We find significant changes in the distribution by MPI and PU. On the other hand, subtracted cumulants in FIG. 2, remains almost same except for large PU (PU 50 curve). For the large PU, p_T is also shifted by the SUEs and to improved the results, a modified definition of p_T robust against SUEs can be adopted [12]. The simulation results also agree with theoretical prediction within perturbative uncertainty (blue band). This clearly verifies that our subtracted cumulants greatly mitigate MPI and PU contributions.

3.2 Comparison with experimental data

Here, we compare the theoretical predictions and simulations to LHC data measured by ATLAS collaboration at 7 TeV [13] and 5.02 TeV [14]. FIG. 3 shows the results for the predictions at NLL′+NLO accuracy and PYTHIA simulations with and without MPI. The data points are for ATLAS results. Note that in 7 TeV data [13] , the jet mass $s = m_J^2$ is given rather than $\tau = m_J^2/p_T^2$, we rescale the definition of Δ and denote it to Δ_s

$$\Delta_s^{jk} = \langle s\rangle^{[j]} - \langle s\rangle^{[k]} \frac{\langle p_T\rangle^{[j]}}{\langle p_T\rangle^{[k]}} \,. \tag{6}$$

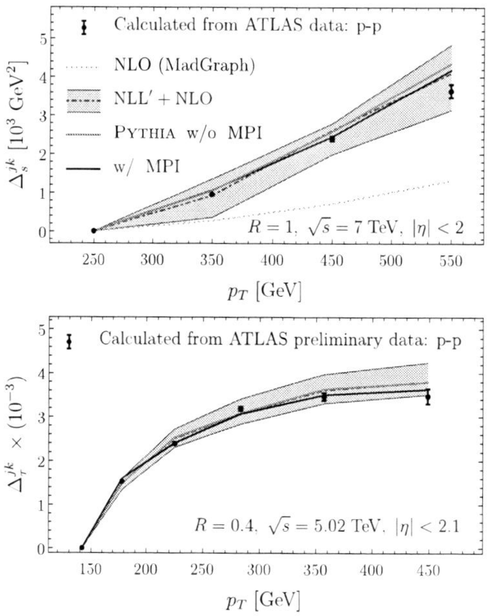

Figure 3. subtracted cumulant results: ATLAS results (data points) computed using data in [13, 14], NLL′+NLO calculation (blue band), PYTHIA simulations with and without MPI (black and red curves). The top and bottom panels are for $\sqrt{s} = 7$ TeV and $\sqrt{s} = 5.02$ TeV, respectively.

We find that both of 5 TeV and 7 TeV data points are in good agreement with our predictions as well as simulations.

4 Summary

In this proceeding, we review and summarize our results in [11, 12]. We introduce a new method for jet substructure observables called a subtracted cumulant. This quantity is insensitive to soft emissions uncorrelated with hard process hence, robust against multiparton interactions and pile-up effects. The is independent way of studying jet substructure from conventional techniques which either correct jets by removing soft constituents or use a model for soft emissions. In order to verify this method, we computed the subtracted cumulant of jet mass at NLL′+NLO accuracy using SCET and compared it to those obtained from PYTHIA simulations. While in theoretical prediction the soft uncorrelated emissions is absent, in simulations those emissions are turned on and they induce significant changes in the jet mass distributions. However subtracted cumulants of theoretical predictions and simulations still remain in a good agreement. This verifies the subtraction method mitigates large background. Finally, we test it with ATLAS measurements and find a similar agreement with theory predictions.

References

[1] M. Cacciari, G.P. Salam, Phys. Lett. **B659**, 119 (2008), 0707.1378
[2] I. Feige, M.D. Schwartz, I.W. Stewart, J. Thaler, Phys. Rev. Lett. **109**, 092001 (2012), 1204.3898
[3] A.J. Larkoski, J. Thaler, Phys. Rev. **D90**, 034010 (2014), 1406.7011

[4] D. Krohn, M.D. Schwartz, M. Low, L.T. Wang, Phys. Rev. **D90**, 065020 (2014), 1309.4777
[5] M. Cacciari, G.P. Salam, G. Soyez, Eur. Phys. J. **C75**, 59 (2015), 1407.0408
[6] D. Bertolini, P. Harris, M. Low, N. Tran, JHEP **10**, 059 (2014), 1407.6013
[7] P. Berta, M. Spousta, D.W. Miller, R. Leitner, JHEP **06**, 092 (2014), 1403.3108
[8] A.J. Larkoski, S. Marzani, G. Soyez, J. Thaler, JHEP **05**, 146 (2014), 1402.2657
[9] P.T. Komiske, E.M. Metodiev, B. Nachman, M.D. Schwartz, JHEP **12**, 051 (2017), 1707.08600
[10] G. Soyez, Phys. Rept. **803**, 1 (2019), 1801.09721
[11] D. Kang, Y. Makris, T. Mehen, JHEP **09**, 055 (2018), 1803.04413
[12] Y.T. Chien, D. Kang, K. Lee, Y. Makris, Phys. Rev. **D100**, 074030 (2019), 1812.06977
[13] G. Aad et al. (ATLAS), JHEP **05**, 128 (2012), 1203.4606
[14] The ATLAS collaboration (ATLAS), ATLAS-CONF-2018-014 (2018)

EPJ Web of Conferences **235**, 05003 (2020) https://doi.org/10.1051/epjconf/202023505003
ISMD 2019

Probing BFKL dynamics at hadronic colliders in jet gap jet events

Christophe Royon[1,⋆] *Federico* Deganutti[1,⋆⋆]

[1] *Department of Physics and Astronomy, The University of Kansas, Lawrence, USA*

Abstract. In this report, we give the Balitsky Fadin Kuraev Lipton formalism for jet gap jet events at hadronic colliders. We also discuss the case where in addition at least one proton is intact in the final state in diffractive events.

1 Looking for BFKL resummation effects at hadronic colliders: forward jets and Mueller-Navelet jets

The experimental search for resummation effects at low x where x is the momentum fraction of the interacting parton inside the proton, as predicted by the Balitsky Fadin Kuraev Lipatov (BFKL) [1] equation has been an important topic of research at recent colliders The starting point to look for BFKL resummation effects at colliders was to study forward jet production at HERA, the ep collider located at DESY, Hamburg. When Q, the transferred energy between the electron and the interacting quark, is of the same order of the forward jet transverse momentum, and when the interval in rapidity between the forward jet and the scattered electron is large, the cross section predicted by usual Quantum Chromodynamics (QCD) as given by the Dokhitzer Gribov Lipatov Altarelli Parisi (DGLAP) [2] evolution equation is small because of the k_T-ordering of the different emitted gluons on the ladder. On the contrary, the BFKL dynamics predicts a non-negligible cross section. As an example, the H1 collaboration measured the triple differential forward jet cross section as a function of Q^2, jet p_T and jet fractional energy [3]. The BFKL NLL formalism leads to a good description of data whereas the NLO QCD calculation has difficulties to describe the forward jet cross section [4]. However this measurement was not completely conclusive concerning the observation of low x resummation effects since higher order QCD corrections lead to a good description of data.

The second observable that was proposed to look for BFKL resummation effects at hadronic colliders is the Mueller-Navelet jet production [5]. The idea is to look for jets separated will a large interval in rapidity. When the transverse momentum of the two jets is similar, the BFKL dynamics predicts a larger cross section than the DGLAP one because of the k_T ordering of the different gluons along the ladder as before [6, 7]. One possible observable is the measurement of dijet azimuthal decorrelation which is larger for BFKL dynamics because of the multiple gluon emission [6]. Unfortunately, this prediction suffers from high order corrections to BFKL dynamics that has a tendency to bring both predictions to be closer to each other. As an example, the most recent measurements by the

⋆e-mail: christophe.royon@ku.edu
⋆⋆e-mail: fedeganutti@gmail.com

CMS and ATLAS collaborations [8] are shown in Fig. 1. Data can be described by BFKL dynamics as expected but also by NLO DGLAP QCD calculations. It is thus needed to define new possible variables or going to larger interval of rapidity between jets, using the CASTOR [9] very forward calorimeter in CMS as an example, in order to enhance further the BFKL contribution.

New variables were recently proposed to look for BFKL dynamics [10]. The idea is to look for mini-jet production between the two Mueller-Navelet jets that are predicted by DGLAP or BFKL dynamics as illustrated in Fig. 2. The pattern for gluon emission is different between both dynamics because of the k_T-ordering of the gluons in the ladder for the DGLAP prediction. As an example, it would be possible to measure the mini-jet production with transerve momentum above 20 or 25 GeV between the two Mueller Navelet jets with p_T >40 or 45 GeV.

In the next section, we will discuss another method to look for BFKL dynamics, namely the measurement of jet gap jet events.

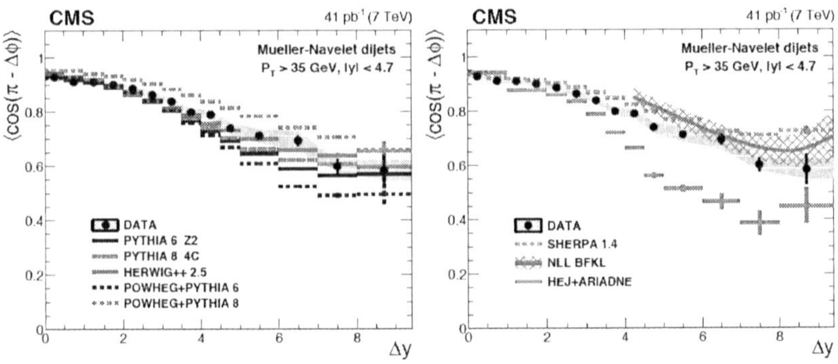

Figure 1. Measurement of the azimuthal decorrelation between Mueller Navelet jets measured by the CMS collaboration and compared to DGLAP Monte Carlo and BFKL calculations.

2 Jet gap jet production

2.1 Gap between jets at the Tevatron and the LHC

The measurement of gap between jets or the Mueller-Tang process [11] was performed at the Tevatron [12] and is also being done at the LHC [13]. The idea is to look for two jets separated by a large interval in rapidity $\Delta\eta$ with a gap devoid of any activity or energy (no particle emission) in a central part of the detector $\Delta\eta_{gap}$ (for instance between (-1) and (1) in rapidity). This means that a colorless object, or a Pomeron, is emitted between the two jets. The natural description of this object is the BFKL Pomeron. The prediction of the BFKL cross section to produce these events is not negligible whereas the DGLAP prediction is very small for gap sizes above two units in rapidity. This is why this process is ideal to look for BFKL resummation effects. However, it suffers from potential additional interactions, such as soft gluon exchanges, that can destroy the gap between the jets. This is the so-called the survival probability that can be as low as 0.1 at Tevatron center-of-mass energy (about 2 TeV) and 0.03 at the LHC (about 14 TeV).

$$\langle p_T \rangle = \frac{1}{N}\sum_{i=1}^{N} |p_{Ti}|$$

$$\langle R_y \rangle = \frac{1}{N+1}\sum_{i=1}^{N+1} \frac{y_i}{y_{i-1}}$$

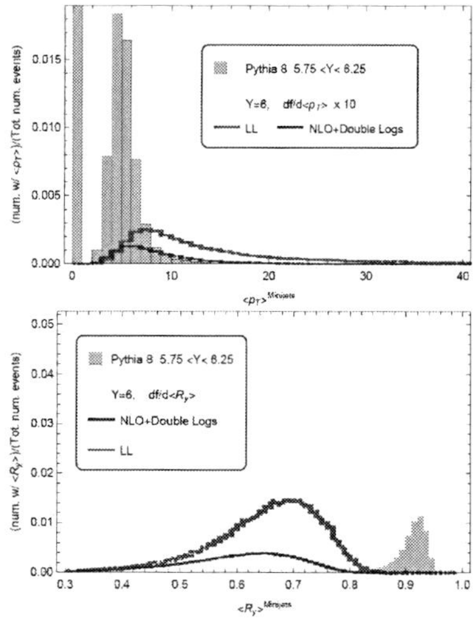

Figure 2. New less inclusive measurements in Mueller-Navelet jet events to look for BFKL resummation effects: average p_T rapidity of mini-jet production.

2.2 The NLL BFKL formalism and comparison with data

In order to describe the measurements of gap between jets at the Tevatron and the LHC, the following BFKL formalism was used. The production cross section of two jets with a gap in rapidity between them reads

$$\frac{d\sigma^{pp\to XJJY}}{dx_1 dx_2 dE_T^2} = \mathcal{S} f_{eff}(x_1, E_T^2) f_{eff}(x_2, E_T^2) \frac{d\sigma^{gg\to gg}}{dE_T^2}, \tag{1}$$

where $\sqrt{s}$ is the total energy of the collision, E_T the transverse momentum of the two jets, x_1 and x_2 their longitudinal fraction of momentum with respect to the incident hadrons, S the survival probability, and f the effective parton density functions. The rapidity gap between the two jets is $\Delta\eta = \ln(x_1 x_2 s/p_T^2)$.

The cross section is given by

$$\frac{d\sigma^{gg\to gg}}{dE_T^2} = \frac{1}{16\pi}\left|A(\Delta\eta, E_T^2)\right|^2 \tag{2}$$

in terms of the $gg \to gg$ scattering amplitude $A(\Delta\eta, p_T^2)$.

In the following, we consider the high energy limit in which the rapidity gap $\Delta\eta$ is assumed to be very large. The BFKL framework allows to compute the $gg \to gg$ amplitude in this regime, and the result is known up to NLL accuracy (in this calculation, the coupling to the proton, the so-called impact factors, was computed at LL)

$$A(\Delta\eta, E_T^2) = \frac{16 N_c \pi \alpha_s^2}{C_F E_T^2} \sum_{p=-\infty}^{\infty} \int \frac{d\gamma}{2i\pi} \frac{[p^2 - (\gamma - 1/2)^2]}{[(\gamma-1/2)^2 - (p-1/2)^2]} \frac{\exp\left\{\bar{\alpha}(E_T^2)\chi_{eff}[2p, \gamma, \bar{\alpha}(E_T^2)]\Delta\eta\right\}}{[(\gamma-1/2)^2 - (p+1/2)^2]} \tag{3}$$

with the complex integral running along the imaginary axis from $1/2 - i\infty$ to $1/2 + i\infty$, and with only even conformal spins contributing to the sum, and $\bar{\alpha} = \alpha_S N_C/\pi$ the running coupling.

In this study, we performed a parametrised distribution of $d\sigma^{gg\to gg}/dE_T^2$ so that it can be easily implemented in the HERWIG Monte Carlo [15] since performing the integral over γ in particular would be too much time consuming in a Monte Carlo. The implementation of the BFKL cross section in a Monte Carlo is absolutely necessary to make a direct comparison with data. Namely, the measurements are sensitive to the jet size (for instance, experimentally the gap size is different from the rapidity interval between the jets which is not the case by definition in the analytic calculation).

Let us first notice that the sum over all conformal spins is absolutely necessary. Considering only $p = 0$ in the sum of Equation 3 leads to a wrong normalization and a wrong jet E_T dependence, and the effect is more pronounced as $\Delta\eta$ diminishes.

The DØ Collaboration measured the jet gap jet cross section ratio with respect to the total dijet cross section, requesting for a gap between -1 and 1 in rapidity, as a function of the second leading jet E_T, and $\Delta\eta$ between the two leading jets for two different low and high E_T samples (15< E_T <20 GeV and E_T >30 GeV). To compare with theory, we compute the following quantity

$$\text{Ratio} = \frac{\text{NLL HERWIG}}{\text{LO Herwig}} \times \frac{\text{LO QCD dijets}}{\text{NLO QCD dijets}} \tag{4}$$

in order to take into account the NLO corrections on the dijet cross sections, where BFKL NLL HERWIG and Dijet HERWIG denote the BFKL NLL and the dijet cross section implemented in HERWIG [15]. The NLO QCD cross section was computed using the NLOJet++ program [16].

The comparison with DØ data [12] is shown in Fig. 3. We find a good agreement between the data and the BFKL calculation. It is worth noticing that the BFKL NLL calculation leads to a better result than the BFKL LL one (note that most studies in the literature considered only the $p = 0$ component which is not a valid assumption). The comparison with CDF data [12] leads to a similar conclusion and data are well described by the BFKL NLL formalism.

The latest results have been provided by the CMS Collaboration using data collected in pp collisions at $\sqrt{s} = 7$ TeV [13]. Each of the leading two jets has transverse momentum of $p_T > 40$ GeV, pseudorapidity of $1.4 < |\eta_{\text{jet}}| < 4.7$, and opposite signed pseudorapidities $\eta_{\text{jet 1}} \cdot \eta_{\text{jet 2}} < 0$. The fraction of jet-gap-jet events were measured as a function of $p_{\text{T, jet 2}}$ and pseudorapidity difference between the leading two jets $\Delta\eta_{\text{jj}}$.

2.3 The NLL BFKL formalism including NLO impact factors

The full NLL BFKL calculation using the NLO impact factor is in progress [18]. Several complications appear when the impact factors are refined to NLO.

- The back-to-back symmetry proper to an elastic interaction is lost since a third parton can appear in the final state.
- Since the NLO impact factor interference diagrams connect in a non trivial way to the gluon-Green function (GGF) their convolution cannot be factorized any more into an amplitude square involving only the integrated GGF. Moreover, the GGF in its full form is written in terms of Gauss-hypergeometric functions in some particularly complicated combinations especially for large conformal spins and it complicates the numerical implementation.
- A formal divergence appears in the real-corrections published in [19] when the proper jet reconstruction algorithm is employed. The problematic behavior appears when besides the two tagged jets, a third parton is emitted in the gap region but with a sub-threshold transverse energy. One would think that such configuration should be suppressed when the additional parton becomes more central. This puzzling behavior motivates an independent cross-check of previously pubished results that is currently a work-in-progress.

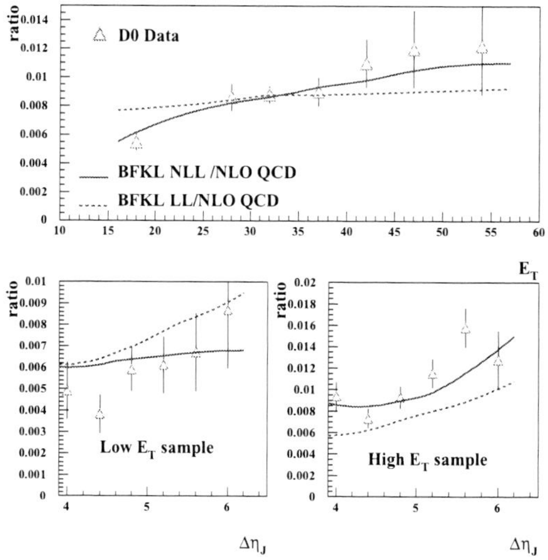

Figure 3. Comparison between the measurement of gap between jets events by the DO collaboration and the BFKL calculation as a function of jet E_T or $\Delta\eta$ between jets for the low and high E_T samples.

It is expected that the full NLO calculation will be available soon, leading to more accurate comparison with data.

2.4 Jet gap jet events in diffraction

In this section, we discuss the observation of gap between jets in single diffractive or double Pomeron exchange events [20]. These processes are similar to the one discussed in the previous section but they require in addition the existence of at least one intact proton after interaction. This process was implemented in the FPMC generator [17].

The DPE jet-gap-jet event ratio is defined as the ratio of the cross section for the DPE jet-gap-jet (JGJ) production to the DPE inclusive dijet (Jets) production $R = \sigma(DPE\ \text{JGJ})/\sigma(DPE\ \text{Jets})$ and is plotted in Fig. 4 as a function of the transverse momenta of the first-leading jet. To take into account NLO QCD effects, absent in the FPMC program, the LO ratio obtained was corrected by the cross section ratio $\sigma(DPE\ LO\ Jet++)/\sigma(DPE\ NLO\ Jet++)$, obtained by the NLO Jet++ program [16].

As far as the gap fraction $\sigma(DPE\ JGJ)/\sigma(DPE\ Jets)$ is concerned, note that we did not consider an additional suppression factor for DPE jet-gap-jet production, on top of the 0.03 of DPE inclusive jet production. Therefore, in the predictions of Fig. 4, all the rapidity gap survival probability cancel. We would like to point out that this last point is an assumption, DPE jet-gap-jet production could be subject to a bigger suppression than DPE inclusive jet production, due to the extra color-singlet exchange in the hard cross section. However, we do not expect this potential additional factor (on top of 0.03) to be large, due to the fact that extra soft interactions with the BFKL pomeron are unlikely: the 2-to-2 hard scattering takes place on much shorter time scale compared to the soft interactions filling the rapidity gaps. In any case, this will be checked at the LHC and if necessary, our numbers can then be adjusted accordingly.

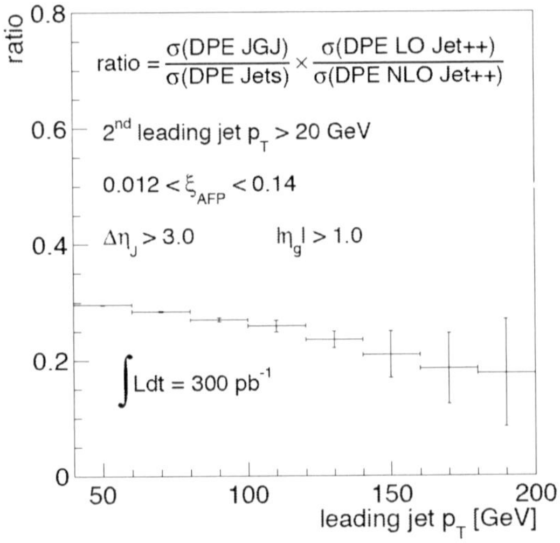

Figure 4. Ratio of the jet-gap-jet to the inclusive jet cross sections at the LHC as a function of jet p_T in double pomeron exchange events where the protons are detected in AFP or CT-PPS.

3 Conclusion

References

[1] V. S. Fadin, E. A. Kuraev, L. N. Lipatov, Phys. Lett. B**60** (1975) 50; L. N. Lipatov, *Sov. J. Nucl. Phys.* **23** (1976) 338; E. A. Kuraev, L. N. Lipatov and V. S. Fadin, *Sov. Phys. JETP* **45** (1977) 199;I. I. Balitsky, L. N. Lipatov, Sov.J.Nucl.Phys. **28** (1978) 822; V.S. Fadin and L.N. Lipatov, *Phys. Lett.* **B429** (1998) 127; M. Ciafaloni, *Phys. Lett.* **B429** (1998) 363; M. Ciafaloni and G. Camici, *Phys. Lett.* **B430** (1998) 349.

[2] G.Altarelli and G.Parisi, *Nucl. Phys.* **B126** 18C (1977) 298; V.N.Gribov and L.N.Lipatov, *Sov. Journ. Nucl. Phys.* (1972) 438 and 675; Yu.L.Dokshitzer, *Sov. Phys. JETP.* **46** (1977) 641.

[3] H1 Coll., Eur. Phys. J. C**46** (2006) 27; ZEUS Coll., Phys. Lett. B **632** (2006) 13.

[4] O. Kepka, C. Royon, C. Marquet, R. Peschanski, Eur. Phys. J. C**55** (2008) 259; O. Kepka, C. Royon, C. Marquet, R. Peschanski, Phys. Lett. B**655** (2007) 236; O. Kepka, C. Royon, C. Marquet, R. Peschanski, Eur. Phys. J. C**55** (2008) 259; C. Marquet, C. Royon, Nucl. Phys. B**739** (2006); H. Navelet, R. Peschanski, C. Royon, S. Wallon, Phys. Lett. B**385** (1996) 357; H. Navelet, R. Peschanski, C. Royon, Phys. Lett. B366 (1996) 329; J.G. Contreras, R. Peschanski and C. Royon, Phys. Rev. D**62** (2000) 034006; C. Marquet, R. Peschanski and C. Royon, Phys. Lett. B**599** (2004) 236.

[5] A.H. Mueller, H. Navelet, Nucl. Phys. B**282** (1987) 727.

[6] C. Marquet. C. Royon, Phys. Rev. D**79** (2009) 034028.

[7] B. Ducloué, L. Szymanowski, S. Wallon, JHEP **1305** (2013) 096; D. Colferai, F. Schwennsen, L. Szymanowski, S. Wallon, JHEP **1012** (2010) 026; F. Caporale, F.G. Celiberto, G. Chachamis, D. Gordo Gomez, A. Sabio Vera, Phys. Rev. D**95** (2017) no.7, 074007; M. Angioni, G. Chachamis, J.D. Madrigal, A. Sabio Vera, Phys. Rev. Lett. **107** (2011) 191601; A. Sabio Vera and F.

Schwennsen, Nucl. Phys. B **776** (2007) 170; A. Sabio Vera and F. Schwennsen, Phys. Rev. D**77** (2008) 014001; B. Ducloue, L. Szymanowski, S. Wallon, Phys. Rev. Lett. **112** (2014) 082003.

[8] CMS Coll., JHEP **08** (2016) 139; ATLAS Coll. Eur. Phys. J. C**74** (2014) 3117.

[9] CMS Coll. Eur. Phys. J. C**67** (2010) 601-615.

[10] G. Chachamis, A. Sabio Vera, JHEP **1602** (2016) 064; G. Chachamis, A. Sabio Vera, Phys.Rev. D**93** (2016) no.7, 074004; C. Baldenegro, F. Deganutti, D. Gordo, M. Klasen, C. Royon, J. Salomon, in preparation

[11] A. H. Mueller, W. K. Tang, Phys. Lett. B **284** (1992) 123.

[12] B. Abbott *et al.*, Phys. Lett. B **440**, 189 (1998); F. Abe *et al.*, Phys. Rev. Lett. **80**, 1156 (1998).

[13] CMS Coll., Eur, Phys. J. C**78** (2018) 242.

[14] O. Kepka, C. Marquet, C. Royon, Phys. Rev. D**83** (2011) 034036; F. Chevallier, O. Kepka, C. Marquet, C. Royon, Phys. Rev. D**79** (2009) 094019.

[15] G. Marchesini *et al.*, Comp. Phys. Comm. **67**, 465 (1992).

[16] Z. Nagy and Z. Trocsanyi, Phys. Rev. Lett. **87**, 082001 (2001).

[17] M. Boonekamp, A. Dechambre, V. Juranek, O. Kepka, M. Rangel, C. Royon, R. Staszewski, e-Print: arXiv:1102.2531; M. Boonekamp, V. Juranek, O. Kepka, C. Royon "Forward Physics Monte Carlo", "Proceedings of the workshop: HERA and the LHC workshop series on the implications of HERA for LHC physics," arXiv:0903.3861 [hep-ph].

[18] D. Colferai, F. Deganutti, C. Royon, T, Raben, in preparation.

[19] M. Hentschinski, J. D. Madrigal Martinez, B. Murdaca, A. Sabio Vera, Nucl. Phys. B**887** (2014) 309.

[20] C. Marquet, C. Royon, M. Trzebinski, R. Zlebcik, Phys.Rev. D**87** (2013) 3, 034010; O. Kepka, C. Marquet, C. Royon, Phys. Rev. D**83** (2011) 034036.

Jet charge in heavy-ion collisions

Hai Tao Li[1,⋆]

[1] *Theoretical Division, Los Alamos National Laboratory, Los Alamos, NM, 87545, USA*

Abstract.
Jet quenching effects have been widely used to study the properties of strongly-interacting matter, quark-gluon plasma, in heavy-ion collisions. Flavor tagging in heavy-ion collisions plays an important role to reveal the medium parton showers for quark and gluon evolution. Combining with kinematic information, the average jet charge can be used to separate the contribution of different jet flavors, which is defined as the momentum-weighted sum of the charges of hadrons inside a given jet. Using soft-collinear effective theory with medium interactions, we investigate the factorization of the jet charge in QCD medium. We provide predictions for jet charge distributions and their modifications compared to the ones in proton-proton collisions.

1 Introduction

The jet charge is defined as the transverse momentum-weighted sum of the charges of the hadrons in jet cone [1]

$$Q_{\kappa,\mathrm{jet}} = \sum_{i\in\mathrm{jet}} Q_i \left(\frac{p_T^i}{p_T^{\mathrm{jet}}}\right)^{\kappa} , \qquad (1)$$

where p_T^{jet} and p_T^i is the transverse momenta of the jet and hadron i. Q_i is the electric charge of particle i and κ is a free parameter.

In proton-proton collisions, the works [2, 3] studied the jet charge distribution using soft-collinear effective theory, where this observable can be written as the product of perturbative calculable jet functions and the non-perturbative fragmentation functions. The average charge of gluon jet is approximately zero if we ignore the soft correlations. Ideally the charges of different jet flavors are different which makes it is possible to tag the flavor origin of the jet. At the LHC, the ATLAS collaboration in Ref. [4] extracted the average charges of up- and down-quark jet as a function of jet p_T, where the charges are well separated. Furthermore, they also measured the scale violation of the jet charge distribution, for which the prediction is

$$\frac{p_T}{\langle Q_{\kappa,q}\rangle}\frac{d}{dp_T}\langle Q_{\kappa,q}\rangle = \frac{\alpha_s}{\pi}\tilde{P}_{qq}(\kappa), \qquad (2)$$

where $\tilde{P}_{qq}(\kappa)$ is the moment of the QCD splitting function.

⋆e-mail: haitaoli@lanl.gov

In heavy-ion collisions, jet charge is a promising observable, which can be used to study the initial-state and final-state effects. For the initial state effects, the parton densities of up-quark and down-quark are very different in heavy nucleus resulting in different fractions of up-quark and down-quark jets in comparison with proton-proton collisions. The charge distributions of untagged jets are significantly modified. For the final state effects, the gluon tends to lose more energy due to the larger color factor. Therefore, more quark jets are expected in heavy-ion collisions. Furthermore, the hot QCD medium produced in the collisions also modifies the jet and fragmentation functions. The evolution of energetic partons medium can be tested uniquely.

In these proceedings, we present our recent work [5] about jet charge in heavy-ion collisions, where we introduce a framework for perturbative calculations of the average jet charge in analogy to the study [2, 3] in vacuum. The global-fit nuclear PDFs are used for the initial state effects. The final state effects are constructed by the medium-induced collinear splitting kernels [6–10]. For similar applications see Refs. [11–14]. And we will show that the initial state isospin effects and final state jet quenching effects can be disentangled utilizing jet charge distributions.

2 Framework

2.1 Jet charge in proton-proton collisions

According to the factorization of the cross section for finding a hadron inside a given jet in SCET, the average jet charge for a quark jet can be written as [2, 3]

$$\langle Q_{\kappa,q}\rangle = \frac{\tilde{\mathcal{J}}_{qq}(E,R,\kappa,\mu)}{J_q(E,R,\mu)}\tilde{D}_q^Q(\kappa,\mu)\,, \tag{3}$$

with $\tilde{\mathcal{J}}_{qq}(E,R,\kappa,\mu)$ the Wilson coefficient matching the fragmenting jet function onto a fragmentation function. E, R and μ are the jet energy, the jet radius, and the factorization scale, respectively. The $(\kappa+1)$-th Mellin moments are defined as

$$\tilde{\mathcal{J}}_{qq}(E,R,\kappa,\mu) = \int_0^1 dz\, z^\kappa \mathcal{J}_{qq}(E,R,z,\mu)\,, \qquad \tilde{D}_q^Q(\kappa,\mu) = \int_0^1 dz\, z^\kappa \sum_h Q_h D_q^h(z,\mu)\,. \tag{4}$$

The detail discussions about the jet function and the jet matching coefficients can be found in Refs. [3, 15, 16]. For a given flavor, jet charge only depends on one nonperturbative parameter, the fragmentation function D_j^Q, which is the momentum weighted charge for parton j evolving into all hadrons. The fragmentation function obeys a DGLAP-like evolution. Due to no contribution from gluon jet, the evolution equation of $\tilde{D}_q^Q(\kappa,\mu)$ can be written as

$$\mu\frac{d}{d\mu}\tilde{D}_q^Q(\kappa,\mu) = \frac{\alpha_s(\mu)}{\pi}\tilde{P}_{qq}(\kappa)\tilde{D}_q^Q(\kappa,\mu)\,. \tag{5}$$

2.2 Jet charge in heavy-ion collisions

In analogy to the case of the vacuum jet function which describes the physics in the collinear limit, the medium modifications are introduced by replacing the splitting functions by the in-medium splitting ones. And the medium-induced splitting functions can be constructed using the splitting kernels induced by QCD medium as discussed in Refs. [5, 11, 12]. Similar to the case of semi-inclusive jet functions [13, 14], the medium corrections to the matching coefficients are

$$\mathcal{J}_{qq}^{\rm med}(E,R,x,\mu) = \frac{\alpha_s(\mu)}{2\pi^2}\int_0^{2Ex(1-x)\tan R/2}\frac{d^2\mathbf{k}_\perp}{\mathbf{k}_\perp^2}P_{q\to qg}^{\rm med}(x,\mathbf{k}_\perp)\,,$$
$$\mathcal{J}_{qg}^{\rm med}(E,R,x,\mu) = \frac{\alpha_s(\mu)}{2\pi^2}\int_0^{2Ex(1-x)\tan R/2}\frac{d^2\mathbf{k}_\perp}{\mathbf{k}_\perp^2}P_{q\to gq}^{\rm med}(x,\mathbf{k}_\perp)\,, \tag{6}$$

where $P_{i\to jk}^{\rm med}$ is the medium induced corrections to splitting functions. The medium correction to the total quark jet function is [5]

$$J_q^{\rm med}(E,R,\mu) = \frac{\alpha_s(\mu)}{2\pi^2}\int_0^1 dx\int_0^{2Ex(1-x)\tan R/2}\frac{d^2\mathbf{k}_\perp}{\mathbf{k}_\perp^2}P_{q\to qg}^{\rm med,real}(x,\mathbf{k}_\perp)\,. \tag{7}$$

The fragmentation functions in medium evolve according to a medium-modified DGLAP evolution. The evolution in medium has been used quite widely in literature [8, 11, 12, 17]. In our work, the evolution of the fragmentation function is given by

$$\frac{d}{d\ln\mu}\tilde{D}_q^{Q,\rm full}(\kappa,\mu) = \frac{\alpha_s(\mu)}{\pi}\left(\tilde{P}_{qq}(\kappa)+\tilde{P}_{qq}^{\rm med}(\kappa,\mu)\right)\tilde{D}_q^{Q,\rm full}(\kappa,\mu). \tag{8}$$

The average jet charge in heavy-ion collisions up to first order in QCD and first order in opacity can be written as

$$\langle Q_{q,\kappa}^{\rm AA}\rangle = \langle Q_{q,\kappa}^{\rm pp}\rangle\left(1+\tilde{\mathcal{J}}_{qq}^{\rm med}-J_q^{\rm med}\right)\exp\left[\int_{\mu_0}^{\mu}\frac{d\bar{\mu}}{\bar{\mu}}\frac{\alpha_s(\bar{\mu})}{\pi}\tilde{P}_{qq}^{\rm med}\right]+O(\alpha_s^2,\chi^2)\,. \tag{9}$$

If we choose the scale $\mu = p_T R$, the p_T dependence of the jet charge is

$$\frac{d}{d\ln p_T}\ln\langle Q_{q,\kappa}^{\rm AA}\rangle = \frac{\alpha_s(p_TR)}{\pi}\left[\tilde{P}_{qq}(\kappa)+\tilde{P}_{qq}^{\rm med}(\kappa,p_TR)+\int_0^1 dx(x^\kappa-1)P_{qq}^{\rm med}(\kappa,k_\perp=x(1-x)p_TR)\right]. \tag{10}$$

In principle the scale violation in heavy-ion collisions can be used to extract the information of the medium splitting functions.

3 Numerical results

Only one nonperturbative parameter for a given jet flavor is required for jet charge distribution, which is obtained by fitting our predictions using PYTHIA8 [18, 19] simulations or ALTAS measurements [4]. In QCD medium we evolve the fragmentation functions from $\Lambda_{\rm QCD} = 0.2$ GeV to the jet scale p_TR. The scale of the QCD coupling in medium is set to be $\sqrt{k_\perp^2+\mu_D^2}$ where $\mu_D = 0.75$ GeV. The flavor fractions of jet types are simulated by LO matrix elements with CT14NLO PDF sets [20] for proton and nCTEQ15 [21] PDF sets for nuclei. In heavy-ion collisions for the jet fractions, we choose to use the energy loss approach [11].

In Fig. 1, we present the up- and down-quark jet charges in p+p collisions with $\sqrt{s_{\rm NN}} = 8$ TeV compared with ATLAS measurements [4]. The bands of the predictions are generated by combining the scale uncertainties and 1-σ deviation of the nonperturbative parameter. We find that these scale uncertainties are small. As indicated in the jet charge definition, the absolute value is smaller for larger

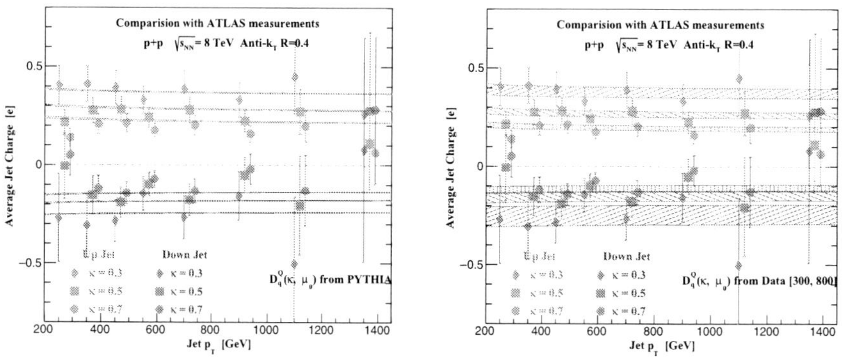

Figure 1. Average charge of up and down-quark jet as a function of jet p_T with κ =0.3, 0.5 and 0.7 in $\sqrt{s_{\rm NN}}$ = 8 TeV p+p collisions. The dots with error bars represent the measurements by ATLAS [4]. The red and blue bands are the predictions with the non-perturbative parameters obtained through fitting to PYTHIA simulations (left) and ATLAS measurements (right).

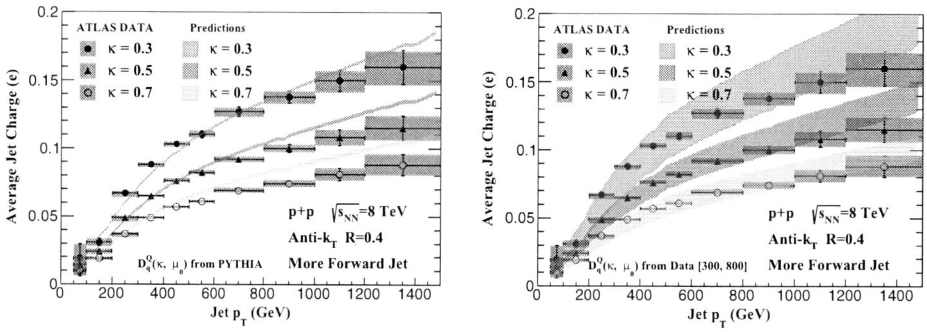

Figure 2. Transverse momentum dependence of the average jet charge distributions for more forward jets in dijet events at $\sqrt{s_{\rm NN}}$ = 8 TeV in p+p collisions. The non-perturbative parameters are obtained by fitting to PYTHIA simulations (left) and ATLAS measurements (right).

κ. The up-quark and down-quark jets are well separated with positive and negative electric charges, respectively. The predictions are consistent with the measurements by ATLAS.

Considering valence quark contribution only, the charge for more forward jets is $\langle Q_\kappa^f \rangle = (f_u^f - f_{\bar{u}}^f)\langle Q_\kappa^u \rangle + (f_d^f - f_{\bar{d}}^f)\langle Q_\kappa^d \rangle$, where the more forward jet is the one with a larger η in dijet events. f_q^f represents the fraction of jet flavors. Q_{κ,q_i} is the q_i-jet's charge. For more forward jets, the valence quark contribution is enhanced, especially in the high transverse momentum region. In Fig. 2 we show the average jet charge distribution for the more forward jet. Our predictions can describe the experimental data very well.

In the rest of this section we will use the fragmentation functions obtained from PYTHIA simulations. The left panel in Fig. 3 shows the average jet charge distribution and its modifications with $\sqrt{s_{\rm NN}}$ = 5.02 TeV in 0-10% central Pb+Pb collisions. The uncertainties are calculated by varying the effective coupling $g \in [1.8, 2.0]$ between the energetic parton and QCD medium. The charges of up- and down-quark jets remain well separated. The right panel in Fig. 3 shows the modification of up-quark jet charge. Because the QCD interaction is flavor-blind, the medium modifications for up-

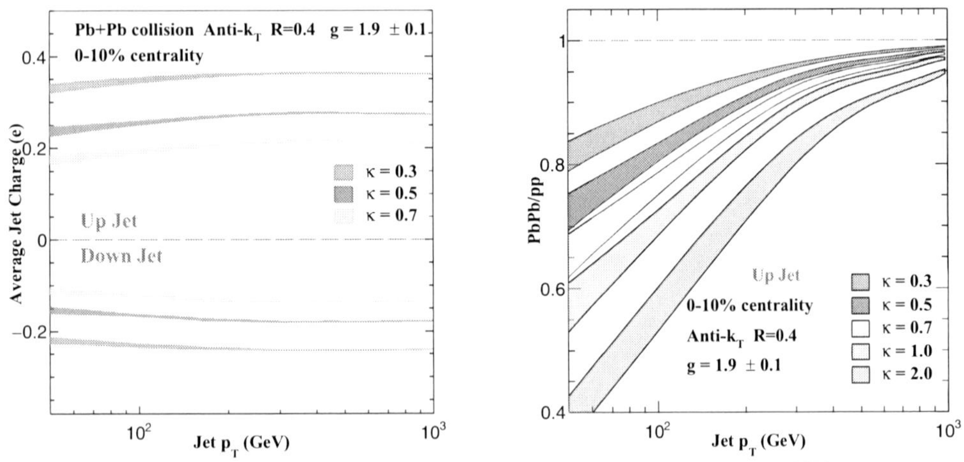

Figure 3. Average quark jet charges and their modifications as a function of jet p_T in $\sqrt{s_{NN}}$ = 5.02 TeV 0-10% central Pb+Pb collisions. The coupling between the jet and QCD medium is set to be g=1.9±0.1.

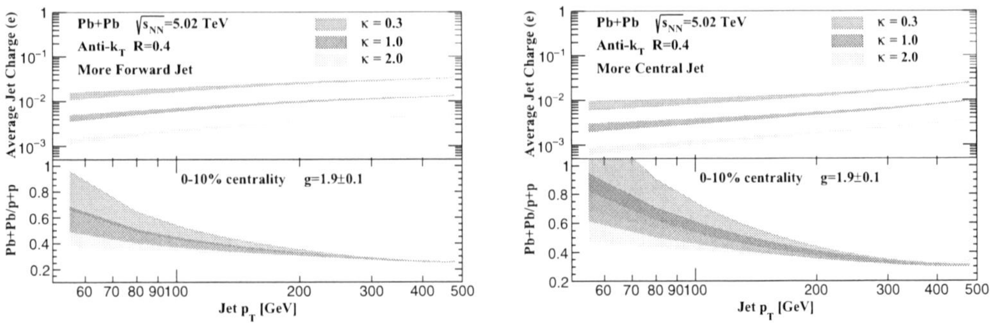

Figure 4. Average jet charge distribution for more forward (right) and more central (left) jet with κ =0.3, 1.0 and 2.0 in Pb+Pb collisions with $\sqrt{s_{NN}}$ =5.02 TeV with nonperturbative parameters from PYTHIA. The bottom panels show the modification of the average jet charge in 0-10% central Pb+Pb collisions relative to the ones in p+p collisions.

and down-quark jet are the same. The modifications are larger for smaller energy jet or larger κ, which arises from the medium-induced parton shower on jet evolution.

Figure 4 presents our predictions for the average charge of more forward and more central jets in Pb+Pb collision with $\sqrt{s_{NN}}$ = 5.02 TeV, which can be measured straightforwardly in heavy-ion collisions. Due to isospin effects, there is a huge cancellation between the contributions from up- and down-quark jets. In results, average charge is smaller in Pb+Pb collisions, especially for very large p_T. At low p_T the modification is sensitive to medium induced radiations, which at high p_T the initial stat effects are more important.

4 Conclusion

In these proceedings, we reported a theoretical framework to investigate average jet charge in heavy-ion collisions, which is build upon the SCET$_G$ approach. The jet charge observable can be written as the product of the jet matching function and fragmentation functions. In heavy-ion collisions the modifications of the collinear objects are constructed with the help of the medium-induced splitting kernels derived with $SCET_G$. We presented the average jet charge distributions for specific jet flavor, as well as the predictions for more forward and central jets at hadronic colliders. The jet charge does not depend on the hard process, which can also be measured at RHIC and a future EIC.

References

[1] R.D. Field, R.P. Feynman, Nucl. Phys. **B136**, 1 (1978), [763(1977)]
[2] D. Krohn, M.D. Schwartz, T. Lin, W.J. Waalewijn, Phys. Rev. Lett. **110**, 212001 (2013), 1209.2421
[3] W.J. Waalewijn, Phys. Rev. **D86**, 094030 (2012), 1209.3019
[4] G. Aad et al. (ATLAS), Phys. Rev. **D93**, 052003 (2016), 1509.05190
[5] H.T. Li, I. Vitev (2019), 1908.06979
[6] G. Ovanesyan, I. Vitev, JHEP **06**, 080 (2011), 1103.1074
[7] G. Ovanesyan, I. Vitev, Phys. Lett. **B706**, 371 (2012), 1109.5619
[8] Z.B. Kang, F. Ringer, I. Vitev, JHEP **03**, 146 (2017), 1610.02043
[9] M.D. Sievert, I. Vitev, Phys. Rev. **D98**, 094010 (2018), 1807.03799
[10] M.D. Sievert, I. Vitev, B. Yoon, Phys. Lett. **B795**, 502 (2019), 1903.06170
[11] Z.B. Kang, R. Lashof-Regas, G. Ovanesyan, P. Saad, I. Vitev, Phys. Rev. Lett. **114**, 092002 (2015), 1405.2612
[12] Y.T. Chien, A. Emerman, Z.B. Kang, G. Ovanesyan, I. Vitev, Phys. Rev. **D93**, 074030 (2016), 1509.02936
[13] Z.B. Kang, F. Ringer, I. Vitev, Phys. Lett. **B769**, 242 (2017), 1701.05839
[14] H.T. Li, I. Vitev, JHEP **07**, 148 (2019), 1811.07905
[15] A. Jain, M. Procura, W.J. Waalewijn, JHEP **05**, 035 (2011), 1101.4953
[16] S.D. Ellis, C.K. Vermilion, J.R. Walsh, A. Hornig, C. Lee, JHEP **11**, 101 (2010), 1001.0014
[17] H.T. Li, I. Vitev, Phys. Lett. **B793**, 259 (2019), 1801.00008
[18] T. Sjostrand, S. Mrenna, P.Z. Skands, Comput. Phys. Commun. **178**, 852 (2008), 0710.3820
[19] T. Sjöstrand, S. Ask, J.R. Christiansen, R. Corke, N. Desai, P. Ilten, S. Mrenna, S. Prestel, C.O. Rasmussen, P.Z. Skands, Comput. Phys. Commun. **191**, 159 (2015), 1410.3012
[20] S. Dulat, T.J. Hou, J. Gao, M. Guzzi, J. Huston, P. Nadolsky, J. Pumplin, C. Schmidt, D. Stump, C.P. Yuan, Phys. Rev. **D93**, 033006 (2016), 1506.07443
[21] K. Kovarik et al., Phys. Rev. **D93**, 085037 (2016), 1509.00792

Reconstruction of neutral-triggered charged recoil jets in $\sqrt{s} =$ 200 GeV p+p collisions at the STAR experiment

Derek Anderson[1,2,*]

[1] *for the STAR Collaboration*
[2] *Cyclotron Institute, Texas A&M University*

Abstract. Jets — collimated sprays of hadrons — are produced by hard-scattered partons during the early stages of heavy ion collisions. Hence, they provide a valuable probe of the complex multi-particle dynamics within the hot, dense medium produced in such collisions. In particular, the study of jets recoiling from direct photons (γ_{dir}+jet) and those recoiling from energetic π^0 (π^0+jet) may shed light on the path-length and initial flavor (quark vs. gluon) dependence of the energy-loss experienced by a parton as it traverses the medium. We present here measurements of the yields of charged recoil jets tagged by γ_{dir} and π^0 in p+p collisions at $\sqrt{s} = 200$ GeV. These measurements will serve as a vacuum fragmentation reference for an upcoming measurement in Au+Au collisions.

1 Introduction

The collisions of ultra-relativistic heavy ions such as the Au+Au collisions studied at the Relativistic Heavy-Ion Collider (RHIC) produce an extremely hot, dense medium with properties consistent with those expected of a liquid of deconfined partons, the Quark-Gluon Plasma (QGP). One of the primary goals of RHIC is to obtain a quantitative and precise understanding of the complex multi-particle dynamics in this extreme state of matter.

An important signature of the formation of this medium is "jet quenching" [1]. As partons move through the medium, they lose energy via collisional and radiative interactions. This in-medium partonic energy loss results in a suppression of high energy particles relative to a collision system where no medium is expected to be produced (e.g. p+p).

Jets — collimated sprays of hadrons produced by the fragmentation of hard-scattered partons -— are a natural observable for probing this phenomenon. Jets are powerful probes due to the large momentum-transfers involved in their production, which makes them amenable to a perturbative QCD description. Additionally, they are produced in the earliest stages of the collisions and so are expected to be sensitive to the evolution of the medium.

A so-called "golden channel" to study jet quenching is jets tagged by "prompt photons" (γ_{prompt}), photons scattered from energetic partons. Since γ_{prompt} are color-neutral, they will not interact with the medium. It follows that the measured transverse energy of a γ_{prompt} will closely approximate the initial transverse energy of the recoiling parton. Thus, the recoiling parton and the ensuing jet serves as a well-calibrated probe of the in-medium energy loss [2]. In practice, however, γ_{prompt} can not be

*e-mail: dmawxc@physics.tamu.edu

measured in isolation. Rather, a mixture of prompt, thermal, and fragmentation photons is measured. These are collectively referred to as "direct photons" ($\gamma_{\rm dir}$). For the energies considered here, the thermal contribution is negligible [3], but there is a sizable background due to hadronic decays which must be corrected for.

In order to gain insight into the path-length and flavor (quark vs. gluon) dependence of in-medium energy loss, $\gamma_{\rm dir}$-tagged recoil jets are compared to recoil jets tagged by an energetic π^0. Energetic π^0, being produced as part of a jet, are subject to jet-quenching, and thus triggering on them biases their production towards the surface of the medium [4]. On the other hand, $\gamma_{\rm dir}$ have no such bias. This implies that on average the recoil jets tagged by π^0 have a longer path-length through the medium than do recoil jets tagged by $\gamma_{\rm dir}$. Moreover, the production of $\gamma_{\rm dir}$ is dominated by (anti-)quark-gluon Compton scattering. So, the recoil jets tagged $\gamma_{\rm dir}$ are predominantly from (anti-)quarks.

These two observations, that energetic π^0 are biased towards surface production and that $\gamma_{\rm dir}$-tagged jets predominantly originate from (anti-)quarks, suggest that energy loss experienced by the recoil jets may differ between the two systems. Thus, it might be that the observed suppression of the recoil jets also differs. This suppression is quantified by $I_{\rm AA} \equiv D^{\rm Au+Au}/D^{\rm p+p}$, where $D^{\rm Au+Au}$ and $D^{\rm p+p}$ are the per-trigger yields of recoil jets in Au+Au and p+p, respectively. Presented in these proceedings are measurements of the $\gamma_{\rm dir}$- and π^0-tagged per-trigger charged recoil jet spectra in p+p collisions ($D^{\rm p+p}$), which will ultimately serve as vacuum-fragmentation (no medium) baselines for future measurements of I_{AA}.

2 The STAR Detector

The data were obtained by analyzing p+p collisions at $\sqrt{s}$ = 200 GeV recorded by the Solenoidal Tracker at RHIC (STAR) detector during the running year 2009. The two sub-systems of the STAR detector most relevant to jet measurements are the Time Projection Chamber (TPC) [5] and the Barrel (i.e. mid-rapidity) Electromagnetic Calorimeter (BEMC) [6]. The TPC is used to measure the transverse momentum ($p_{\rm T}$), azimuth (φ), and pseudorapidity (η) of charged particles, and the BEMC is used to trigger on p+p collisions which may contain an energetic π^0 or $\gamma_{\rm dir}$.

At approximately 5.6 radiation lengths into the BEMC at $\eta = 0$ sits the Barrel Shower Maximum Detector (BSMD). The BSMD is a grid of readout wires which is used to distinguish hadronic decays from single photons, and has a granularity of approximately 0.007×0.007 in the η-φ plane. Candidate triggers are identified as one to two BEMC tower clusters with transverse energy in the range of $E_{\rm T}^{\rm trg} \in (9, 20)$ GeV and a centroid whose pseudorapidity lies in the range $|\eta^{\rm trg}| < 0.9$. Single photons and π^0 are identified from these candidates based on their shower shape, which is quantified by the so-called "Transverse Shower Profile" (TSP):

$$\text{TSP} \equiv \frac{E_{\rm clust}}{\Sigma_i e_i r_i^{1.5}}$$

where $E_{\rm clust}$ is the total energy of the cluster, e_i is the energy of the $i^{\rm th}$ BSMD strip of the cluster, and r_i is the distance from the $i^{\rm th}$ BSMD strip to the centroid of the cluster [7]. The exponent of 1.5 in the denominator of the TSP was tuned via simulations to give maximal separation in TSP between $\pi^0 \rightarrow \gamma\gamma$ decays (which produce broader showers in the BEMC) and isolated $\gamma_{\rm dir}$ (which produce narrower showers in the BEMC). It should be noted that the opening angle of a 9 - 20 GeV π^0 decaying to two photons is small enough to be contained within one to two BEMC towers.

From the candidate triggers, the TSP is used to select a 95% pure sample of π^0 triggers and to select a sample of triggers with an enhanced fraction of $\gamma_{\rm dir}$, referred to as $\gamma_{\rm rich}$ triggers. The identified π^0 triggers have TSP < 0.08, and the $\gamma_{\rm rich}$ triggers have TSP $\in (0.2, 0.6)$. A representative TSP distribution is shown in Fig. 1. Finally, events tagged by π^0 and $\gamma_{\rm rich}$ triggers which have transverse energies of 9 - 11, 11 - 15, or 15 - 20 GeV are analyzed.

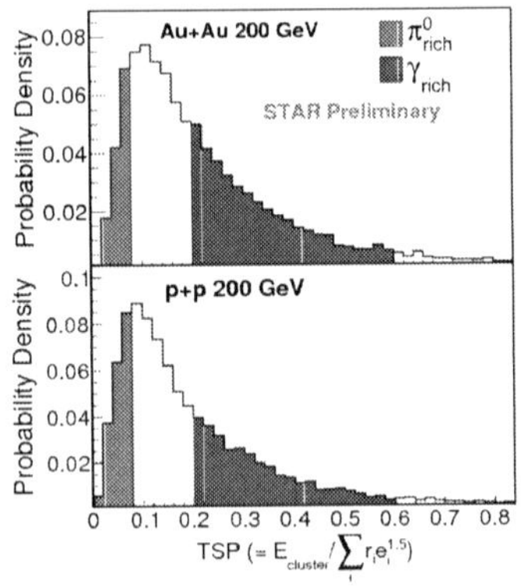

Figure 1: The TSP distribution in Au+Au and p+p collisions. The blue and red regions indicate the cuts used to identify the π^0 and $\gamma_{\rm rich}$ samples of triggers respectively. The figure is from [8].

The $\gamma_{\rm rich}$ background due to hadronic decays is estimated in a data-driven manner by comparing the near-side per-trigger azimuthal yields of the π^0 and $\gamma^{\rm rich}$ samples. For $E_{\rm T}^{\rm trg} \in (9, 11)$ GeV, the fraction of $\gamma_{\rm rich}$ triggers due to hadronic decays ($\mathfrak{B}$) is measured to be $\mathfrak{B} \approx 0.57$.

3 Corrected Recoil Jet Spectra

Jets are reconstructed from TPC tracks using the anti-$k_{\rm T}$ algorithm as implemented in FastJet 3.0.6 [9]. A resolution parameter of $R_{\rm jet} = 0.2$ was used for the data shown in these proceedings. The TPC tracks utilized in jet reconstruction were required to have $p_{\rm T}^{\rm trk} \in (0.2, 30)$ GeV/c and $|\eta^{\rm trk}| < 1$. Only the reconstructed jets which satisfy the following conditions are considered:

(i) the raw reconstructed ($p_{\rm T}^{\rm jet,raw}$) satisfies $p_{\rm T}^{\rm jet,raw} \in (0.2, 30)$ GeV/c;

(ii) the jet area ($A^{\rm jet}$) satisfies $A^{\rm jet} > 0.05$;

(iii) the jet axis satisfies $|\eta^{\rm jet}| < 1 - R_{\rm jet}$;

(iv) and the jet axis satisfies $\Delta\varphi^{\rm jet} \in (3\pi/4, 5\pi/4)$.

Here, $\Delta\varphi^{\rm jet}$ indicates the difference between the azimuthal angles of the trigger and the jet axis. A "recoil jet" is one whose axis satisfies condition (iv). The reconstructed jets are then corrected for the median background energy density via the standard ρ prescription:

$$p_{\rm T}^{\rm reco} = p_{\rm T}^{\rm jet,raw} - \rho \cdot A^{\rm jet}$$

where ρ is the median background energy density. The procedure to calculate ρ is described in detail in [9]. For the range of trigger energies presented in these proceedings, the correction $\rho \cdot A^{\rm jet}$ removes no energy on average for $R_{\rm jet} = 0.2$ jets. However, increasing $R_{\rm jet}$ increases the jet area, and for $R_{\rm jet} = 0.5$ jets, the procedure removes 20 MeV/c on average for the presented range of trigger energies.

The data are corrected for detector effects following the procedure utilized in a recent measurement of hadron-jet correlations by the STAR Collaboration [10]. The jet energy resolution is corrected for

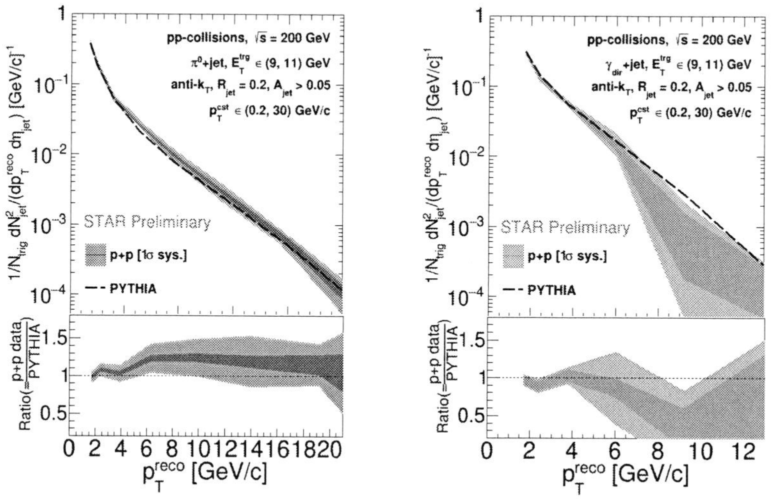

Figure 2: Corrected charged recoil jet spectra compared against PYTHIA 8. The darker bands indicate statistical uncertainty and the lighter bands indicate total uncertainty (statistical and systematic).

at the event-averaged ensemble level via an iterative unfolding algorithm based on Bayes' theorem as implemented in the RooUnfold software package [11]. The jet reconstruction efficiency is then corrected bin-by-bin. The fully corrected π^0- and $\gamma_{\rm dir}$-triggered charged recoil jet spectra for $E_{\rm T}^{\rm trg} \in (9, 11)$ GeV are compared against PYTHIA 8.1.85 [12] in Fig. 2.

The dominant systematic uncertainties are the unfolding process (which includes the choice of prior, number of iterations, etc.), and the tracking efficiency. However, in the case of the $\gamma_{\rm dir}$-tagged recoil jets, there is an additional systematic uncertainty associated with the subtraction of the background due to hadronic decays; this is the largest uncertainty of the $\gamma_{\rm dir}$-tagged recoil jets.

Within uncertainty, the corrected data are largely consistent with PYTHIA 8, though there is some tension between the two. It is known that at small $R_{\rm jet}$, PYTHIA 8 has difficulty describing jet spectra. Moreover, the shown PYTHIA 8 distributions were generated "out-of-the-box" with no additional tuning of the PYTHIA 8 simulations.

4 Summary and Future Work

Charged recoil jets with $R_{\rm jet} = 0.2$ tagged by energetic π^0 and $\gamma_{\rm dir}$ with $E_{\rm T}^{\rm trg} \in (9, 11)$ GeV have been reconstructed in p+p collisions at $\sqrt{s} = 200$ GeV. The fully corrected data compare favorably to PYTHIA 8 simulations. However, there is tension, particularly between the $\gamma_{\rm dir}$-triggered data and simulation. Corresponding measurements of charged recoil jet spectra for different values of $R_{\rm jet}$ and $E_{\rm T}^{\rm trg}$ ranges of 11 - 15 and 15 - 20 GeV are underway. This analysis will also be extended to full jets, i.e. jets consisting of TPC tracks and BEMC towers. The impact on the recoil jet spectra from incorporating the neutral component of the jet is striking: Figure 3 illustrates this by comparing a charged and full recoil jet spectrum generated in PYTHIA 8. The addition of the neutral component will enable more precise measurements of the recoil jets' energy and enable an easier comparison to theoretical calculations which will lead to a more complete picture of the dynamics of in-medium energy loss.

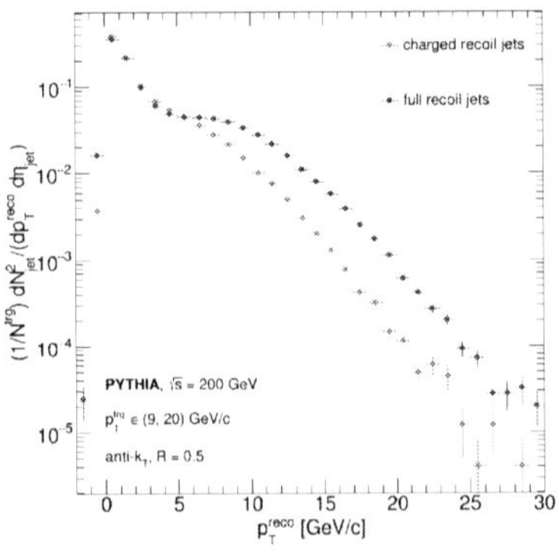

Figure 3: Charged and full γ_{dir} tagged recoil jet spectra for $R_{jet} = 0.5$ generated in PYTHIA 8.

References

[1] M. Connors, C. Nattrass, R. Reed, and S. Salur, RMP **90**, 025005 (2018)
[2] Wang et al., PRL **77**, 231 (1996)
[3] G. David, (2019) arXiv:1907.08893 [nucl-ex]
[4] T. Renk, PRC **88**, 054902 (2013)
[5] M. Anderson et al., NIM A **499**, 659 (2003)
[6] M. Beddo et al., NIM A **499**, 725 (2003)
[7] B. I. Abelev et al., PRC **82**, 034909 (2010)
[8] N. Sahoo and the STAR collaboration, NPA **956**, 621 (2016)
[9] M. Cacciari, G. P. Salam, and G. Soyez, EPJ C **72**, 1896 (2012)
[10] L. Adamcyzk et al., PRC **96**, 024905 (2017)
[11] T. Adye, (2011) arXiv:1105.1160 [physics.data-an]
[12] T. sjöstrand, S. Mrenna, and P. Z. Skands, Comput. Phys. Commun. **178**, 852 (2008)

The dijet mass distribution in heavy ion collisions

Jared Reiten[1,2,3,⋆]

[1] *Department of Physics and Astronomy, University of California, Los Angeles, California 90095, USA*
[2] *Mani L. Bhaumik Institute for Theoretical Physics, University of California, Los Angeles, California 90095, USA*
[3] *Theoretical Division, Los Alamos National Laboratory, Los Alamos, New Mexico 87545, USA*

Abstract. In these proceedings, we review the production of both light and heavy flavor dijets in heavy ion collisions and highlight a promising observable to expose their distinct signatures. We propose the modification of dijet invariant mass distributions in heavy ion collisions as a new observable that exhibits striking sensitivity to the heavy quark mass dependence of in-medium parton showers. This observable has the advantage of amplifying the effects of jet quenching in contrast to conventional observables, such as the dijet momentum imbalance shift, which involve cancellations of such effects and, hence, result in less pronounced signals. Predictions are presented for Au+Au collisions at $\sqrt{s_{NN}} = 200$ GeV to guide the future sPHENIX program at the Relativistic Heavy Ion Collider.

1 Introduction

Heavy flavor jets are a new frontier in the hard probes thrust of the ultra-relativistic nuclear collisions program. Back-to-back dijet measurements, in particular, are exciting experimental channels that are well-suited to probe the physics of jet production and subsequent evolution through the quark-gluon plasma (QGP) formed by colliding heavy ions. Such measurements provide valuable insight into the transport properties of the medium. For our present purposes, we are interested in the mass dependence of quark and gluon energy loss. A clear advantage that heavy flavor dijets have over single inclusive heavy flavor jets is that they are predominately initiated by $Q + \bar{Q}$ pairs in the final partonic state [1, 2], whereas single jets receive large contributions from prompt gluons – splitting to heavy quarks only in later stages of the showering process [3, 4]. This means that the study of the modification of dijet observables provides direct information about heavy flavor energy loss mechanisms in the medium, while single jet observables reflect those of light partons, as has been experimentally verified by the CMS collaboration [5].

In performing such measurements, experimentalists typically focus on the most energetic jet, the so-called "leading" jet, and the second most energetic jet, the "subleading" jet, of a given event [6–8]. This correlated pair then provides information on energy loss mechanisms in the QGP through the enhancement of their asymmetry, namely, the development of an imbalance between their respective momenta. Such imbalances develop as a result of the differing expanses of medium each jet must traverse before reaching detectors.

⋆e-mail: jdreiten@physics.ucla.edu

In this review, we will focus on two dijet observables to be utilized at the future sPHENIX experiment at RHIC: the traditional modification of the dijet imbalance distribution and the newly proposed modification of the dijet mass distribution [1].

2 Dijet production in heavy ion collisions

We begin by discussing the calculation of the dijet production cross section in heavy ion collisions. To do so, we first generate the dijet production cross section in elementary p+p collisions using Pythia 8 [9], utilizing the kinematic cuts to be used by the sPHENIX collaboration [10]. To map this to the case of heavy ions, we must take into account nuclear geometry as well as the effects of medium-induced energy loss. The former induces a dependence on the impact parameter $|\mathbf{b}_\perp|$ in the cross section, while the latter accounts for the alteration of phase space that comes from radiative [11–18] and collisional [19–24] energy loss. Altogether we have:

$$\begin{aligned}\frac{d\sigma^{AA}(|\mathbf{b}_\perp|)}{dp_{1T}dp_{2T}} &= \int d^2\mathbf{s}_\perp T_A\left(\mathbf{s}_\perp - \frac{\mathbf{b}_\perp}{2}\right)T_A\left(\mathbf{s}_\perp + \frac{\mathbf{b}_\perp}{2}\right) \\ &\times \sum_{q,g}\int_0^1 d\epsilon \frac{P^1_{q,g}(\epsilon;\mathbf{s}_\perp,|\mathbf{b}_\perp|)}{1-f^{1\,\mathrm{loss}}_{q,g}(R;\mathbf{s}_\perp,|\mathbf{b}_\perp|)\,\epsilon}\int_0^1 d\epsilon' \frac{P^2_{q,g}(\epsilon';\mathbf{s}_\perp,|\mathbf{b}_\perp|)}{1-f^{2\,\mathrm{loss}}_{q,g}(R;\mathbf{s}_\perp,|\mathbf{b}_\perp|)\,\epsilon'} \\ &\times \frac{d\sigma^{NN}_{q,g}\left(p_{1T}\big/\left[1-f^{1\,\mathrm{loss}}_{q,g}(R;\mathbf{s}_\perp,|\mathbf{b}_\perp|)\,\epsilon\right],\ p_{2T}\big/\left[1-f^{2\,\mathrm{loss}}_{q,g}(R;\mathbf{s}_\perp,|\mathbf{b}_\perp|)\,\epsilon'\right]\right)}{dp_{1T}dp_{2T}}, \qquad (1)\end{aligned}$$

where the T_A is known as the thickness function of the optical Glauber model [25], $P^i_{q,g}(\epsilon)$ is the probability for partons $i = 1,2$ to lose energy fraction ϵ to medium-induced gluon bremsstrahlung, and $f^{i,\mathrm{loss}}_{q,g}$ is the fraction of energy lost outside each jet cone. Energy may be lost to both radiative and collisional effects, and is determined via

$$f^{\mathrm{loss}}_{q,g}(R;\mathrm{rad}+\mathrm{coll}) = 1 - \left(\int_0^R dr\int_{\omega_{\min}}^E d\omega \frac{dN^g_{q,g}(\omega,r)}{d\omega dr}\right)\Bigg/\left(\int_0^{R_{\max}} dr\int_0^E d\omega \frac{dN^g_{q,g}(\omega,r)}{d\omega dr}\right), \qquad (2)$$

where $\omega_{\min}$ parameterizes collisional energy loss. The medium-induced splitting functions are calculated within the formalism of $\mathrm{SCET_{M,G}}$ [26], and may be obtained equivalently through the use of lightcone wavefunctions [27]. These splitting functions are evaluated in the hydrodynamically evolved medium simulated by the iEBE-VISHNU package [28].

Once we have Eq. (1), we may then calculate the so-called nuclear modification factor through

$$R_{AA}(p_{1T},p_{2T},|\mathbf{b}_\perp|) = \frac{1}{\langle N_{\mathrm{bin}}\rangle}\frac{d\sigma^{AA}(|\mathbf{b}_\perp|)/dp_{1T}dp_{2T}}{d\sigma^{pp}/dp_{1T}dp_{2T}}, \qquad (3)$$

where, in order to make a meaningful comparison to the cross section obtained in the p+p baseline, we have normalized by the average number of binary collisions for the given centrality, $\langle N_{\mathrm{bin}}\rangle$. The R_{AA} for light and heavy flavor dijets is displayed in Fig. 1. Note that $R_{AA} = 1$ denotes no modification relative to p+p, while $R_{AA} > 1$ signals enhancement, and $R_{AA} < 1$ marks suppression. Further note that Fig. 1 shows the greatest suppression existing along the main diagonal – a characteristic feature of the aforementioned asymmetry, which develops for dijets produced in heavy ion collisions.

3 Phenomenological results at RHIC

The dijet production cross section Eq. (1), differential in the transverse momenta of both the leading and subleading jets, provides us with the physical information required to construct the dijet imbalance and mass distributions. This allows us to study their respective modifications induced by the medium.

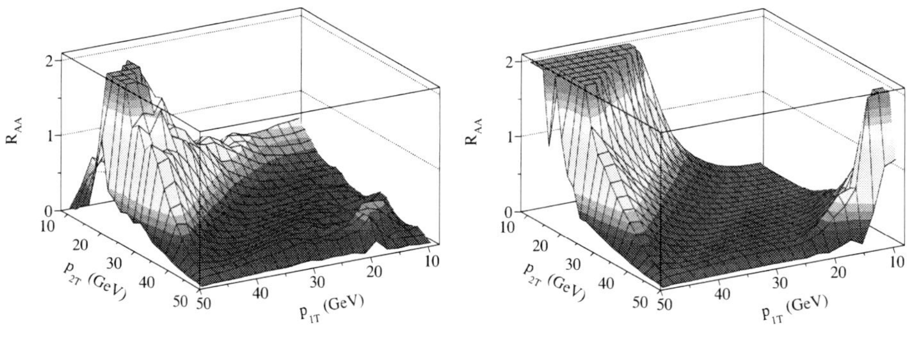

Figure 1. Nuclear modification factor for b-tagged (left) and inclusive (right) dijet production in Au+Au collisions at $\sqrt{s_{NN}} = 200$ GeV with sPHENIX kinematics [10].

3.1 Dijet imbalance distribution

The dijet imbalance distribution in heavy ion collisions may be calculated according to:

$$\frac{d\sigma^{AA}}{dz_J} = \int dp_{1T}\, dp_{2T}\, \frac{d\sigma^{AA}}{dp_{1T}dp_{2T}}\, \delta\left(z_J - \frac{p_{2T}}{p_{1T}}\right), \tag{4}$$

where we have defined the dijet imbalance z_J to be the ratio of the subleading to leading jet transverse momenta, making the allowed values for z_J lie between zero and unity.

Results for this distribution with sPHENIX kinematics are displayed in Fig. 2. In this figure, the black histogram is the distribution formed in p+p collisions, while the green is that of A+A. The relative shift between the two is the mark of modification due to the QGP. What is important to note is that the difference between the shift for heavy flavor dijets and that of inclusive (light flavor) is quite subtle. In fact, defining $\Delta\langle z_J\rangle \equiv \Delta\langle z_J\rangle_{pp} - \Delta\langle z_J\rangle_{AA}$, we find that $\Delta\langle z_J\rangle_{bb} = 0.065 \pm 0.012$ for heavy flavor dijet and $\Delta\langle z_J\rangle_{jj} = 0.100 \pm 0.005$ for light flavor dijet. Such a small effect offers little in differentiation between light and heavy flavors. This leads us to look for other dijet observables that amplify the mass dependence of in-medium parton propagation – the dijet mass distribution is such an observable.

3.2 Dijet mass distribution

The dijet mass distribution in heavy ion collisions can be obtained through use of Eq. (1) in a fashion analogous to the way Eq. (4) is obtained:

$$\frac{d\sigma^{AA}}{dm_{12}} = \int dp_{1T}\, dp_{2T}\, \frac{d\sigma^{AA}}{dp_{1T}dp_{2T}}\, \delta\left(m_{12} - \sqrt{\langle m_1^2\rangle + \langle m_2^2\rangle + 2p_{1T}\, p_{2T}\langle\cosh(\Delta\eta) - \cos(\Delta\phi)\rangle}\right). \tag{5}$$

Letting p_1 and p_2 denote the four-momenta of the jets forming the dijet pair, the dijet invariant mass m_{12} is obtained through $m_{12} = \sqrt{(p_1 + p_2)^2}$. The above expression, Eq. (5), is expressed in the high p_T limit where $p_{iT} \gg m_i$ for $i = 1, 2$ as well as in terms of the differences in rapidity, $\Delta\eta$, and azimuthal angle, $\Delta\phi$, of the jets forming the dijet pair. Furthermore, the brackets denote averages over each bracketed quantity in p+p collisions, whose modifications in A+A are insignificant, as observed by the ALICE collaboration at the LHC [29].

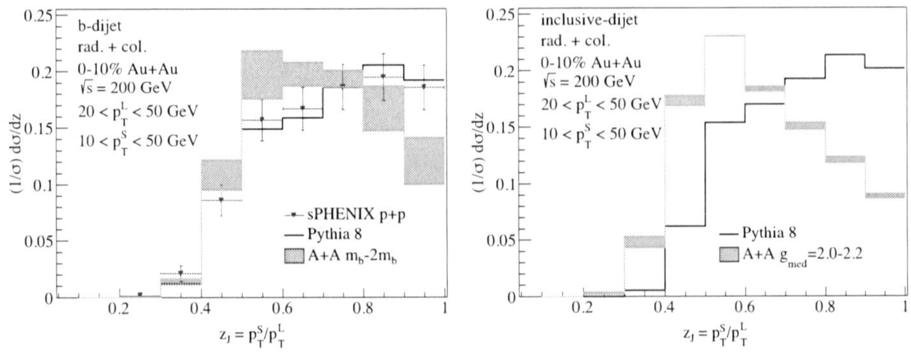

Figure 2. The dijet imbalance z_J distributions for b-tagged (left) and inclusive (right) dijet production at $\sqrt{s_{NN}}$ = 200 GeV with sPHENIX kinematics. The black histogram marks the distribution for p+p collisions, while the green marks that of Au+Au collisions. The blue "data" points come from preliminary simulations by the sPHENIX collaboration [10]. Left: the coupling of the jet to the medium is fixed at $g_{\rm med} = 2.0$ and the band corresponds to varying the range of masses of the propagating system between m_b and $2m_b$. Right: the band corresponds to varying the medium coupling $g_{\rm med} = 2.0 - 2.2$.

With the dijet mass distribution in hand, we may then define a new observable, namely, the nuclear modification factor for the dijet mass distribution:

$$R_{AA}\left(m_{12}, |\mathbf{b}_\perp|\right) = \frac{1}{\langle N_{\rm bin}\rangle} \frac{d\sigma^{AA}\left(|\mathbf{b}_\perp|\right)/dm_{12}}{d\sigma^{pp}/dm_{12}}. \tag{6}$$

Predictions for this new R_{AA}, differential in the dijet invariant mass, are displayed in Fig. 3 – these are to guide the future sPHENIX experiment.

In contrast to the traditional dijet momentum imbalance shift, the dijet mass modification displays a strong and clear dependence on the mass of the partons initiating the dijet. This is exemplified by the factor of ~ 10 or more suppression of inclusive dijets up to a mass of about 100 GeV. Note that this suppression is markedly different from that of b-tagged dijets through most of the mass range covered. This point is highlighted in Fig. 4, where we plot the ratio of b-tagged R_{AA} to that of inclusive. Indeed, in the low invariant mass range around 20 GeV, the suppressions experienced by light and heavy flavor dijets differ by almost an order of magnitude. This means that the kinematic range of sPHENIX is particularly well-suited to reveal the mass dependence of quark and gluon energy loss in the QGP – an important open question in the heavy ion community.

The amplification of quenching effects exhibited in the modification of dijet invariant mass distributions can be understood in the following simple way. Reconstructed jets emerging from A+A collisions have less energy than their counterparts in p+p, hence, in the reverse engineering of A+A cross sections from those of p+p, a jet with a given p_T in A+A is mapped to a jet of a higher value, say $p_T + \delta p_T$, in p+p. This is the qualitative content of Eq. (1). Now, the dijet imbalance z_J is defined as a ratio of transverse momenta in Eq. (4), while the dijet mass m_{12} involves a product of transverse momenta in Eq. (5). Thus, the effects of the medium between the leading and subleading jets work to cancel one another out in the modification of z_J, while they strictly combine to amplify the effects of one another in their modification of m_{12}. Therefore, the overall alteration of z_J is more minute, resulting in minimal modification. At the same time, the change to m_{12} is substantial and leads to a sizable plunge down the p+p cross section to obtain that of A+A, resulting in significant modification.

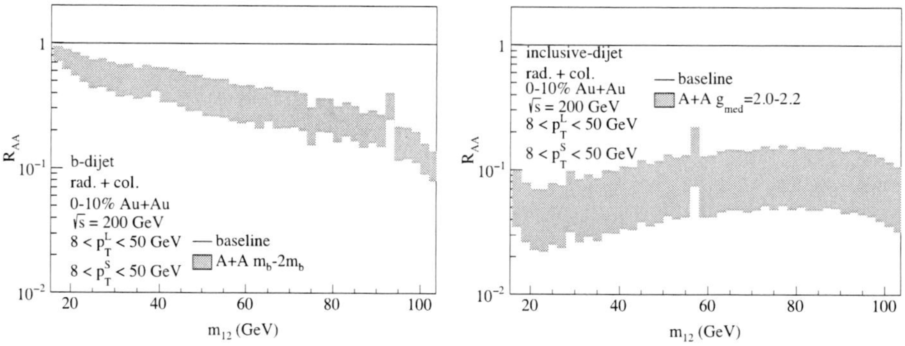

Figure 3. Nuclear modification factor R_{AA} with respect to dijet invariant mass m_{12} for b-tagged (left) and inclusive (right) dijet production in Au+Au collisions at $\sqrt{s_{NN}}$ = 200 GeV for sPHENIX at RHIC. Left: we fix $g_{\rm med}$ = 2.0 and the band corresponds to varying the mass of the propagating system from m_b to $2m_b$. Right: the band corresponds to the range of medium couplings $g_{\rm med}$ = 2.0 − 2.2.

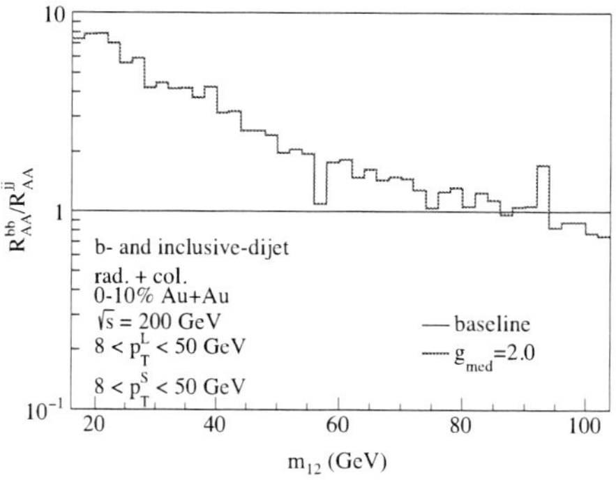

Figure 4. Ratio of the nuclear modification factors for b-tagged $\left(R^{bb}_{AA}\right)$ to inclusive $\left(R^{jj}_{AA}\right)$ dijet production vs. dijet invariant mass m_{12} for sPHENIX at RHIC. For b-tagged dijets, the mass of the propagating system is held fixed at m_b. For both b-tagged and inclusive dijets, we fix $g_{\rm med}$ = 2.0.

4 Conclusion

In these proceedings, we have reviewed the recent predictions for the modification of the dijet mass distribution in heavy ion collisions [1]. In particular, we have examined the behavior of this observable within the kinematic range of the future sPHENIX experiment at RHIC. The dramatic difference between the modification patterns exhibited by light and heavy flavor dijets makes this observable a promising tool in the growing arsenal of hard probes of the QCD medium. The dijet mass distribution has the advantage of amplifying the effects of jet quenching in heavy ion collisions – making it stand out among traditional dijet observables, such as the dijet imbalance shift. This is due to its dependence on the product of the transverse momenta of the jets forming the dijet pair, whereas the dijet imbalance distribution depends on their quotient. By amplifying the effects of the medium, the differences in energy loss mechanisms governing the propagation of light and heavy flavor partons within the QGP become more apparent, as displayed in Figs. 3 and 4. We look forward to the first experimental measurements of this novel observable, as they are sure to enrich our knowledge of the QGP.

5 Acknowledgements

I thank Theoretical Division of Los Alamos National Laboratory for its hospitality during the completion of part of this work. This work is supported by the UC Office of the President through the UC Laboratory Fees Research Program under Grant No. LGF-19-601097.

References

[1] Z.-B. Kang, J. Reiten, I. Vitev, and B. Yoon, Phys. Rev. D. **99**, 034006 (2019), arXiv:1810.10007.
[2] J. Huang, Z. B. Kang, I. Vitev and H. Xing, Phys. Lett. B. **750**, 287 (2015), arXiv:1505.03517.
[3] J. Huang, Z.-B. Kang, and I. Vitev, Phys. Lett. **B726**, 251 (2013), arXiv:1306.0909.
[4] H. T. Li and I. Vitev, (2018), arXiv:1811.07905.
[5] CMS, S. Chatrchyan *et al.*, Phys. Rev. Lett. **113**, 132301 (2014), arXiv:1312.4198, [Erratum: Phys. Rev. Lett.115,no.2,029903(2015)].
[6] CMS, V. Khachatryan *et al.*, JHEP **01**, 006 (2016), arXiv:1509.09029.
[7] STAR, L. Adamczyk *et al.*, Phys. Rev. Lett. **119**, 062301 (2017), arXiv:1609.03878.
[8] CMS, A. M. Sirunyan *et al.*, JHEP **03**, 181 (2018), arXiv:1802.00707.
[9] T. Sjostrand, S. Mrenna, and P. Z. Skands, Comput. Phys. Commun. **178**, 852 (2008), arXiv:0710.3820.
[10] sPHENIX: Heavy Flavor Jet Simulation and Analysis, https://www.phenix.bnl.gov/WWW/publish/jinhuang/sPHENIX/sPH-HF-2017-001-v1.1.pdf.
[11] B. G. Zakharov, JETP Lett. **65**, 615 (1997), arXiv:hep-ph/9704255.
[12] R. Baier, Y. L. Dokshitzer, A. H. Mueller, S. Peigne, and D. Schiff, Nucl. Phys. **B483**, 291 (1997), arXiv:hep-ph/9607355.
[13] M. Gyulassy, P. Levai, and I. Vitev, Phys. Rev. Lett. **85**, 5535 (2000), arXiv:nucl-th/0005032.
[14] U. A. Wiedemann, Nucl. Phys. **B588**, 303 (2000), arXiv:hep-ph/0005129.
[15] X.-N. Wang and X.-f. Guo, Nucl. Phys. **A696**, 788 (2001), arXiv:hep-ph/0102230.
[16] P. B. Arnold, G. D. Moore, and L. G. Yaffe, JHEP **06**, 030 (2002), arXiv:hep-ph/0204343.
[17] B.-W. Zhang, E. Wang, and X.-N. Wang, Phys. Rev. Lett. **93**, 072301 (2004), arXiv:nucl-th/0309040.
[18] Y. L. Dokshitzer and D. E. Kharzeev, Phys. Lett. **B519**, 199 (2001), arXiv:hep-ph/0106202.
[19] E. Braaten and M. H. Thoma, Phys. Rev. **D44**, R2625 (1991).
[20] S. Wicks, W. Horowitz, M. Djordjevic, and M. Gyulassy, Nucl. Phys. **A784**, 426 (2007), arXiv:nucl-th/0512076.
[21] A. Adil, M. Gyulassy, W. A. Horowitz, and S. Wicks, Phys. Rev. **C75**, 044906 (2007), arXiv:nucl-th/0606010.
[22] M. H. Thoma, Rev. Mod. Phys. **81**, 959 (2009), arXiv:0801.0956.
[23] H. Berrehrah *et al.*, Phys. Rev. **C89**, 054901 (2014), arXiv:1308.5148.
[24] R. B. Neufeld, I. Vitev, and H. Xing, Phys. Rev. **D89**, 096003 (2014), arXiv:1401.5101.
[25] M. L. Miller, K. Reygers, S. J. Sanders, and P. Steinberg, Ann. Rev. Nucl. Part. Sci. **57**, 205 (2007), arXiv:nucl-ex/0701025.
[26] Z.-B. Kang, F. Ringer, and I. Vitev, JHEP **03**, 146 (2017), arXiv:1610.02043.
[27] M. D. Sievert and I. Vitev, Phys. Rev. **D98**, 094010 (2018), arXiv:1807.03799.
[28] C. Shen *et al.*, Comput. Phys. Commun. **199**, 61 (2016), arXiv:1409.8164.
[29] ALICE, S. Acharya *et al.*, Phys. Lett. **B776**, 249 (2018), arXiv:1702.00804.

EPJ Web of Conferences **235**, 06001 (2020) https://doi.org/10.1051/epjconf/202023506001
ISMD 2019

Exclusive Diffraction at the LHC
On behalf of the ATLAS, LHCb, CMS and TOTEM experiments

Christophe Royon[1,3,⋆]

[1] *Department of Physics and Astronomy, The University of Kansas, Lawrence, USA*

Abstract. In this report, we describe the most recent results on exclusive diffraction from the ATLAS, CMS, LHCb, TOTEM experiments at the LHC concerning exclusive pions, J/Ψ, $\Psi(2S)$, dilepton, diphoton, WW productions and prospects concerning the search for anomalous couplings and axion-like particle production.

1 Introduction: Definition of exclusive diffraction

We will start this report by defining what we call "Exclusive" diffraction. The first left diagram of Fig. 1 corresponds to Double Pomeron Exchange in inclusive diffraction. In this event, both protons are intact in the final state and two Pomerons are exchanged. Gluons are extracted from each Pomeron in order to produce jets (or diphotons, Ws...). Some energy is "lost" in Pomeron remnants. The three other diagrams in Fig. 1 are exclusive in the sense that the full energy is used to produce the diffractive object. In other word, there is no energy loss in Pomeron remnants. The second diagram corresponds to exclusive diffraction [1], the third one to photon exchanges and the last one to photon Pomeron exchanges that produce vector mesons. In the following, we will discuss the measurements at the LHC corresponding to these different diagrams.

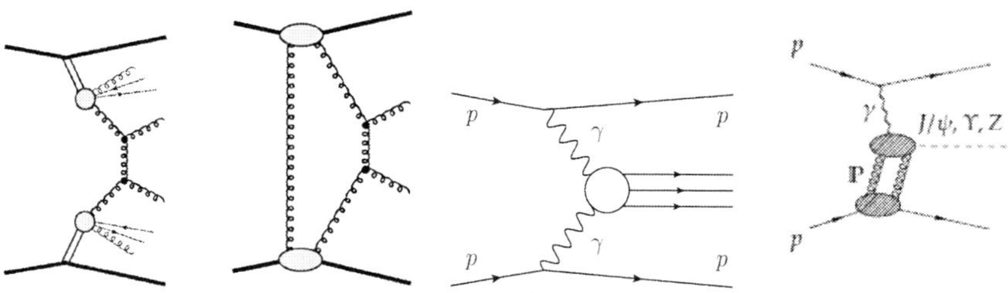

Figure 1. Diagrams of inclusive and exclusive diffractive processes.

⋆e-mail: christophe.royon@ku.edu

2 Low mass exclusive diffraction

In this section, we will focus on low mass exclusive diffraction results from LHCb and CMS, namely on vector meson and pion production.

2.1 Measurement of central exclusive production in LHCb

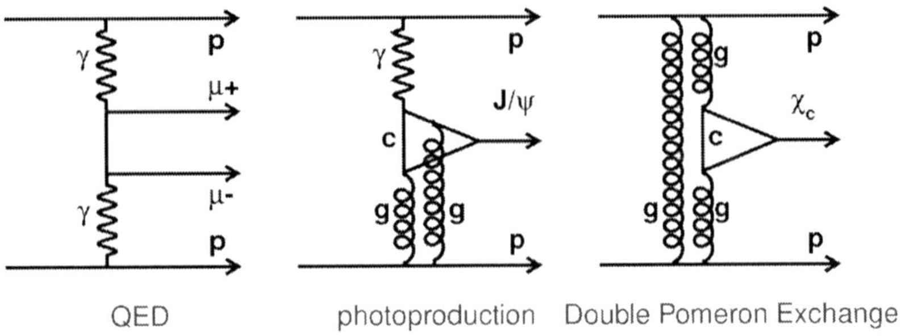

Figure 2. Central exclusive production in LHCb.

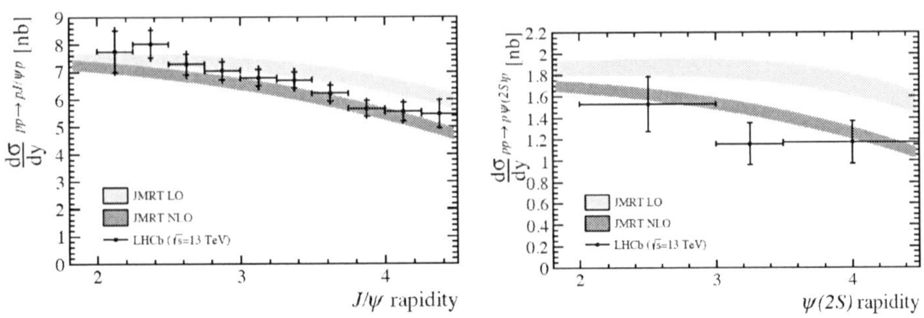

Figure 3. Measurement of the J/Ψ and $\Psi(2S)$ cross sections as a function of rapidity using 13 TeV data including the new HERSCHEL detector.

In Fig. 2, we display three diagrams that correspond to exclusive productions in LHCb. The first left diagram corresponds to the QED production of di-muon, the middle one the photoproduction of J/Ψ, and the right one the inclusive Double Pomeron Exchange production of χ_C mesons. The exclusive production of J/Ψ is strongly sensitive to the gluon distribution in the proton since it appears as a square dependence in the amplitude. The signal is to measure the J/Ψ in the central part of LHCb and the background is due to inclusive diffractive processes where the Pomeron remnants are not detected since exclusive processes are selected via the rapidity gap method. The issue is the gap detection since, for some events, the Pomeron remnants are outside detector acceptance.

In order to improve the coverage at high rapidities, the LHCb collaboration installed a new detector called HERSCHEL (High Rapidity Shower Counter for LHCb). HERSCHEL is installed on both

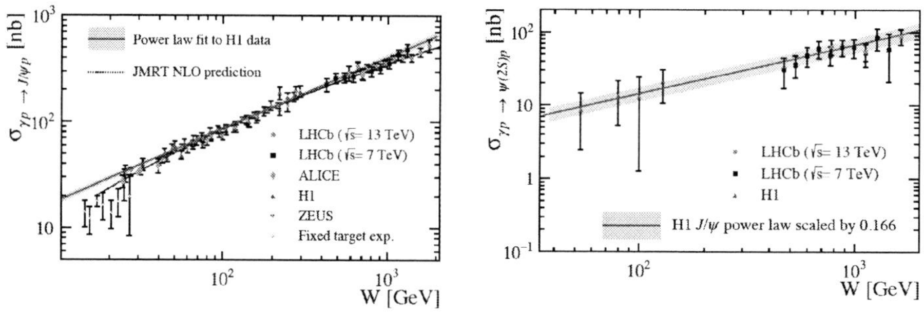

Figure 4. Measurement of the $\gamma p \to J/\Psi p$ and $\gamma p \to \Psi(2S)p$ cross sections as a function of energy using 7 and 13 TeV data compared to the JMRT NLO parametrization. Data cannot be described by a simple power law.

sides of LHCb and allows a coverage in pseudo-rapidity of $-10.0 < \eta < -3.5$ and $5.0 < \eta < 10.0$ which leads to a coverage of LHCb of $-10.0 < \eta < -1.5$ and $2.0 < \eta < 10.0$. HERSCHEL was used to select exclusive data at 13 TeV, and about 14753 J/Ψ and 440 $\Psi(2S)$ were selected in 204 pb^{-1} of data with a respective purity of 75.5±1.5% and 72.6±6.1%. The measured cross sections are $\sigma(J/\Psi \to \mu^+\mu^-)(2 < \eta < 4.5) = 435 \pm 18 \pm 11 \pm 17$ pb, and $\sigma(\Psi(2S) \to \mu^+\mu^-)(2 < \eta < 4.5) = 11.1 \pm 1.1 \pm 0.3 \pm 0.4$ pb. The differential cross sections for J/Ψ and $\Psi(2S)$ compared between 7 and 13 TeV data are shown in Fig. 3 [2]. The results are in agreement with NLO calculations and the best PDF parametrisation is from JMRT at NLO [3].

This measurement allows LHCb to compare with previous measurements performed by the H1 and ZEUS collaborations at HERA using the following simple relation between HERA and LHCb measurements

$$\frac{d\sigma}{dy_{pp\to pVp}} = r(y)\left[k_+\frac{dn}{dk_+}\sigma_{\gamma p\to Vp}(W_+) + k_-\frac{dn}{dk_-}\sigma_{\gamma p\to Vp}(W_-)\right] \tag{1}$$

where the $pp \to pVp$ is measured by LHCb, $\sigma(W_-)$ (resp. $\sigma(W_+)$) at HERA, and $\sigma(W_+)$ (resp. $\sigma(W_-)$) is extracted from the previous formula. The results are shown in Fig. 4 [2] and a simple power law does not lead to a good description of data.

2.2 Measurement of exclusive pion production in CMS

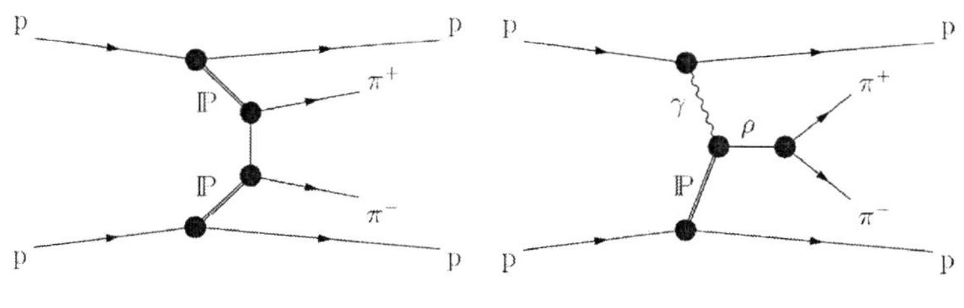

Figure 5. Leading diagrams leading to exclusive pion production.

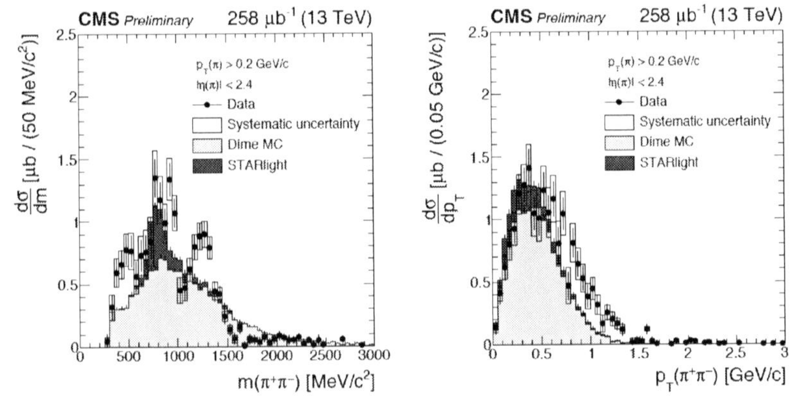

Figure 6. Measurement of exclusive pion cross section by the CMS collaboration as a function of the di-pion mass in linear and logarithmic scales.

In this section, we will describe the measurement of exclusive pion production by the CMS collaboration [4]. The two diagrams that are dominant at low masses are shown in Fig. 5. The left diagram corresponds to double Pomeron exchange and is implemented in the DIME Monte carlo [5] whereas the right diagram corresponds to γ-Pomeron exchanges producing a ρ meson that decays into a pion pair as implemented in the STARLIGHT Monte Carlo [6]. The measurement was performed using low pile up runs since the production cross section is high and not much accumulated luminosity is needed. The same measurement in a high pile up environment would be challenging because of the low p_T track requirement. The experimental signature is to request two opposite tracks from the same primary vertex and no additional signal in the calorimeter outside the two pions with $p_T(\pi) > 0.2$ GeV and $|y(\pi)| < 2$. The background is computed using same sign events directly in data.

The measurement of the di-pion cross section is shown in Fig. 6 as a function of $m_{\pi\pi}$ and $p_T(\pi\pi)$ [4]. We see a discrepancy between the measurements and the sum of the DIME and STARLIGHT Monte Carlo expectations both in shape and normalization. Both Monte Carlos do not contain proton dissociation and it might be a reason for most of the discrepancy (since the measurements require the existence of a rapidity gap in the CMS detector, it is impossible to distinguish experimentally between the cases when the protons are intact or dissociate). The exclusive pion cross section was measured to be $\sigma_{\pi^+\pi^-}(13TeV) = 19.0 \pm 0.6(stat) \pm 3.2(syst) \pm 0.01(lumi)$ μb; $\sigma_{\pi^+\pi^-}(5.02TeV) = 19.6 \pm 0.4(stat) \pm 3.3(syst) \pm 0.01(lumi)$ μb.

3 Exclusive measurements at high mass and photon exchange processes

In this section, we will describe the results on exclusive production at high mass together with the ones on photon exclusive exchange processes.

3.1 ATLAS and CMS results on exclusive WW production

ATLAS and CMS measured the exclusive production of a pair of W bosons $pp \rightarrow pWWp$ using the rapidity gap technique. This measurement is motivated by the search for quartic anomalous $\gamma\gamma WW$

couplings defined as additional terms in the SM lagrangian

$$\mathcal{L}_6^0 \sim \frac{-e^2}{8}\frac{a_0^W}{\Lambda^2}F_{\mu\nu}F^{\mu\nu}W^{+\alpha}W^-_\alpha - \frac{e^2}{16\cos^2(\theta_W)}\frac{a_0^Z}{\Lambda^2}F_{\mu\nu}F^{\mu\nu}Z^\alpha Z_\alpha$$

$$\mathcal{L}_6^C \sim \frac{-e^2}{16}\frac{a_C^W}{\Lambda^2}F_{\mu\alpha}F^{\mu\beta}(W^{+\alpha}W^-_\beta + W^{-\alpha}W^+_\beta) - \frac{e^2}{16\cos^2(\theta_W)}\frac{a_C^Z}{\Lambda^2}F_{\mu\alpha}F^{\mu\beta}Z^\alpha Z_\beta$$

All parameters are equal to 0 in the SM.

This search has to be performed using the highest possible integrated luminosity at the LHC since exclusive WW events are rare which means at high pile up at the LHC. The CMS and ATLAS collaborations used respectively 19.7 fb^{-1} and 20.2 fb^{-1} at a center-of-mass of 8 TeV (and CMS an additional 5.05 fb^{-1} at 7 TeV). The exclusivity selection requires opposite sign electron and muon originating from the common primary vertex and no extra track from that vertex. In addition, $M_{e\mu} >$ 20 GeV to avoid the low mass resonance region and $P_T^{e\mu} >$ 30 GeV to remove Drell-Yan and $\gamma \to \tau\tau$ events.

CMS and ATLAS measured the exclusive WW production and obtained $\sigma(pp \to pWWp \to p\mu ep) = 2.2^{+3.3}_{-2.0}$ fb at 7 TeV (SM: 4.0 ± 0.7 fb) , $\sigma(pp \to pWWp \to p\mu ep) = 10.8^{+5.1}_{-4.1}$ fb at 8 TeV (SM: 6.2 ± 0.5 fb) for CMS after correction for proton dissociation, and $\sigma = 6.9 \pm 2.2(stat) \pm 1.4(syst)$ fb) for ATLAS (SM: 4.4 ± 0.3 fb). The observed significance for 7 and 8 TeV combination: is 3.4 σ for CMS and 3.0 σ for ATLAS [7]

Since data are in agreement with the standard model, it is possible to obtain the most stringent limits to date on $\gamma\gamma WW$ quartic anomalous coupling and the results are given in Fig. 7. It is worth noticing that these limits can potentially be improved by about two orders of magnitude measuring the intact protons in the final state using the CT-PPS and AFP detectors [8].

Dimension-6 AQGC parameter	7 TeV ($\times 10^{-4}$ GeV^{-2})	8 TeV ($\times 10^{-4}$ GeV^{-2})	7+8 TeV ($\times 10^{-4}$ GeV^{-2})
$a_0^W/\Lambda^2(\Lambda_{cutoff} = 500\,\mathrm{GeV})$	$-1.5 < a_0^W/\Lambda^2 < 1.5$	$-1.1 < a_0^W/\Lambda^2 < 1.0$	$-0.9 < a_0^W/\Lambda^2 < 0.9$
$a_C^W/\Lambda^2(\Lambda_{cutoff} = 500\,\mathrm{GeV})$	$-5 < a_C^W/\Lambda^2 < 5$	$-4.2 < a_C^W/\Lambda^2 < 3.4$	$-3.6 < a_C^W/\Lambda^2 < 3.0$

Coupling	Λ_{cutoff}	Observed allowed range [GeV $^{-2}$]	Expected allowed range [GeV $^{-2}$]
a_0^W/Λ^2	500 GeV	$[-0.96 \times 10^{-4}, 0.93 \times 10^{-4}]$	$[-0.90 \times 10^{-4}, 0.87 \times 10^{-4}]$
a_C^W/Λ^2	500 GeV	$[-3.5 \times 10^{-4}, 3.3 \times 10^{-4}]$	$[-3.3 \times 10^{-4}, 3.1 \times 10^{-4}]$
a_0^W/Λ^2	∞	$[-1.7 \times 10^{-6}, 1.7 \times 10^{-6}]$	$[-1.5 \times 10^{-6}, 1.6 \times 10^{-6}]$
a_C^W/Λ^2	∞	$[-6.4 \times 10^{-6}, 6.3 \times 10^{-6}]$	$[-5.9 \times 10^{-6}, 5.8 \times 10^{-6}]$

Figure 7. Limits on $\gamma\gamma WW$ quartic anomalous couplings by the CMS and ATLAS collaborations..

3.2 Exclusive di-lepton production in ATLAS and in CMS-TOTEM (PPS)

The ATLAS Forward Proton [9] (AFP) and CMS-TOTEM Precision Proton Spectrometer [10] (PPS) detectors were installed recently in order to detect and measure intact protons in the final state that leads to a possible better identification of exclusive events. The LHC magnets bend the scattered protons outside the beam envelope. The roman pots detectors are located at about 210-220 m from the center part of the CMS or ATLAS detectors and cover a region in diffractive mass between typically 350 and 2000 GeV depending on the beam lattice. The PPS detector started taking data in 2016 and could accumulate about 15 fb^{-1} in 2016, and about 115 fb^{-1} between 2016 and 2018. While the full data set is still being analyzed, we will describe the first observation of exclusive di-leptons using 9.4 fb^{-1}.

The ATLAS and CMS collaborations measured in 2016 the exclusive di-lepton production with two different strategies. The corresponding diagrams are shown in Fig. 8. The LHC is turned into a $\gamma\gamma$ collider and the flux of quasi-real photons is computed using the Equivalent Photon Approximation (EPA). ATLAS did not have the possibility of detecting intact protons at the time of this measurement, and they selected exclusive $\mu\mu$ production using the rapidity gap technique. The exclusivity requirement in presence of pile up is that zero additional track exists within 1 mm of the $\mu^+\mu^-$ vertex. From the rapidity gap technique, it is impossible to know if the protons are dissociate or not and the measurement corresponds to the three diagrams shown in Fig. 8. The ATLAS collaboration fitted the $\mu\mu$ acoplanarity distribution at 13 TeV using 3.2 fb^{-1} of data in order to extract the fiducial cross section for $p_T^\mu > 6$ GeV ($12 < m_{\mu\mu} < 30$ GeV) and for $p_T^\mu > 10$ GeV ($30 < m_{\mu\mu} < 70$ GeV) and found $\sigma_{\gamma\gamma\rightarrow\mu\mu}^{excl.fid} = 3.12 \pm 0.07(stat) \pm 0.10(syst)$ pb [11]. The comparison with the Superchic Monte Carlo [12] indicates an insufficient suppression in Monte Carlo due to absorptive effects.

The PPS collaboration measured exclusive di-lepton production by tagging one intact proton in the final state. This is the first time the semi-exclusive di-lepton processes are measured with proton tag. In Fig. 8, the two left diagrams correspond to the signal whereas the rightmost diagram is part of the background. The reason that only one proton is requested to be tagged is that less than one event is expected for double tagged events with about 10 fb^{-1} of data due to the mass acceptance above about 350 GeV for the forward proton detectors. A pair of opposite sign muons or electrons with $p_T > 50$ GeV and $M_{ll} > 110$ GeV above the Z boson peak is requested. In order to suppress background, there is a veto on additional tracks around the di-lepton vertex (within 0.5 mm) and leptons are required to be back-to-back, $|1 - \Delta\Phi/\pi| < 0.006$ for electrons (0.009 for muons). The main background is due to Drell-Yan di-lepton production with the intact proton originating from pile up events. This background is estimated using Drell-Yan Z events in data and extrapolating from the Z peak region to our exclusive di-lepton signal region. 40 events (17 $\mu\mu$ and 23 ee) are found with protons in the PPS acceptance and 20 (12 $\mu\mu$ and 8 ee) show a less than 2σ matching between the values of ξ computed using the TOTEM roman pots and using the di-lepton measured in CMS as shown in Fig. 9 [13]. This leads to a significance larger than 5σ to observe 20 events for a background of 3.85 ($1.49 \pm 0.07(stat) \pm 0.53(syst)$ for $\mu\mu$ and $2.36 \pm 0.09(stat) \pm 0.47(syst)$ for ee). As expected, no event was double tagged with an intact proton on each side.

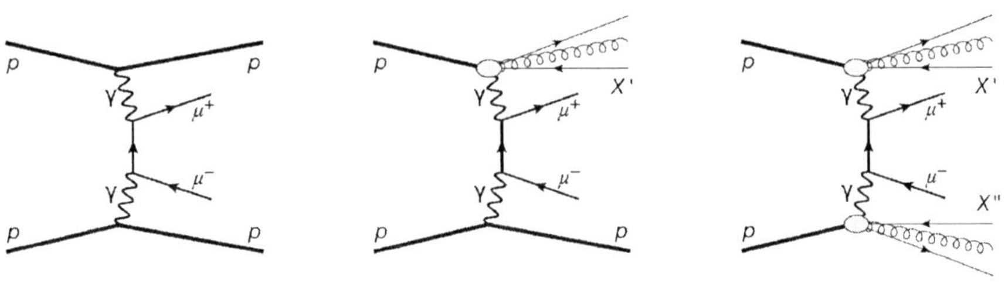

Figure 8. Diagrams leading to di-muon production via photon exchanges..

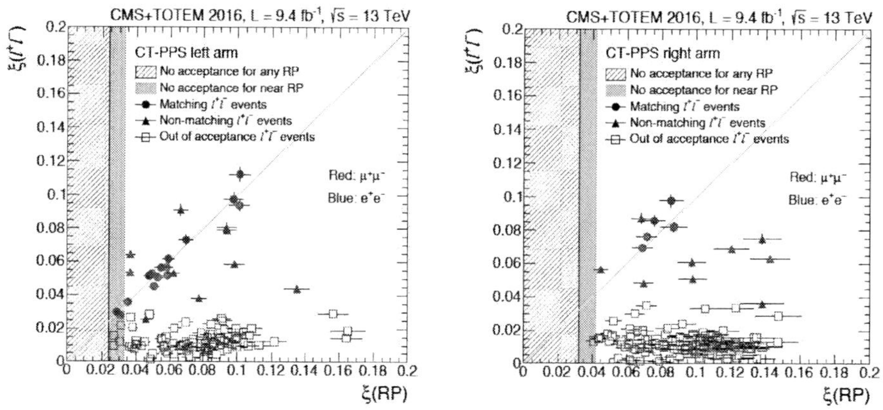

Figure 9. Correlation between the ξ values computed using the TOTEM roman pots and the di-lepton measured in CMS. The 20 semi-exclusive events are indicated in red. The left (right) plot displays the left (right) arm of TOTEM.

3.3 Exclusive di-photon production via photon exchanges

The ATLAS collaboration recently measured exclusive di-photon production in *PbPb* production at the LHC. Exclusive di-photons can be produced via photon exchanges and the cross section is enhanced by a factor Z^4 with respect to the pp case. In 480 μb^{-1} of *PbPb* data at $\sqrt{s}$ =5.02 TeV in 2015, 13 events were observed for an estimated background of 2.6±0.7. Uisng 1.7 nb^{-1} of data in 2018, 59 events were observed for a background of 12±3. For photon $E_T > 3$ GeV, $|\eta| < 2.37$, $M_{\gamma\gamma}$ >6 GeV, $p_T^{\gamma\gamma} < 2$ GeV:, ATLAS measured $\sigma = 78 \pm 13(stat) \pm 8(syst)$ nb in agreement with SM (49 ± 5 nb) [14].

Using PPS and AFP, it is also possible to search for $\gamma\gamma\gamma\gamma$ anomalous couplings in a very clean way. Within the acceptance of the PPS and AFP detectors during standard high luminosity runs at the LHC (basically for a di-photon mass above 350 GeV), it is possible to show that the exclusive production of di-photons is completely dominated by photon exchange processes and gluon exchanges can be neglected. Since the signal only shows two photons and two intact protons in the final state, we measure all final state particles. That allows us to obtain a negligible background for 300 fb^{-1} at the LHC. The basic idea is to compare the proton missing mass and the di-photon mass as shown in Fig. 10, left [15]. The signal peaks around 1.0 and the gaussian width is due to the detector resolution whereas the pile-up background leads to a much flatter distribution since the two protons are not related with the two photons. The same requirement can be performed using the difference in rapidity between the di-photon and di-proton systems, as shown in Fig. 10, right. This leads to a sensitivity better by two orders of magnitude to $\gamma\gamma\gamma\gamma$ anomalous couplings compared to other methods at the LHC without detecting intact protons [15]. It is possible to show that the gain of two orders of magnitude is also valid for $\gamma\gamma WW$ and $\gamma\gamma ZZ$ anomalous couplings whereas the gain reaches three orders of magnitude for $\gamma\gamma\gamma Z$ anomalous couplings [8, 17]. The search for anomalous couplings with tagged protons is now being pursued by both CMS-TOTEM and ATLAS collaborations.

Looking for exclusive di-photon events with tagged protons can be directly applied to the search for axion-like particles at high mass [18]. The ALP would be produced by $\gamma\gamma$ interactions and would decay into two photons. The sensitivity in the coupling versus ALP mass is shown in Fig. 11 and we see the gain of about two orders of magnitude in coupling at high mass using this method. We also note that this is complementary to looking for exclusive di-photons in pPb, $PbPb$, and $ArAr$ collisions at lower masses of ALPs. This is due to the fact that the cross section is enhanced by a factor Z^4 in heavy ion collisions but the sensitivity at high mass is reduced to a large suppression at small impact parameter due to the size of the heavy ion [19]

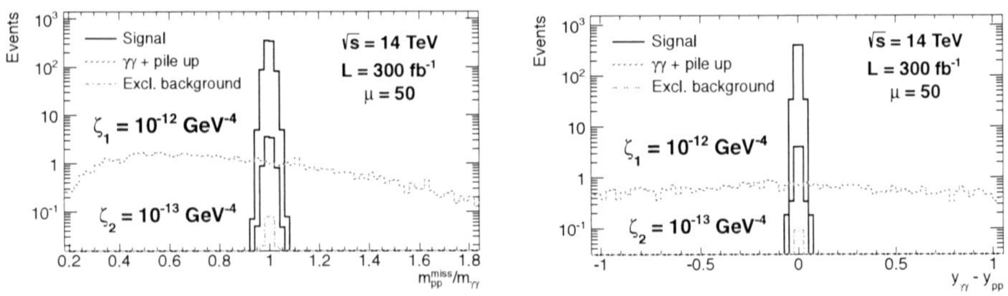

Figure 10. Left: Ratio between the proton missing mass and di-photon mass for exclusive di-photon signal events and background. Right: Difference between the di-photon and di-proton rapidity for exclusive di-photon signal and background.

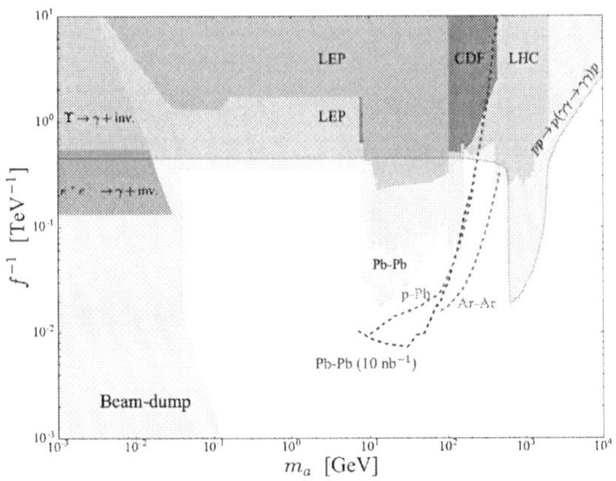

Figure 11. Exclusion plot on axion like particles in the coupling versus mass plane and sensitivity at the LHC in pp collisions with 300 fb^{-1} (grey band) and in $PbPb$ (blue dashed line), pPb (green dashed line), $ArAr$ (red dashed lines) collisions.

4 Conclusion

We discussed many complementary measurements concerning exclusive production at the LHC from the ATLAS, CMS, LHCb and TOTEM experiments either using the "rapidity gap" technique or the proton tags. The LHCb collaboration measured the J/Ψ and $\Psi(2S)$ production and the preferred model is JMRT at NLO. The CMS exclusive pion measurement is in disagreement with theoretical expectations which is probably due to the fact that proton dissociation is not included in models.

The exclusive production of W pair production was observed and led to the best limits to date on $\gamma\gamma WW$ anomalous couplings. For the first time, the CMS and TOTEM collaborations observed the exclusive production of di-leptons at high mass with a proton tag. Last but not least, AFP and PPS open a completely new search for quartic anomalous couplings between γ, W and Z bosons with a gain in reach by two or three orders of magnitude compared to more standard methods at the LHC, reaching some of the values predicted by extra-dimension, composite Higgs models and enhancing the sensitivity to axion-like-particle at high masses.

References

[1] V. Khoze, A. Martin, M. Ryskin, Eur. Phys. J. C**23** (2002) 311.
[2] LHCb Coll., JHEP **1810** (2018) 167; LHCb Coll., J. Phys. G **41** (2014) 055002.
[3] P. Jones, A. Martin, G. Ryskin, T. Teubner, JHEP **11** (2013) 085.
[4] CMS Coll., ArXiv:1706.08310.
[5] L. Harland Lang, V. A. Khoze, M. G. Rykin, ArXiv:1312.4553.
[6] S. R. Klein, J. Nystrand, J. Seger, Y. Gorbunv, J. Butterworth, Comp. Phys. Comm. **212** (2017) 258.
[7] CMS Coll. JHEP08 (2016) 119; ATLAS Coll., Phys Rev D94 (2016) 032011.
[8] E. Chapon, O. Kepka, C. Royon, Phys. Rev. D**81** (2010) 074003; O. Kepka, C. Royon, Phys. Rev. D**78** (2008) 073005.
[9] ATLAS Coll., CERN-LHCC-2011-012.
[10] CT-PPS project, see https://cds.cern.ch/record/1753795; CMS and TOTEM coll., CERN-LHCC-2014-021.
[11] ATLAS Coll., preprint ArXiv:1708.04503.
[12] L. Harland Lang, V.A. Khoze, M.G. Ryskin, preprint ArXiv:1508.02718.
[13] CMS and TOTEM Coll., JHEP **1807** (2018) 153.
[14] ATLAS Coll., Nature Physics **13** (2017) 852; ATLAS Coll., Phys. Rev. Lett. **123** (2019) 052001.
[15] S. Fichet, G. von Gersdorff, B. Lenzi, C. Royon, M. Saimpert, JHEP **1502** (2015) 165; S. Fichet, G. von Gersdorff, O. Kepka, B. Lenzi, C. Royon, M. Saimpert, Phys. Rev. D**89** (2014) 114004. S. Fichet, G. von Gersdorff, C. Royon, Phys. Rev. Lett. **116** (2016) no 23, 231801; S. Fichet, G. von Gersdorff, C. Royon, Phys. Rev. D**93** (2016) no 7, 075031;
[16] M. Boonekamp et al., preprint ArXiv:1102.2631.
[17] C. Baldenegro, S. Fichet, G. von Gersdorff, C. Royon, JHEP **1706** (2017) 142; O. Kepka, C. Royon, Phys. Rev. D**78** (2008) 073005.
[18] C. Baldenegro, S. Fichet, G. von Gersdorff, C. Royon, JHEP **1806** (2018) 131.
[19] C. Baldenegro, S. Hassani, C. Royon, L. Schoeffel, Phys. Lett. B**795** (2019) 339.

Proton Holography

Discovering Odderon from Scaling Properties of Elastic Scattering

T. Csörgő[1,2,*], T. Novák[2,**], R. Pasechnik[3,4,***], A. Ster[1,****], and I. Szanyi[1,5,†]

[1]Wigner RCP, H-1525 Budapest 114, P.O.Box 49, Hungary
[2]Eszterházy University KRC, H-3200 Gyöngyös, Mátrai út 36, Hungary
[3]Department of Astronomy and Theoretical Physics, Lund University, SE-223 62 Lund, Sweden
[4]Nuclear Physics Institute ASCR, 25068 Řež, Czech Republic
[5]Eötvös University, H - 1117 Budapest XI, Pázmány P. s. 1/A, Hungary

Abstract. We investigate the scaling properties of elastic scattering data at ISR and LHC energies, and find that the significance of an Odderon observation is larger than the discovery threshold of 5σ. As an unexpected by-product of these investigations, for certain experimentally relevant cases, we also conjecture the possibility of proton holography with the help of elastic proton-proton scattering.

1 Introduction

This work summarizes two related topics: the discovery of a significant Odderon effect at LHC energies and the investigation of the possibility of four-momentum transfer t dependent phase measurement in elastic proton-proton (pp) and proton-antiproton ($p\bar{p}$) scattering.

The Odderon corresponds to a crossing-odd contribution to the scattering amplitude of elastic pp and $p\bar{p}$ scattering at asymptotically high energies, proposed by Lukaszuk and Nicolescu in 1973 [1]. In QCD, the quantum field theory of the strong interactions, such an Odderon exchange corresponds to the t-channel exchange of a color-neutral gluonic compound state consisting of an odd number of gluons, as noted by Bartels, Vacca and Lipatov in 1999 [2].

As the modulus square of the elastic scattering amplitude is proportional to the differential cross-section of elastic scattering, a phase reconstruction is equivalent to the possibility of proton holography. It is well known that such a t-dependent phase reconstruction from the measurements of modulus squared amplitudes, without further information, is simply impossible. We thus ask a different question here: what are those specific additional, and experimentally testable conditions, that actually allow for a t-dependent phase reconstruction in elastic pp collisions at LHC energies?

Holography of light by now is well developed technique that has several sub-topics and applications not only in science but also in arts, banking, programming, interferometry and security, including holograms on bank-notes, vehicles, credit and identity cards as well as passports. The essential point of holography was highlighted in D. Gabor's Nobel Lecture [3]: Holography is based on the wave nature of light, and corresponds to phase level reconstruction. As every quantum field has a dual wave and quantum property, holography is possible not only using the wave properties of light quanta, but also that of other particles like electrons [4], atoms [5] as well as neutrons [6]. The key concept of holography is to produce a coherent source of these quanta, and have parts of this field diffractively scatter on some scattering center. Recording the resulting interference pattern corresponds to recording a holographic picture. Illuminating this picture with the original beam allows for reconstruction of the scattered wave including both the modulus and the phase of the amplitude. Actually, two different holographic images are created, the object beam that carries the original phase and modulus, and the conjugated object beam that reconstructs the modulus but conjugates the complex phase of the object beam.

The title of this work is proton holography, as we discuss here possibilities for a phase-level reconstruction using the elastically scattered protons at the TeV energy scale. Although such a title may look a bit fancy, it turns out that our idea is not unprecedented, as Ref. [7] has already considered four-momentum transfer dependent phase reconstruction in elastic scattering of protons at the ISR energies of $\sqrt{s} = 23.5 - 62.5$ GeV, but without introducing such a term.

A related topic is the search for a crossing-odd contribution in elastic proton-proton collisions at CERN LHC energies, the so called Odderon effect. As the Odderon effect is not significant at the ISR energy range but is found to be significant at LHC energies [8], the methodology of phase reconstruction at these two energies are also slightly different. For example, the phase-level reconstruction of

*e-mail: tcsorgo@cern.ch
**e-mail: tamas.novak@uni-eszterhazy.hu
***e-mail: Roman.Pasechnik@thep.lu.se
****e-mail: Ster.Andras@wigner.mta.hu
†e-mail: Istvan.Szanyi@cern.ch

Ref. [7] disregarded any Odderon contribution, so that method cannot be readily applied at TeV energies. Due to this reason, let us first briefly summarize some of our new results related to the Odderon discovery at the CERN LHC, and this way also prepare the ground for the idea of proton holography at high energies, as detailed in the second part of this work.

2 Formalism

Let us consider the elastic scattering of particles a and b with incoming four-momenta (p_1, p_2), and outgoing four-momenta (p_3, p_4), respectively. The Mandelstam variables s and t are defined as $s = (p_1 + p_2)^2$ and $t = (p_1 - p_3)^2$. The differential cross-section of such $(a, b) \to (a, b)$ scattering, $\frac{d\sigma(s,t)}{dt}$ can be expressed in terms of the scattering amplitude $T_{\rm el}(s, t)$ as

$$\frac{d\sigma(s,t)}{dt} = \frac{1}{4\pi}|T_{\rm el}(s,t)|^2 . \qquad (1)$$

The elastic cross-section is given as an integral of the differential cross-section as

$$\sigma_{\rm el}(s) = \int_0^\infty dt\, \frac{d\sigma(s,t)}{dt}. \qquad (2)$$

At a given s, the t-dependent slope parameter $B(s, t)$ is the logarithmic slope of the differential cross-section:

$$B(s,t) = \frac{d}{dt} \ln \frac{d\sigma(s,t)}{dt} . \qquad (3)$$

In the low-t region, corresponding to the diffractive cone, this function is frequently assumed or found within experimental errors to be a constant. To characterize this diffractive cone, a t-independent slope parameter $B(s)$ can be introduced as

$$B(s) \equiv B_0(s) = \lim_{t\to 0} B(s,t) , \qquad (4)$$

where the $t \to 0$ limit, the so called optical point is measured with a finite experimental resolution, so in general $B(s)$ is resolution, or $-t$ fit-range, dependent. The lowest values of $|-t|$ that we analyze in this work correspond to $-t \geq 0.00515$ GeV^2 at $\sqrt{s} = 7$ TeV, which is outside of the so-called Coulomb-Nuclear Interference (CNI) region, see Fig. 2 of Ref. [9]. This is why $B(s)$ is called hadronic or nuclear slope, as it is defined outside of the range of the CNI effects.

The optical point is also found by extrapolations from the measurements performed in the diffractive cone, in the $-t > 0$, experimentally accessible regions. The total cross-section is determined as

$$\sigma_{\rm tot}(s) \equiv 2\,{\rm Im}\, T_{\rm el}(t = 0, s) . \qquad (5)$$

In general, the (s, t) dependent ratio of the real to imaginary parts of the elastic amplitude is defined as

$$\rho(s,t) \equiv \frac{{\rm Re}\, T_{\rm el}(s,t)}{{\rm Im}\, T_{\rm el}(s,t)} . \qquad (6)$$

The $t \to 0$ limit of this ratio is given by

$$\rho(s) \equiv \rho_0(s) = \lim_{t\to 0} \rho(s,t) . \qquad (7)$$

Technically, ρ_0 is measured in the CNI region [9], with the help of the interference of the well understood Coulomb wave with the nuclear (or strong) amplitude. The $t \to 0$ limit is understood in terms of the finite experimental resolution and refers to ρ of the nuclear phase, without including the CNI effects. Then the differential cross section at the optical point can be written as

$$\left.\frac{d\sigma(s,t)}{dt}\right|_{t\to 0} = \frac{1+\rho_0^2(s)}{16\pi}\, \sigma_{\rm tot}^2(s) . \qquad (8)$$

In the impact-parameter b-space, we have

$$t_{\rm el}(s,b) = \int \frac{d^2\mathbf{\Delta}}{(2\pi)^2}\, e^{-i\mathbf{\Delta} \mathbf{b}}\, T_{\rm el}(s,t) ,$$
$$\Delta \equiv |\mathbf{\Delta}| , \quad b \equiv |\mathbf{b}| , \qquad \Delta = \sqrt{|t|} . \qquad (9)$$

Here, the Fourier-transformed elastic amplitude $t_{el}(s, b)$ can be cast in the eikonal form as follows

$$t_{\rm el}(s,b) = i\left[1 - e^{-\Omega(s,b)}\right] , \qquad (10)$$

where $\Omega(s, b)$ is the so-called opacity function (known also as the eikonal function), which is generically complex. The shadow profile function is then defined as

$$P(s,b) = 1 - \left|e^{-\Omega(s,b)}\right|^2 . \qquad (11)$$

When the real part of the scattering amplitude is neglected, $P(b, s)$ is frequently denoted as $G_{\rm inel}(s, b)$, see for example Refs. [10–14] for more details.

3 Odderon

The pp or $p\bar{p}$ elastic scattering amplitude can be written as a sum or a difference of crossing-even and crossing-odd contributions, respectively,

$$T_{\rm el}^{pp}(s,t) = T_{\rm el}^{+}(s,t) + T_{\rm el}^{-}(s,t), \qquad (12)$$
$$T_{\rm el}^{p\bar{p}}(s,t) = T_{\rm el}^{+}(s,t) - T_{\rm el}^{-}(s,t), \qquad (13)$$
$$T_{\rm el}^{+}(s,t) = T_{\rm el}^{P}(s,t) + T_{\rm el}^{f}(s,t), \qquad (14)$$
$$T_{\rm el}^{-}(s,t) = T_{\rm el}^{O}(s,t) + T_{\rm el}^{\omega}(s,t) . \qquad (15)$$

The even-under-crossing part consists of the Pomeron and the f Reggeon trajectory, while the odd-under-crossing part contains the Odderon and a contribution from the ω Reggeon. A direct way to "see" the Odderon in the data is to compare the differential cross-section of in elastic pp and $p\bar{p}$ scattering at TeV energies [15, 16] since at sufficiently high $\sqrt{s}$ the Reggeon contributions decrease below the experimental errors. In this case, the Odderon and Pomeron contributions, $T_{\rm el}^{O}(s, t)$ and $T_{\rm el}^{P}(s, t)$, respectively, are found as follows

$$T_{\rm el}^{P}(s,t) = \frac{T_{\rm el}^{pp} + T_{\rm el}^{p\bar{p}}}{2} \quad \text{if } \sqrt{s} \geq 1 \text{ TeV} , \qquad (16)$$
$$T_{\rm el}^{O}(s,t) = \frac{T_{\rm el}^{pp} - T_{\rm el}^{p\bar{p}}}{2} \quad \text{if } \sqrt{s} \geq 1 \text{ TeV} , \qquad (17)$$

where we have suppressed the (s, t) dependence of the pp and $p\bar{p}$ scattering amplitudes, for the sake of brevity. Indeed, if the pp differential cross sections differ from that amplitude cannot be equal to zero [8], i.e.

$$\frac{d\sigma^{pp}}{dt} \neq \frac{d\sigma^{p\bar{p}}}{dt} \quad \text{for } \sqrt{s} \geq 1 \text{ TeV} \implies T_{\rm el}^{O}(s,t) \neq 0\,, \tag{18}$$

which is considered to be the main Odderon signal. The Odderon contribution can be particularly strong around $t_{\rm dip}$, as a pronounced diffractive dip (minimum) is seen in elastic pp collisions, while this structure is apparently missing in $p\bar{p}$ collisions in the TeV energy range [17, 18].

3.1 Scaling of the differential cross-section

The low-$|t|$ part of the measured distribution can be frequently approximated with an exponential shape, if the experimental errors are sufficiently large:

$$\frac{d\sigma}{dt}(s,t) = A(s)\, \exp\left[B(s)t\right]\,. \tag{19}$$

The normalization parameter is denoted by $A(s) = \frac{d\sigma}{dt}(s, t=0)$ and the nuclear slope parameter by $B(s)$ as discussed above. In the diffractive cone, $A(s) = B(s)\sigma_{\rm el}(s)$, and the nearly exponential differential cross-sections can be scaled to a universal scaling function defined as [8]

$$H(x) \equiv \frac{1}{B(s)\sigma_{\rm el}(s)}\frac{d\sigma}{dt}\,, \tag{20}$$

$$x = -tB(s)\,. \tag{21}$$

At low-$|t|$ this scaling function is approximately written as $H(x) = \exp(-x)$. Thus, $H(x)$ scales out the trivial s dependencies of $B(s)$ and $\sigma_{\rm el}(s)$ from the differential cross-sections in the diffractive cone. In the exponential approximation,

$$A(s) = B(s)\,\sigma_{\rm el}(s) = \frac{1+\rho_0^2(s)}{16\pi}\,\sigma_{\rm tot}^2(s)\,, \tag{22}$$

$$B(s) = \frac{1+\rho_0^2(s)}{16\pi}\,\frac{\sigma_{\rm tot}^2(s)}{\sigma_{\rm el}(s)}\,, \tag{23}$$

thus $H(x)$ scales out the s-dependence, if it arises only due to the s-dependence of the total cross-section $\sigma_{\rm tot}(s)$ and the real-to-imaginary ratio $\rho(s)$.

In Ref. [8] we have shown that our $H(x)$ scaling, defined as above, is valid not only in the diffractive cone but, surprisingly, also at the diffractive minimum and maximum of elastic proton-proton collisions: the $H(x, s)$ is s-independent in the ISR energy range of $\sqrt{s}$ = 23.5 - 62.5 GeV. In addition, the $H(x)$ scaling was shown to be also valid, within statistical errors, in the LHC energy range of $\sqrt{s}$ = 2.76 − 7.0 TeV, but with a scaling function that is significantly different from the one at ISR energies [8].

From the s-independence of the $H(x, s)$ at the lower LHC energies as well as from Eqs. (20,21) a new Odderon signature follows [8]: If the $H(x, s)$ scaling function for pp collisions differs from that of $p\bar{p}$ in a given s domain, where $H(x, s) \equiv H(x)$ is s-independent for pp collisions, then the differential cross-sections in that interval of s cannot be equal for pp and $p\bar{p}$ collisions, either. Hence, the

$$\text{if } H(x,s_1) \text{ for } pp \neq H(x,s_2) \text{ for } p\bar{p}$$
$$\text{for} \quad 1 \lessapprox \sqrt{s_1},\ \sqrt{s_2} \lessapprox 8\,\text{TeV}$$
$$\implies T_{\rm el}^{O}(s,t) \neq 0\,. \tag{24}$$

The left panel of Fig. 1 indicates that for elastic proton-proton collisions, the $H(x, s)$ scaling function of the differential cross-section is independent, within statistical errors, of the colliding energy for $\sqrt{s}$ = 2.76 and 7.0 TeV, at the confidence level of CL = 99 %. Similar results are obtained for the TOTEM preliminary large $-t$ differential cross-sections at $\sqrt{s}$ = 8 TeV [19]. Given that in addition to the statistical errors, there are significant systematic errors present as well, this result indicates that the validity of the s-independence of the $H(x) \equiv H(x, s)$ scaling for pp collisions is greater than the $2.76 \leq \sqrt{s} \leq 8.0$ TeV region. Within *systematic* errors, indicated by vertical bars with widths proportional to the horizontal systematic error on $x = -tB$, the confidence level of the agreement of the $H(x, s)$ scaling function for the two datasets at $\sqrt{s}$ = 2.76 and 7.0 TeV is 100 %. In Fig. 1, the regions with smaller and larger boxes correspond to two different datasets, measured in two different acceptance regions [20, 21].

Quantitatively, we have found a statistically significant Odderon signal in the comparison of the $H(x, s)$ scaling functions of elastic pp collisions at $\sqrt{s}$ = 7.0 to that of $p\bar{p}$ collisions at $\sqrt{s}$ = 1.96 TeV. On the right panel of Fig. 1, we compare the $H(x, s)$ scaling functions of elastic pp collisions at $\sqrt{s}$ = 7.0 TeV with that of the elastic $p\bar{p}$ collisions at $\sqrt{s}$ = 1.96 TeV. These scaling functions are statistically significantly different. The confidence level of their agreement is maximum $CL = 3.7 \times 10^{-8}$ %, corresponding to a statistically significant, of at least 6.26 σ Odderon signal. This difference is larger, than the 5σ threshold, required for a discovery in particle physics. The advantages of our method, with respect to comparing the cross sections directly include the scaling out of the s-dependencies of $\sigma_{\rm tot}(s)$, $\sigma_{\rm el}(s)$, $B(s)$ and $\rho(s)$, as well as the normalization of the $H(x)$ scaling function that cancels the point-to-point correlated and t-independent normalization errors.

4 Proton holography

A new possibility for phase reconstruction in elastic proton-proton scattering is based on our detailed analysis of elastic pp collisions data in the dip region at ISR and LHC energies [8, 22–25]. To support this conjecture, we present some qualitative considerations as well as the results of a model-independent study, and also discuss the currently known limitations of this promising new method.

As obvious, the phase reconstruction from the measurement of the differential cross-section of elastic scattering, in general and without additional information, is not possible. For example, if the measurements are limited to the diffractive cone, outside of the CNI region, but stopping well before the diffractive minimum-maximum (dip and bump) structure, the phase reconstruction from a structureless, nearly exponential cone is clearly meaningless and impossible. These limiting cases were discussed

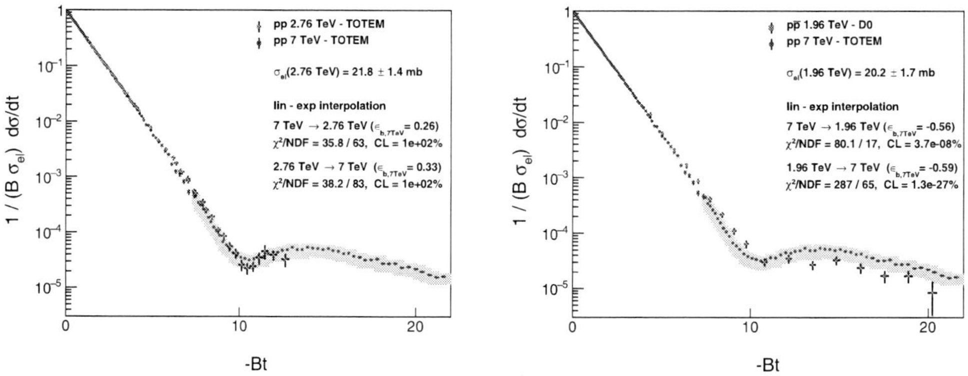

Figure 1. Left panel: For proton-proton elastic scattering, the $H(x, s) = \frac{1}{B\sigma_{el}}\frac{d\sigma}{dt}$ scaling function is energy independent in terms of $x = -tB$ at the TeV energy scale, or specifically, in the $\sqrt{s} = 2.76$ - 7 TeV energy range. TOTEM preliminary data at $\sqrt{s} = 8$ TeV [19] also follow this scaling law, while scaling violations are observed at $\sqrt{s} = 13$ TeV [8]. Right panel: A statistically significant difference is found between the $H(x, s)$ scaling functions for elastic pp collisions at $\sqrt{s} = 7.0$ TeV and that of $p\bar{p}$ collisions at $\sqrt{s} = 1.96$ TeV. On both panels, vertical lines stand for the type A (point-to-point fluctuating) errors, while the horizontal lines indicate the point-to-point fluctuating uncertainty of the horizontal errors. Grey vertical bars stand for the type B (point-to-point changing, but overall correlated) errors and the width of this grey bars correspond to the type B errors in the horizontal (x) direction. Larger and smaller boxes correspond to two different measurements at $\sqrt{s} = 7$ TeV, detailed in Refs. [20, 21], respectively.

for example in Appendices C and D of Ref. [22]. However, the observation of a diffractive minimum, followed by a diffractive maximum corresponds to the observation of an interference pattern, one of the conditions of phase reconstruction and holography. Such an interference pattern is clearly visible as a function of the deflection angle in Fig. 1 of Ref. [26], originally from Ref. [27].

Recently, we have discussed model-independent, whole $-t$ range fits to the differential cross-section in Appendix A and B of Ref. [22] for pp and for $p\bar{p}$ elastic collisions, respectively. A closer inspection of these fits indicate that at the ISR energy range, the total cross-section is reproduced reasonably well, while the ρ_0 measurements are reproduced only in one case, at $\sqrt{s} = 44.7$ GeV, within two standard deviations. Given that we did *not* utilize any data in the CNI region, the bad reproduction of several of the measured ρ_0 values were not surprising. As discussed in Ref. [22], in the case of the very well measured TOTEM data at $\sqrt{s} = 13$ TeV, we were able to reproduce within experimental errors also the measured value of ρ_0, and we have obtained similarly good quality preliminary results at $\sqrt{s} = 7$ TeV as well.

We started to wonder, whether or not such a good quality ρ_0 determination from outside the CNI region is a coincidence only: it is possible that there is a deeper reason for it. As a first step, we have shown in Ref. [22], that *within* the technique of the model-independent Lévy series expansion, the reconstruction of the t-dependent phase or equivalently the determination of the $-t$ dependent $\rho(t)$ function is unique, similarly to how the coefficients of a Taylor-series can be uniquely determined if the measurement of the Taylor-expanded function is precise enough. Of course, behind this reconstruction lies a powerful physical assumption about the Lévy processes driving the elastic scattering in QCD which dictates a very specific form of the elastic amplitude, in the form of generalised Lévy series. However, other methods may result in other $\phi(t)$ functions. We are currently working on these subtleties of the $-t$ dependent phase reconstruction by comparing the results of Ref. [22] with other, model-dependent efforts.

In this work, we thus start to explore some of the basic details of t-dependent phase reconstruction. We ask the following question: what are those special experimentally testable, but mathematically well defined conditions that are sufficient for an experimentally validated phase reconstruction in elastic pp or $p\bar{p}$ scattering? Let us recall two examples here about how such a phase reconstruction might work in elastic pp scattering.

The first example deals with phase reconstruction at the ISR energies. Such a reconstruction of the nuclear phase from the ISR data was performed in Ref. [7] utilizing the Phillips-Barger model [28]. This model was shown to be well suited to describe the ISR data in the $\sqrt{s} = 23.5$ - 62.1 GeV region with statistically acceptable, good quality fits in the $1.05 \leq -t \leq 2.5$ GeV^2 interval. Table 3 of Ref. [7] reported the values of $|\epsilon|$, the modulus square of the phase difference from π, with values small but close to zero in the range of $(0.06 \pm 0.06) - (0.34 \pm 0.10)$. These values suggested that the imaginary part of the scattering amplitude approximately vanishes near the diffractive minimum, and the real part is dominant at the dip region, but due to the $\cos(\phi) = \cos(\pi \pm \epsilon) \approx 1 + 0.5\epsilon^2$ type of Taylor expansion, the sign of the phase difference, ϵ cannot be determined from the experimental data. Assuming that a crossing-odd contribution is negligible at the ISR energies, a more detailed, t-dependent reconstruction of $\rho(t) = \mathcal{R}eT/\mathcal{I}mT$ is reported in Fig. 22 of Ref. [7].

The Reggeized Phillips-Barger model [28] gave good quality fits to the ISR and LHC data in larger $-t$ intervals as well, starting from $-t \geq 0.3$ GeV^2 and in 2011 it was used to study the Odderon effect in the elastic pp scattering data at LHC energies [15]. However, this model at small values of $-t \leq 0.3$ GeV^2 did not give a statistically acceptable data description, hence this model cannot be used to connect the phase measurement at the dip region of $-t \approx 0.5$ GeV^2 with CNI measurements at the optical point $t \approx 0$ GeV^2, so this model cannot be used to resolve the Cul-de-Sac of elastic scattering [14]. However, the model-independent Lévy series [22] is able to do this job. As far as we know, this is the only method so far, that is able to describe the elastic pp collisions at the currently highest energy of $\sqrt{s} = 13$ TeV [27] with a statistically acceptable, 2% confidence level. Within this model-independent approach the t-dependent phase reconstruction appears to be unique [22]. Increasing the order of the series, the domain of the convergence starts from $t \approx 0$ and is increasing gradually. The convergence region for the phase of the fourth order Lévy series is $0 \leq -t \lessapprox 0.5$ and 0.9 GeV^2 at $\sqrt{s} = 7$ and 13 TeV, respectively, as shown on Fig. 2.

The left panel of Fig. 2 indicates that for proton-proton elastic scattering at $\sqrt{s} = 13.0$ TeV, the nuclear phase reconstruction based on a 4th-order Levy expansion converges up to $-t \lessapprox 0.9$ GeV^2. The right panel of the same figure indicates similar results but at $\sqrt{s} = 7.0$ TeV. These data are less precise as compared to the measurements at $\sqrt{s} = 13.0$ TeV, correspondingly the domain of convergence of the phase reconstruction is smaller. Generically, the greater the experimental precision, the larger the domain of convergence of this method. These plots suggest that the reconstruction of the t-dependent phase is possible with the help of this Lévy series technique, providing a new kind of model-independent proton holography. This is our second example, detailed below.

One of the most intriguing properties of elastic pp collisions at the ISR energy range of $\sqrt{s} = 23.5 - 62.1$ GeV and at the LHC energy range of $\sqrt{s} = 2.76 - 13.0$ TeV is the existence of a unique, single diffractive minimum, and the existence of a nearly exponential diffractive cone not only at very low $|t|$ but also a similar, secondary diffractive cone that follows after the diffractive maximum [25]. From measurements of the CNI region, reviewed recently for example in Fig. 15 of Ref. [29], we also know that in each of these measurements, $\rho_0 \leq 0.15$ that implies that close to $t \approx 0$, the real part of the scattering amplitude is, as compared to the imaginary part, relatively small.

By means of the optical theorem, a total cross section measurement uniquely determines the imaginary part of the elastic amplitude at $t = 0$, as indicated by Eq. (5). This implies that not only the imaginary but also the ρ_0^2 or the squared value of real part of the forward scattering amplitude can also be uniquely determined at the optical point, $t = 0$. Indeed, using the diffractive cone approximation and Eqs. (22) and (23), from the values of $A(s)$, $B(s)$ and $\sigma_{\rm tot}(s)$, the values of $\sigma_{\rm el}(s)$ and $\rho_0^2(s)$ are determined uniquely. From such a determination of $\rho_0^2(s)$, the value of the phase of the scattering amplitude at the optical point, $\phi(t = 0)$ is fixed to two possible values by measurements. However, to select one of these values unambiguously, further information about the CNI interference has to be utilized. This dual degeneracy is related to the fact that the sign of ρ_0 cannot be determined from measurements of ρ_0^2, such that both possible values, corresponding to $\pm\sqrt{\rho_0^2(s)}$, are allowed. These two different allowed phases correspond to the object and the conjugated object wave reconstruction in holography.

One can determine the nuclear phase at the dip $t_{\rm dip}$, too, where the differential cross-section of elastic proton-proton collisions has a diffractive minimum. At this point, the imaginary part of the elastic scattering amplitude approximately vanishes, and the differential cross-section is dominated in the dip region by the real part of the forward scattering amplitude. Thus, the nuclear phase is approximately an integer multiple of π close to the dip region. This is a degeneracy similar to the ρ_0 sign problem, mentioned in the previous paragraph, and to the object beam and the conjugated object beam reconstruction in holography, mentioned in the Introduction. This problem has been called the Cul-de-Sac of elastic scattering in Ref. [14]: at the diffractive minimum, only the modulus but not the sign of the real part can be uniquely determined, hence the nuclear phase is either 0 or π, up to small corrections.

The fact that at $\sqrt{s} = 13$ TeV, the phase at the dip connects analytically to the phase measured in the CNI region indicates, that with the help of the Lévy expansion method, the nuclear phase of the elastic scattering amplitude can be realistically determined, at least in certain specific cases. A posteriori, our method is validated by the excellent reproduction of the ρ_0 value measured in the CNI region in elastic pp collisions at $\sqrt{s} = 13$ TeV by the TOTEM Collaboration [29]. Fits detailed in Appendices A and B of Ref. [22] indicate that indeed the 4th and 3rd order Lévy expansions reproduce well measured values of ρ_0 not only at $\sqrt{s} = 13$ TeV but at lower LHC energies as well, if the fit quality is satisfactory. However, one should put a pull on ρ_0 in the definition of χ^2 as the sign problem at the dip cannot be resolved otherwise. The two different signs of the phase near the dip can lead to two different analytic continuation of the phase to $t = 0$ but only one of these continuations lead to the correct value of ρ_0.

5 Summary

We have described a new Odderon signal [8], the validity of the $H(x)$ scaling for pp collisions and its violation in $p\bar{p}$ collisions in the few TeV energy range shown and quantified in the right panel of Fig. 1. The statistical significance of this Odderon signal is found to be at least 6.26σ.

We have also discussed the related topic of proton holography or the t-dependent reconstruction of the nuclear phase. If we assume that this nuclear phase $\phi(t)$ is an analytic and continuous function of t, and the differential cross-section of elastic scattering data is determined with great precision, the two different approximate values of the phases at the dip, 0 and π can be analytically and continuously extrapolated to $t \to 0$. Furthermore, our analysis with Lévy series performed in Ref. [22] suggested that if these two extrapolated values are within the errors of the extrapolations significantly different from one another, and only one of them is consistent with the determination of

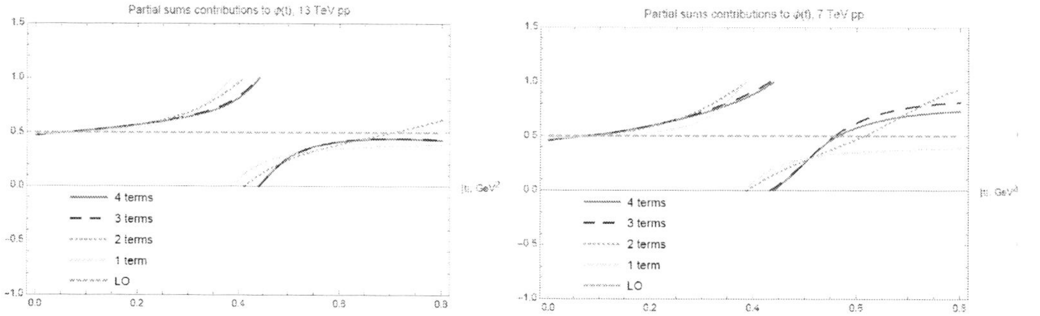

Figure 2. Left: For *pp* elastic scattering at $\sqrt{s}$ = 13.0 TeV, the nuclear phase reconstruction based on a 4th order Levy expansion converges up to $-t \lessapprox 0.9$ GeV^{-2}. Right: The phase reconstruction at $\sqrt{s}$ = 7.0 TeV converges up to $-t \lessapprox 0.5$ GeV^{-2}, as detailed in [24].

the ρ_0 using the CNI methods, then the t-dependence of the nuclear phase can be reconstructed not only at a given value of $t = 0$ or $t = t_{\rm dip}$ but in the entire t region, where such an extrapolation is possible and convergent.

The search for a sufficient and necessary condition that would make the t-dependent phase reconstruction unique not only within, but also outside the Lévy series method is still ongoing at the time of closing this manuscript.

Acknowledgments

We thank I. Vitev and co-organizers for an inspiring and successful meeting. We acknowledge stimulating discussions with G. Gustafson, Y. Hatta, L. Lönnblad, and M. Šumbera as well as with members of the TOTEM Collaboration. T. Cs. and A. S. were partially supported by the NKFIH grants No. FK-123842, FK-123959 and K-133046, and by the EFOP 3.6.1-16-2016-00001 grant (Hungary). R. P. is partially supported by Swedish Research Council grants No. 621-2013-4287 and 2016-05996, by ERC H2020 grant No 668679, as well as by the Ministry of Education, Youth and Sports of the Czech Republic project LTT17018. Our collaboration was supported by THOR, the EU COST Action CA15213.

References

[1] L. Lukaszuk, B. Nicolescu, Lett. Nuovo Cim. **8**, 405 (1973)
[2] J. Bartels, L.N. Lipatov, G.P. Vacca, Phys. Lett. **B477**, 178 (2000), hep-ph/9912423
[3] D. Gabor (World Scientific Publishing Co., 1992), https://www.nobelprize.org/prizes/physics/1971/gabor/lecture/
[4] R.E. Dunin-Borkowski et al., Microscopy Res. and Tech. **64**, 390 (2004)
[5] F. Shimizu, J. Fujita, Phys. Rev. Lett. **88**, 123201 (2002)
[6] D. Sarenac et al., Opt. Express **24**, 22528 (2016)
[7] E. Nagy et al., Nucl. Phys. **B150**, 221 (1979)
[8] T. Csörgő, T. Novák, R. Pasechnik, A. Ster, I. Szanyi (2019), 1912.11968
[9] G. Antchev et al. (TOTEM), Eur. Phys. J. **C76**, 661 (2016), 1610.00603
[10] V.A. Petrov, A.P. Samokhin, Int. J. Mod. Phys. Conf. Ser. **47**, 1860097 (2018), 1801.03809
[11] I.M. Dremin, V.A. Nechitailo, Nucl. Phys. **A916**, 241 (2013), 1306.5384
[12] I.M. Dremin, Phys. Usp. **58**, 61 (2015), 1406.2153
[13] I.M. Dremin, V.A. Nechitailo, Eur. Phys. J. **C78**, 913 (2018)
[14] I. Dremin, MDPI Physics **1**, 33 (2019)
[15] L.L. Jenkovszky, A.I. Lengyel, D.I. Lontkovskyi, Int. J. Mod. Phys. **A26**, 4755 (2011), 1105.1202
[16] A. Ster, L. Jenkovszky, T. Csörgő, Phys. Rev. **D91**, 074018 (2015), 1501.03860
[17] V.M. Abazov et al. (D0), Phys. Rev. **D86**, 012009 (2012), 1206.0687
[18] G. Antchev et al. (TOTEM) (2018), 1812.08610
[19] J. Kaspar, EPJ Web of Conf. **172**, 06005 (2018)
[20] G. Antchev et al. (TOTEM), EPL **95**, 41001 (2011)
[21] G. Antchev et al. (TOTEM), EPL **101**, 21002 (2013)
[22] T. Csörgő, R. Pasechnik, A. Ster, Eur. Phys. J. **C79**, 62 (2019), 1807.02897
[23] T. Csörgő, R. Pasechnik, A. Ster (2018), 1811.08913
[24] T. Csörgő, R. Pasechnik, A. Ster, EPJ Web Conf. **206**, 06007 (2019), 1903.08235
[25] T. Csörgő, R. Pasechnik, A. Ster (2019), 1910.08817
[26] T. Csörgő (TOTEM), EPJ Web Conf. **206**, 06004 (2019), 1903.06992
[27] G. Antchev et al. (TOTEM), Eur. Phys. J. C **79**, 861 (2019), 1812.08283
[28] R. Phillips, V. Barger, Phys. Lett. **46B**, 412 (1973)
[29] G. Antchev et al. (TOTEM), Eur. Phys. J. C **79**, 785 (2019), 1812.04732

EPJ Web of Conferences **235**, 07001 (2020) https://doi.org/10.1051/epjconf/202023507001
ISMD 2019

Recent Results from the High Altitude Water Cherenkov Observatory

Kelly Malone for the HAWC Collaboration[1,⋆]

[1] *Physics Division, Los Alamos National Laboratory, Los Alamos, NM, USA*

Abstract. The High Altitude Water Cherenkov (HAWC) Gamma-Ray Observatory, an extensive air shower detector consisting of 300 water Cherenkov tanks located at 4100m in Puebla, Mexico, has been surveying the TeV gamma-ray sky for almost five years. HAWC can observe steady sources, variable sources, transients, which allows for probes of both astrophysical and particle physics phenomena. This includes the production and propagation of cosmic rays, studies of Lorentz invariance violation, and dark matter searches. I will discuss recent HAWC results as well as the future of the observatory.

1 Introduction: Cosmic-ray origins and propagation

It is a known fact that cosmic rays (CRs) are accelerated to $\sim 10^{20}$ eV in nature [1], but which sources are responsible for this and the exact mechanisms are still largely open questions. CR acceleration is inherently interesting since the energies involved are much higher than can be reached by accelerators on Earth, but since cosmic rays are charged they bend in magnetic fields on their way to the Earth and it is somewhat difficult to pinpoint their origins.

There are advantages in studying gamma rays instead. CRs interact with their environment (such as the interstellar medium); neutral pions are created in this process. These neutral pions decay to gamma rays. Since gamma rays are not charged, they point back to sites of CR acceleration. This is the hadronic production mechanism of gamma rays.

Gamma rays are also produced via leptonic mechanisms. In the TeV range (discussed in this proceeding), the dominant mechanism is Inverse Compton scattering, in which a lower-energy photon scatters off an electron and has its energy boosted in the process.

Distinguishing between hadronic and leptonic gamma-ray sources (and thus identifiying which sources may be associated with cosmic rays) is a challenge in the field. For leptonic sources, the Klein Nishina component causes the gamma-ray emission to become suppressed starting at a few tens of TeV, resulting in an energy-dependent spectral index and curvature in the spectra [2].

2 Introduction to HAWC

The High Altitude Water Cherenkov (HAWC) Observatory is an extensive air shower experiment with sensitivity to gamma rays with energies between a few hundred GeV and a few hundred TeV [3, 4]. It

⋆e-mail: kmalone@lanl.gov

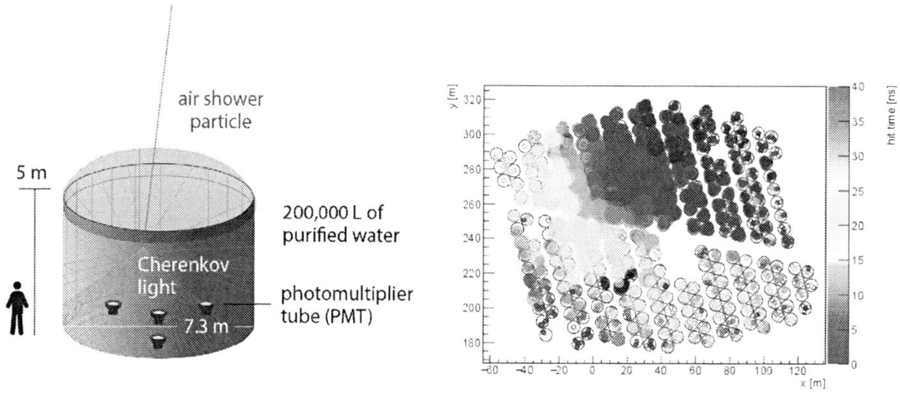

Figure 1: Left: A schematic of a tank. Right: The footprint of a detector as a high-energy event hits it. The colors denote time (with dark blue hitting earliest) and the size of the circles denotes the charge detected by each PMT. The shower core is denoted by a star.

is located at 19° North on the saddle point between Pico de Orizaba and Sierra Negra in the state of Puebla, Mexico. The altitude is 4100 meters. HAWC has a duty cycle of $> 95\%$ and can observe $\sim 2/3$ of the overhead sky per day with an angular resolution of $\geq 0.1°$. Partial science operations began in 2013, with full operations beginning in March 2015.

Extensive air shower (EAS) experiments work by detecting byproducts of the electromagnetic cascade induced when a gamma ray enters the Earth atmosphere and interacts with molecules in the air. This cascade consists of electrons, positrons, and lower-energy photons. Note that protons and other hadronic particles induce showers with a similar electromagnetic component, resulting in a very large backround for EAS arrays.

HAWC consists of an array of 300 tanks of water, each instrumented with four photomultiplier tubes (PMTs). These PMTs detect the Cherenkov light produced when the charged particles from the air shower pass through the tanks of water. Each tank is 5 meters high and 7.3 meters in diameter, and has 200,000 liters of water in a light-tight bladder. The entire footprint is 22,000 meters^2. Figure 1 shows a schematic of a HAWC tank.

As the EAS sweeps across the array, the timing of the PMT hits is used to reconstruct the direction and the charge is used to reconstruct the shower core and energy. The charge is also used for background rejection. Hadronic showers will have large depositions of charge far from the shower core due to the presence of muons, which have high transverse momenta. For more information on how HAWC operates, see [5].

3 Recent Results

HAWC has a far-reaching science agenda covering both Galactic and extragalactic observations as well as studies of fundamental physics phenomena and cosmic ray observations. Transients such as flaring sources and gamma-ray bursts are monitored. Additionally, there is an active follow-up program of multi-wavelength and multi-messenger phenomena.

One of the advantages of an EAS is the ability to perform an unbiased sky survey. The 2HWC (second HAWC) catalog was published in 2017 using 507 days of early HAWC data [6]. 39 sources

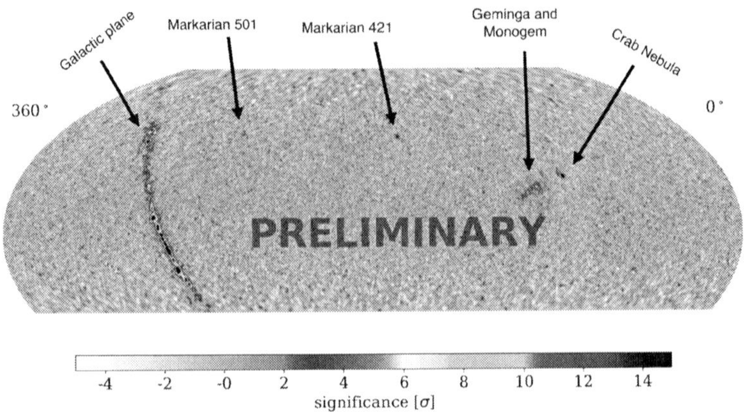

Figure 2: Significance map of the sky as seen by HAWC. 1346 days of data are used. Some prominent features are labeled.

were identified; 16 of them were new TeV discoveries. A handful have since been confirmed by other experiments [7]. Roughly half of the sources are unidentified (meaning that their nature cannot be determined using data from other wavelengths) while the rest are pulsar wind nebulae (PWNe), supernova remnants, and blazars.

Since the publication of the 2HWC catalog, additional data has been collected. Figure 2 shows an updated skymap with 1346 days of data (~2.6 times that data published in the second catalog). Most of the emission is concentrated along the Galactic plane, although a few other sources are also visible. A few selected sources are discussed in the sections below.

3.1 Galactic observations: Steady sources

HAWC's wide field-of-view enables the observation of very extended sources (>1° across). An example of this is HAWC's observations of the nearby (250-300 parsec) middle-aged (~10^5 year) PWNe Geminga and Monogem [8]. It has been previously hypothesized that contributions from nearby pulsars can explain the positron excess (originally observed by PAMELA [9]) beyond what is expected from CR interactions with the interstellar medium. In this scenario, gamma rays are produced when electrons and positrons diffuse and scatter on the cosmic microwave background.

The radial profiles of these sources show that Geminga and Monogem are indeed local sources of leptons, but that diffusion away from these sources is occurring on a slower timescale than previously was thought. Assuming that this slow diffusion constant persists over the entire distance between the Earth and the Geminga/Monogem region, pulsars are unlikely to be the origin of the positron excess. However, in the case of variable diffusion (as suggested by [10–12]), local pulsars are capable of producing the excess positrons. More studies of the region are needed.

HAWC has also studied less extended Galactic sources, such as the binary system SS 433 [13]. This source consists of a supergiant star accreting onto a compact object (either a black hole or a neutron star). This results in two perpendicular jets of plasma moving with a bulk velocity of ~$0.26c$ [14]. The jets terminate inside the supernova remnant W50. Interactions inside this nebula accelerate charged particles, creating gamma rays. This is the first detection of spatially resolved

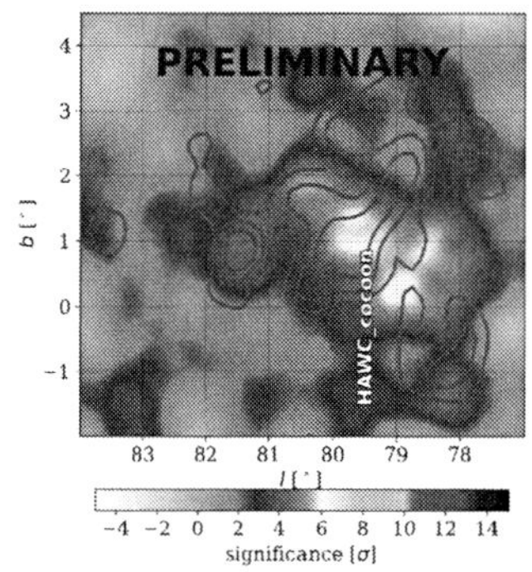

Figure 3: Significance map of the Cocoon region, after the nearby PWN has been removed. From [17]. A large-scale residual emission, roughly 2 degrees across, can be clearly seen.

TeV-emitting lobes, and therefore the first direct proof of particle acceleration above 100 TeV in jets. A leptonic origin of the TeV emission is favored over pure hadronic models [13]. Assuming proton-proton collisions, the predicted gamma-ray flux is much less than what has been detected by HAWC. However, there are still open questions about how the acceleration occurs.

This is an important development because the jets studied here are thought to be scale models of more powerful jets in active Galactic nuclei (AGN), which are likely the sources of the highest-energy CRs. Since AGN are extragalactic objects, we are not able to spatially resolve them at this time.

Recent improvements to HAWC's data analysis chain have allowed the extension of the dynamic range of the experiment to above 100 TeV, resulting in the highest energy source catalog to date [15]. There are nine Galactic sources that emit at very high significance ($>5\sigma$) above 56 TeV, eight of which are within a degree of the Galactic plane (the ninth source is the Crab Nebula). Curiously, all of the eight inner Galactic plane sources are coincident with high spin-down pulsars ($\dot{E} > 10^{36}$ erg/s). Each of these sources also remains extended in apparent size above 56 TeV. This is unexpected, as their spatial coincidence with pulsars implies leptonic emission mechanisms. The high-energy electrons that create >56 TeV gamma rays are expected to cool fairly quickly, so assuming an injection of electrons from the center of each source, it is puzzling to see how they have diffused such distances before cooling. Three of the nine >56 TeV sources have significant emission continuing above 100 TeV. These are the highest-energy collection of sources ever detected in any astrophysical messenger.

One of the most interesting regions that emits above 56 TeV is the Cygnus Cocoon region. The emission seen by HAWC is coincident with a lower-energy region detected by the Fermi-LAT telescope [16]. The Fermi observation was attributed to a cocoon of freshly accelerated cosmic rays. HAWC observations extend this cocoon to higher energies. The Cocoon region is fairly complicated in the TeV regime, as there is also the PWN of PSR J2032+4127 and the γ-Cygni SNR. Once these two sources are subtracted, a large-scale (~2°) residual emission (the Cocoon) is visible [17]. Various possible hadronic models for this emission are currently being explored. See Figure 3 for an image of the region.

3.2 Extragalactic observations: Bursts and transients

HAWC also monitors the extragalactic sky. To date, two sources have been detected: Markarian 421 and Markarian 501 [6].

HAWC is continuously scanning the sky for new extragalactic sources (both steady ones and transients). The flare monitor program provides real-time daily monitoring of known blazars for flares [18]. In the event of a detection, observatories capable of performing in-depth followups are automatically notified, which can give multi-wavelength pictures of flaring sources. It is designed to detect flares with timescales of minutes to days.

HAWC also monitors the sky for short-term bursts. This includes the gamma ray burst (GRB) program. The sky is monitored in real-time for bursts with durations of less than 100 seconds. HAWC is also capable of performing archival data analysis of GRBs detected by other observatories.

HAWC also has several active programs that emphasize multi-messenger observations. HAWC receives alerts from IceCUBE about PeV neutrino events and from LIGO about gravitational wave events. Both time-dependent (around the time of the alert) and time-integrated (using archival data) analyses are possible. In the event that an alert covers a region that is, at the time of the event, out of HAWC field-of-view, an analysis is completed as soon as the source moves into HAWC's field of view. HAWC has placed limits on both electromagnetic counterparts to gravitational waves [19] and astrophysical neutrinos.

3.3 Fundamental physics

In addition to observing Galactic and extragalactic astrophysical objects, HAWC probes fundamental physics phenomena such as Lorentz invariance violation (LIV) and dark matter.

Detections of extremely-high-energy photons (> 100 TeV, like the ones discussed in Section 3.1), lead naturally to studies of LIV. LIV can modify the particle dispersion relation. In some superluminal LIV models, this modification is predicted to lead to a decay of photons above some critical energy over short distances. Over astrophysical distances, this decay probability is predicted to be 100%, which would lead to hard cutoffs in astrophysical spectra. HAWC observations above 100 TeV limit the LIV energy scale to $>10^{31}$ eV, or 800 times the Planck energy scale [20, 21].

This analysis is performed by finding the energy above which the spectrum of an astrophysical source is inconsistent with both background and events who have mis-reconstructed energies by inserting a hard cutoff (step function) into the spectrum. The likelihood value for the nominal spectral fit is compared to the likelihood of the version with the hard cutoff inserted to determine the highest energy that that gamma rays are significantly detected at. The most significant high-energy source is 2HWC J1825-134. At the 95% confidence level, photons of ~250 TeV are detected [21].

HAWC's wide field-of-view also means that there are many targets to search for indirect signatures of dark matter annihilation and decay. Optimal sensitivity is achieved for dark matter masses between 1 and 100 TeV. Examples of targets include the Andromeda galaxy, dwarf spheroidal galaxies, the Sun, and the Galactic halo [22].

4 HAWC upgrade and future

HAWC is on its way to achieving many of its stated science goals. An "outrigger array" has recently been installed [23]. This consists of smaller WCD tanks (each with only one PMT) placed around the main array, with coarser spacing. This breaks the degeneracy between large showers with cores far from the array and smaller showers falling closer to the array. With an increase in effective area of approximately a factor of four, this will lead to a predicted increase in sensitivity of at least a factor of two above 10 TeV.

Planning has also begun for a new wide-field of view TeV experiment in the Southern Hemisphere [24]. This would allow for a better view of the Galactic center as well as provide full-sky coverage for transients. The Southern Wide-field-of-view Gamma Observatory (SWGO) Collaboration (www.swgo.org) has recently been formed, and sites in Chile and Argentina, among others, are under active consideration.

References

[1] M. Tanabashi et al. (Particle Data Group), Phys. Rev. D **98**, 030001 (2018)
[2] R. Moderski, M. Sikora, P.S. Coppi, F. Aharonian, Monthly Notices of the Royal Astronomical Society **363**, 954 (2005)
[3] A.U. Abeysekara, A. Albert, R. Alfaro, C. Alvarez, J.D. Álvarez et al., The Astrophysical Journal **843**, 39 (2017)
[4] A.U. Abeysekara, A. Albert, R. Alfaro, C. Alvarez, J.D. Álvarez et al., The Astrophysical Jorunal **881**, 134 (2019)
[5] A.J. Smith, Proceedings of Science (34th International Cosmic Ray Conference) **966** (2015)
[6] A.U. Abeysekara, A. Albert, R. Alfaro, C. Alvarez, J.D. Álvarez et al., The Astrophysical Journal **843**, 40 (2019)
[7] A.U. Abeysekara, A. Archer, W. Benbow, R. Bird, R. Brose et al., The Astrophysical Journal **866**, 24 (2018)
[8] A.U. Abeysekara, A. Albert, R. Alfaro, C. Alvarez, J.D. Álvarez et al., Science **358**, 911 (2017)
[9] O. Adriani, G.C. Barbarino, G.A. Bazilevskaya, R. Bellotti, M. Boezio et al., Nature **458**, 607 (2009)
[10] D. Hooper, I. Cholis, T. Linden, K. Fang, Physical Review D **96**, 103013 (2017)
[11] S. Profumo, J. Reynoso-Cordova, N. Kaaz, M. Silverman, Physical Review D **97**, 123008 (2018)
[12] K. Fang, X.J. Bi, P.F. Yin, Q. Yuan, The Astrophysical Journal **863**, 30 (2018)
[13] A.U. Abeysekara, A. Albert, R. Alfaro, C. Alvarez, J.D. Álvarez et al., Nature **562**, 82 (2018)
[14] S. Fabrika, Astrophysics and Space Physics Reviews **12**, 1 (2004), `astro-ph/0603390`
[15] A.U. Abeysekara, A. Albert, R. Alfaro, J.R. Angeles Camacho, J.C. Arteaga-Velázquez et al., Phys. Rev. Lett. **124**, 021102 (2020)
[16] M. Ackermann, M. Ajello, A. Allafort, L. Baldini, J. Ballet et al., Science **334**, 1103 (2011)
[17] B. Hona, H. Fleischhack, P. Huentemeyer, Proceedings of Science (36th International Cosmic Ray Conference) **699** (2019)
[18] A.U. Abeysekara, R. Alfaro, C. Alvarez, J.D. Álvarez, R. Arceo et al., The Astrophysical Journal **843**, 116 (2017), `1704.07411`
[19] I. Martinez-Castellanos, Proceedings of Science (36th International Cosmic Ray Conference) **737** (2019)
[20] H. Martínez-Huerta, A. Pérez-Lorenzana, Physical Review D **95**, 063001 (2017)
[21] H. Martinez-Huerta, S. Marinelli, J.T. Linnemann, J. Lundeen, Proceedings of Science (36th International Cosmic Ray Conference **738** (2019)
[22] J. Lundeen, J.P. Harding, Proceedings of Science (36th International Cosmic Ray Conference) **528** (2019)
[23] V. Joshi, A. Jardin-Blicq, Proceedings of Science (35th International Cosmic Ray Conference) **806** (2017)
[24] P. Abreu, A. Albert, R. Alfaro, C. Alvarez, R. Arceo et al., *The Southern Wide-Field Gamma-Ray Observatory (SWGO): A Next-Generation Ground-Based Survey Instrument*, in *Bulletin of the AAS* (2019), Vol. 51, p. 109, `1907.07737`

EPJ Web of Conferences **235**, 07002 (2020) https://doi.org/10.1051/epjconf/202023507002
ISMD 2019

What can neutron stars reveal about the equation of state of dense matter?

Ingo Tews[1,★]

[1] *Theoretical Division, Los Alamos National Laboratory, Los Alamos, New Mexico 87545, USA*

Abstract. Neutron stars are astrophysical objects of extremes, reaching the highest densities we can observe in the cosmos, and probing matter under conditions that cannot be recreated in terrestrial experiments. In August 2017, the first neutron-star merger has been observed, which provided compelling evidence that these events are an important site for r-process nucleosynthesis. Furthermore, the gravitational-wave signal of such events might shed light upon the nature of strongly interacting matter in the neutron-star core. To understand these remarkable events, reliable nuclear physics input is essential. In this contribution, I explain how to use chiral effective field theory and advanced many-body methods to provide a consistent and systematic approach to strongly interacting systems from nuclei to neutron stars with controlled theoretical uncertainties. I will discuss recent results for the equation of state relevant for the nuclear astrophysics of neutron stars and neutron-star mergers.

1 Introduction

Neutron stars (NSs) are extreme stellar objects. They are born in core collapse supernovae of stars with about 8-20 solar masses [1] and, hence, are one of the final stages of stellar evolution. With typical masses of $1.4\,M_\odot$ and radii of 10-14 km, NSs reach densities comparable to nuclear saturation density, $n_{\rm sat} \approx 2.7 \cdot 10^{14}$ g/cm^3, and maybe up to $\sim 10\,n_{\rm sat}$ in their core. Due to these high densities, typical temperatures are small compared to the Fermi energy, and NSs can be considered as cold, $T = 0$. In addition to very high densities, NSs also explore large magnetic fields, high rotational frequencies, and extreme regimes of gravity.

Neutron stars consist of strongly interacting matter. Their outer crust is made up of a lattice of nuclei that become increasingly neutron-rich with growing density. At the so-called neutron-drip density, approximately $4 \cdot 10^{11}$ g/cm^3, the neutron chemical potential is so high that neutrons can exist outside of the nuclei. In this inner crust, the lattice of nuclei is surrounded by a neutron fluid. At the bottom of the inner crust, exotic pasta phases can appear, and nuclei form structures lie rods and slabs. At densities of roughly $0.5\,n_{\rm sat}$, the crust dissolves and NSs consist of a nuclear fluid in β equilibrium, with roughly 95% neutrons and 5% protons. At even larger densities, above $2\,n_{\rm sat}$, NSs explore strongly-interacting matter that cannot be realized in terrestrial experiments. At these densities, exotic phases of matter, e.g., deconfined quark matter [2], might appear. Hence, NSs are ideal systems to study the dense-matter equation of state (EOS) and the QCD phase diagram at $T = 0$.

★e-mail: itews@lanl.gov

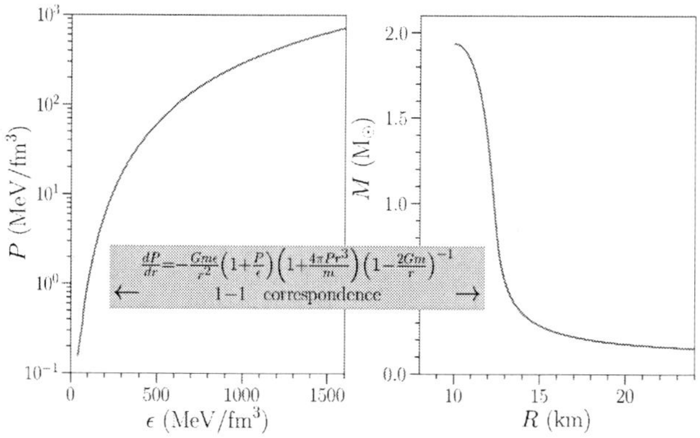

Figure 1. The EOS for NSs (left) and the resulting MR curve (right), for the Skyrme model NRAPR [3]. Figure taken from Ref. [7].

In this contribution, I will explain how we can use state-of-the-art nuclear theory and NS observations to unravel the EOS of dense matter.

2 The equation of state

2.1 General considerations

Neutron stars are described by the EOS, which is a relation among the energy density ϵ, the pressure p, the temperature T and the composition. Because individual NSs can be considered at $T = 0$, the EOS is simplified[1]. With a given composition of NS matter, the EOS reduces to a simple relation $p(\epsilon)$. Given this relation, one can solve the Tolman-Oppenheimer-Volkoff equations and obtain the mass-radius (MR) relation of NSs. The MR relation and the EOS are in 1-to-1 correspondence, see Fig. 1 for an example EOS. As a consequence, NS observations of the MR relation, or even a single point of this curve, allow constraints on the dense-matter EOS itself. While NS masses can be determined quite precisely, radii are very difficult to determine [4]. Currently, the strongest observational constraints come from the existence of two-solar-mass NSs [5, 6], which allows to rule out too soft EOS (with lower pressure) that cannot sustain NSs of this mass against gravitational collapse. Radii, on the other hand, have large uncertainties and range from 10-14 km. This uncertainty range reflects the fact that the EOS at large densities (and even its composition) is unknown.

Theoretical calculations allow us to constrain the EOS and, hence, to improve our understanding of NSs and their properties. At low densities, nuclear theory provides constraints on the EOS and I discuss them in the following section. At very high densities about $50\,n_{\text{sat}}$, well above neutron-star densities, EOS constraints can be obtained by perturbative QCD calculations [8, 9], which are not discussed here. At intermediate densities, there are no robust theoretical EOS constraints, but NS observations of orbital characteristics in binaries, gravitational waves, or in the electromagnetic spectrum might fill this gap.

[1]Note, that NSs explore large temperatures in core-collapse supernovae and in the postmerger phase of NS mergers. In these scenarios, temperature effects cannot be neglected.

2.2 Chiral effective field theory

At low densities, we know that the NS degrees of freedom are nucleonic. In the outer crust, nucleons are contained in increasingly neutron-rich nuclei surrounded by an electron gas, and in the inner crust additionally by a neutron gas. The EOS describing the crust is rather well constrained. Above $0.5\,n_{\rm sat}$, NS matter is composed of a fluid of neutrons, protons, and electrons in β equilibrium. Here, microscopic calculations can provide constraints. Such calculations require nuclear Hamiltonians that describe the interactions among nucleons. There have been very successful phenomenological Hamiltonians describing the interactions as a sum of meson-exchanges and short-range structures [10, 11]. However, nuclear interactions are described by infinitely many processes of different importance involving larger numbers of nucleons, pions, or heavier mesons. It is important to find a way of arranging these processes according to their importance, to enable reliable microscopic calculations of the EOS.

Chiral effective field theory (EFT) is a systematic expansion of nuclear forces in which the most general Lagrangian consistent with all symmetries of QCD, containing all processes among nucleons and pions, is expanded in powers of momenta k [12]. The radius of convergence of this expansion is described by the breakdown scale, $\Lambda_b \sim 600$ MeV [13], which denotes when new d.o.f. need to be explicitly included. In chiral EFT, one- and multi-pion exchanges are explicitly included while short-range interactions are modeled by a series of contact interactions, which are usually parametrized by certain spin-isospin operators together with low-energy couplings (LECs) that determine their strength. Chiral EFT naturally and consistently includes also important many-nucleon forces. The individual terms in the expansion are arranged according to a power-counting scheme, resulting in an order-by-order theory for nuclear forces that is truncated at a certain order. This allows to construct Hamiltonians with desired accuracy that can be systematically improved by going to higher orders. The LECs are typically fit to nucleon-nucleon scattering data as well as properties of light nuclei. However, new ways of constraining the LECs are currently explored [14]. Importantly, the chiral EFT scheme allows to estimate theoretical uncertainties of the nuclear Hamiltonian due to the truncation of the series. These advantages make chiral EFT the current state-of-the-art for nuclear forces, and make it especially suited when studying systems that are not accessible in experiments, like neutron-rich matter. In the next section, I will present results of quantum Monte Carlo (QMC) calculations [17] using the local chiral interactions of Refs. [15, 16]. In particular, the results I show have been obtained using the auxiliary-field diffusion Monte Carlo (AFDMC) method [18]. I will not discuss these methods in detail and refer the reader to the reviews of Refs. [17, 19].

2.3 Results for nuclear systems

The combination of local chiral interactions with QMC methods has lead to an excellent description of various nuclear systems, from atomic nuclei up to ^{16}O [20], neutron-α scattering [16], as well as pure neutron matter [16, 21]. I show results for binding energies of atomic nuclei up to ^{16}O as well as for neutron matter in Fig. 2; details of these calculations can be found in the respective references. Chiral interactions provide a good description of properties of light nuclei and neutron matter with theoretical uncertainties. It is obvious, that uncertainties grow fast with density due to the eventual breakdown of the theory. This makes chiral interactions only applicable below $2\,n_{\rm sat}$ and not over the whole density range explored in NSs.

2.4 Neutron-star equation of state

To address the impact of NS observations on the EOS, we need to extend the microscopic results to all conditions encountered in the NS. First, the neutron-matter results are extrapolated to β equilibrium

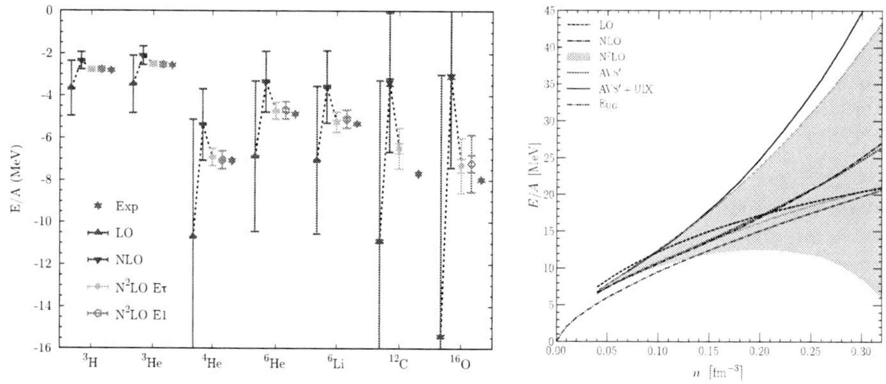

Figure 2. AFDMC results for the binding energies of nuclei up to ^{16}O (left) and for pure neutron matter (right), using local chiral interactions at a cutoff scale $R_0 = 1.0$ fm. Figures taken from Refs. [20, 22].

and below $0.5\,n_{\rm sat}$ a crust EOS is inserted [23]. The result is the neutron-star EOS constrained up to $2\,n_{\rm sat}$. To reliably extract information from NS observations, it is important to extrapolate to larger densities in a model-agnostic fashion, while not assuming a certain set of degrees of freedom, to reduce systematic uncertainties. Instead, it is important to explore all possibilities for the EOS allowed by general physics considerations, i.e., causality and stability of NSs. Such extension schemes can be based on polytropes [24], or the speed of sound [22]. In the following, we will present results obtained using speed-of-sound extension schemes.

3 Neutron-star observations and the equation of state

The EOS extension discussed in the previous section leads to a wide range of NS properties, which can be confronted with observations to improve our knowledge of the EOS.

The first important constraint on the EOS is provided by the observation of heavy NSs. The heaviest NSs known to date have masses of two solar masses [5, 6], which rules out too soft EOS that lead to NSs with small radii and/or small maximum masses. This can be seen in Fig. 3, were the two-solar-mass constraint is indicated by a red-dashed line. This constraint limits the NS radius to be above ~ 9 km for a typical $1.4 M_\odot$ NS, see the light-gray area.

The next constraint I discuss here comes from the first observation of gravitational waves from a binary NS merger and its electromagnetic counterpart, GW170817 and AT2017gfo [27–29]. The observed gravitational wave signal from the inspiral phase of such NS mergers offers the possibility to measure the NS radius, because it is sensitive to the tidal deformation of the stars. A large star shows a stronger deformation and has a larger tidal deformability Λ, while a small star is more difficult to deform and has a small Λ. Because $\Lambda \sim R^6$, the determination of the tidal polarizability from GW170817 allows to constrain the radius of a typical NS to be smaller than 13.6 km [22, 30, 31].

In addition, the properties of the corresponding observed kilonova allows to further constrain the maximum mass. The amount of ejecta from the kilonova allows to rule out a prompt collapse to a black hole, requiring the EOS to be sufficiently stiff. The energetics of the kilonova on the other hand suggests that no longer-lived NS was formed, which requires the maximum mass to be less than $\sim 2.3 M_\odot$ [25]. This produces a narrow range for the maximum mass (indicated in Fig. 3).

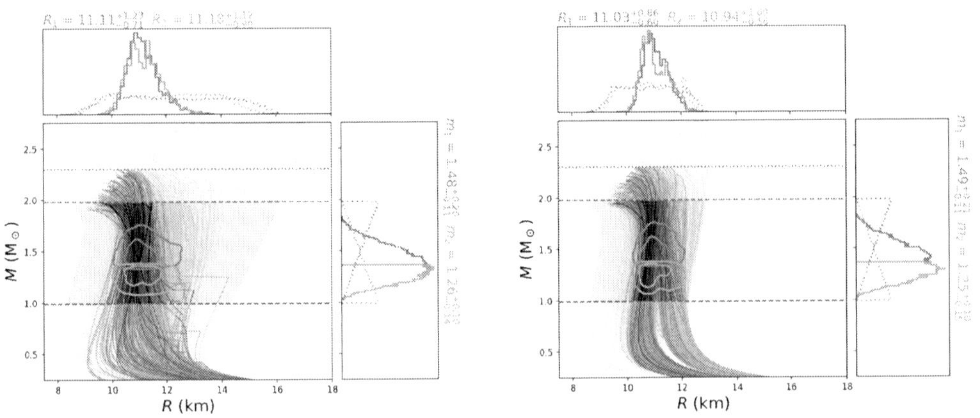

Figure 3. Mass-radius relation of neutron-stars using EOS constrained by chiral EFT up to $n_{\rm sat}$ (left) or $2n_{\rm sat}$ (right) and observational constraints from NS masses, gravitational waves GW170817, and the kilonova observation AT2017gfo. Figure taken from Ref. [26].

A recent reanalysis of the neutron-star merger taking into account EOS constrained by chiral EFT at low densities, the existence of heavy two-solar mass NS, the observed signal GW170817 and the considerations on the maximum mass from the observed kilonova found tight constraints on the NS radius: $R_{1.4} = 11.0^{+0.9}_{-0.6}$ km [26]. We show the corresponding MR relations in Fig. 3 for chiral EFT constraints up to $n_{\rm sat}$ (left) or up to $2n_{\rm sat}$.

4 Summary and Outlook

To extract reliable constraints from NS observations, it is necessary to use EOS with reliable uncertainties at low densities and model-agnostic extensions to higher densities to reduce systematic errors. Together with such EOS sets, NS observations offer a unique possibility to probe the EOS of dense nuclear matter at large densities. Current EOS uncertainties are still sizable but will be reduced by improved nuclear-physics constraints at low densities, future observations of NS mergers, and of radii by the NICER mission [32]. This increased knowledge of the NS mass-radius relation might shed light on the possible existence of phase transitions to exotic forms of matter.

I would like to thank the organizers for the invitation and for delivering a successful workshop. I thank D. Brown, S. Brown, C. Capano, J. Carlson, S. De, S. Gandolfi, A. Gezerlis, K. Hebeler, B. Krishnan, S. Kumar, D. Lonardoni, J. Lynn, B. Margalit, J. Margueron, S. Reddy, A. Schwenk for many insightful discussions and their contributions to the studies presented in this work. This work was supported by the U.S. Department of Energy, Office of Science, Office of Nuclear Physics, under Contract No. DE-AC52-06NA25396, by the Los Alamos National Laboratory (LANL) LDRD program, and by the NUCLEI SciDAC program. This research used resources provided by the LANL Institutional Computing Program, which is supported by the U.S. Department of Energy National Nuclear Security Administration under Contract No. 89233218CNA000001. Computational resources have also been provided by the National Energy Research Scientific Computing Center (NERSC), which is supported by the U.S. Department of Energy, Office of Science, under Contract No. DE-AC02-05CH11231, and by the Jülich Supercomputing Center.

References

[1] C. L. Fryer, Astrophys. J. **522**, 413 (1999).
[2] M. G. Alford, A. Schmitt, K. Rajagopal and T. Schäfer, Rev. Mod. Phys. **80**, 1455 (2008).
[3] A. W. Steiner, M. Prakash, J. M. Lattimer, and P. J.Ellis, Phys. Rep. **411**, 325 (2005).
[4] J. M. Lattimer, Ann. Rev. Nucl. Part. Sci. **62**, 485 (2012).
[5] J. Antoniadis *et al.*, Science **340**, 6131 (2013).
[6] H. T. Cromartie *et al.*, Nat. Astron. **4**, 72 (2019).
[7] S. Gandolfi, J. Lippuner, A. W. Steiner, I. Tews, X. Du and M. Al-Mamun, J. Phys. G **46**, 103001 (2019).
[8] A. Kurkela, P. Romatschke and A. Vuorinen, Phys. Rev. D **81**, 105021 (2010).
[9] A. Kurkela, E. S. Fraga, J. Schaffner-Bielich and A. Vuorinen, Astrophys. J. **789**, 127 (2014).
[10] R. B. Wiringa, V. G. J. Stoks and R. Schiavilla, Phys. Rev. C **51**, 38 (1995).
[11] R. Machleidt, Phys. Rev. C **63**, 024001 (2001).
[12] E. Epelbaum, H. W. Hammer and U. G. Meissner, Rev. Mod. Phys. **81**, 1773 (2009)
[13] J. A. Melendez, S. Wesolowski and R. J. Furnstahl, Phys. Rev. C **96**, no. 2, 024003 (2017)
[14] I. Tews, Z. Davoudi, A. Ekström, J. D. Holt and J. E. Lynn, arXiv:2001.03334 [nucl-th].
[15] A. Gezerlis, I. Tews, E. Epelbaum, M. Freunek, S. Gandolfi, K. Hebeler, A. Nogga and A. Schwenk, Phys. Rev. C **90**, 054323 (2014).
[16] J. E. Lynn, I. Tews, J. Carlson, S. Gandolfi, A. Gezerlis, K. E. Schmidt and A. Schwenk, Phys. Rev. Lett. **116**, 062501 (2016).
[17] J. Carlson, S. Gandolfi, F. Pederiva, S. C. Pieper, R. Schiavilla, K. E. Schmidt and R. B. Wiringa, Rev. Mod. Phys. **87**, 1067 (2015).
[18] K. E. Schmidt and S. Fantoni, Phys. Lett. B **446**, 99 (1999).
[19] J. E. Lynn, I. Tews, S. Gandolfi and A. Lovato, Ann. Rev. Nucl. Part. Sci. **69**, 279 (2019).
[20] D. Lonardoni, S. Gandolfi, J. E. Lynn, C. Petrie, J. Carlson, K. E. Schmidt and A. Schwenk, Phys. Rev. C **97**, 044318 (2018).
[21] I. Tews, J. Carlson, S. Gandolfi and S. Reddy, Astrophys. J. **860**, 149 (2018).
[22] I. Tews, J. Margueron and S. Reddy, Eur. Phys. J. A **55**, 97 (2019).
[23] I. Tews, Phys. Rev. C **95**, 015803 (2017).
[24] K. Hebeler, J. M. Lattimer, C. J. Pethick and A. Schwenk, Astrophys. J. **773**, 11 (2013).
[25] M. Shibata, E. Zhou, K. Kiuchi and S. Fujibayashi, Phys. Rev. D **100**, no. 2, 023015 (2019)
[26] C. D. Capano *et al.*, arXiv:1908.10352 [astro-ph.HE].
[27] B. P. Abbott *et al.* [LIGO Scientific and Virgo Collaborations], Phys. Rev. Lett. **119**, 161101 (2017).
[28] B. P. Abbott *et al.*, Astrophys. J. **848**, 12 (2017).
[29] B. P. Abbott *et al.* [LIGO Scientific and Virgo Collaborations], Phys. Rev. X **9**, 011001 (2019).
[30] E. Annala, T. Gorda, A. Kurkela and A. Vuorinen, Phys. Rev. Lett. **120**, 172703 (2018).
[31] E. R. Most, L. R. Weih, L. Rezzolla and J. Schaffner-Bielich, Phys. Rev. Lett. **120**, 261103 (2018).
[32] G. Raaijmakers *et al.*, Astrophys. J. Lett. **887**, L22 (2019).

EPJ Web of Conferences **235**, 07003 (2020) https://doi.org/10.1051/epjconf/202023507003
ISMD 2019

Studying particle acceleration from driven magnetic reconnection at the termination shock of a relativistic striped wind using particle-in-cell simulations

Yingchao Lu[1,2,⋆], *Fan* Guo[1], *Patrick* Kilian[1], *Hui* Li[1], *Chengkun* Huang[1], and *Edison* Liang[2]

[1]*Theoretical Division, Los Alamos National Laboratory, Los Alamos, New Mexico, 87545, USA*
[2]*Department of Physics and Astronomy, Rice University, Houston, Texas 77005, USA*

Abstract. A rotating pulsar creates a surrounding pulsar wind nebula (PWN) by steadily releasing an energetic wind into the interior of the expanding shockwave of supernova remnant or interstellar medium. At the termination shock of a PWN, the Poynting-flux-dominated relativistic striped wind is compressed. Magnetic reconnection is driven by the compression and converts magnetic energy into particle kinetic energy and accelerating particles to high energies. We carrying out particle-in-cell (PIC) simulations to study the shock structure as well as the energy conversion and particle acceleration mechanism. By analyzing particle trajectories, we find that many particles are accelerated by Fermi-type mechanism. The maximum energy for electrons and positrons can reach hundreds of TeV.

1 Introduction

In recent observations, high-energy emissions have been detected in both young pulsar wind nebulae (PWNe) such as the Crab nebula and mid-age PWNe such as the Geminga nebula[1]. The high-energy emissions is likely due to the high-energy electrons scattering off the cosmic microwave background (CMB) photons[2, 3]. The high-energy electrons and positrons need to be produced by a particle acceleration in the nebula[4]. How pulsar winds efficiently accelerate electrons and positrons to high energies is a major puzzle and holds the key of understanding the near-earth positron anomaly[1, 4–6] as well as gamma rays from the Galactic Center[1, 7, 8]. In the synchrotron spectra of PWNe, the spectral break between the radio and the X-ray cannot be explained by synchrotron cooling and is believed to be attributed to the particle acceleration at the termination shock[9–11]. How the electromagnetic energy is converted into non-thermal particle acceleration at the termination shock is not fully understood.

In the case of an obliquely rotating pulsar, a radially propagating relativistic flow is continuously launched. Asymptoticly near the equatorial plane, such a flow is modeled as a striped wind containing a series of drifting Harries current sheets[12, 13] with opposite polarity magnetic fields in between. Numerical simulations including magnetohydrodynamics[11, 14–16], particle-in-cell (PIC)[17] and test-particle simulations[18] have been used for modeling the termination shock of PWNe. Magnetic reconnection driven by the termination shock may dissipate the magnetic energy and accelerates

⋆e-mail: yclu@lanl.gov

particles[17, 19]. Particle acceleration in relativistic magnetic reconnection has been a recent topic of strong interests. Two candidate mechanisms of particle acceleration[20–24] are direct acceleration surrounding X-points and Fermi acceleration in flows generated within the reconnection layer. While several analyses[20–23] have shown that Fermi acceleration dominates particle acceleration to high energies in a spontaneous reconnection with weak guide field, this has not been studied in a shock-driven reconnection.

In this paper, we carry out two-dimensional particle-in-cell (PIC) simulations to model the relativistic striped wind interacting with the termination shock near the equatorial plane of obliquely rotating pulsars. The magnetic reconnection driven by the precursor perturbation from the shock converts the magnetic energy into particle energy and accelerates particles into a power-law energy spectrum. The maximum energy for electrons and positrons can reach hundreds of TeV if the wind has a bulk Lorentz factor $\gamma_0 \approx 10^6$ and upstream magnetization parameter $\sigma_0 = 10$.

2 Simulation setup

We use the PIC code EPOCH2D[25] to study the kinetic processes in the termination shock of a relativistic striped wind. To ensure that the magnetic reconnection is driven by the physical perturbation from the shock instead of numerical instabilities such as numerical Cherenkov instability (NCI), the code is modified by the authors to use a piecewise polynomial force interpolation scheme with time-step dependency[26]. The relativistic striped wind is a steady magnetized electron-positron flow propagating along $-\hat{x}$. The spatial profile of the electromagnetic field in the downstream rest frame S is

$$\begin{aligned} B_y =& B_0 \tanh\left\{\frac{1}{\delta}\left[\alpha + \cos\left(\frac{2\pi(x+\beta_0 ct)}{\lambda}\right)\right]\right\} \\ E_z =& \beta_0 B_0 \tanh\left\{\frac{1}{\delta}\left[\alpha + \cos\left(\frac{2\pi(x+\beta_0 ct)}{\lambda}\right)\right]\right\} \end{aligned} \tag{1}$$

where β_0 is the velocity of the wind normalized by the speed of light c, and λ is the wavelength of the stripes in the wind. The dimensionless parameters δ and α are such that the half thickness of the current sheet is $\Delta \approx \lambda\delta/(2\pi)$ (actually $\Delta = (\lambda\delta)/(2\pi\sqrt{1-\alpha^2})$ as shown later in this paragraph), and B_y averaged over one wavelength is $\langle B_y \rangle_\lambda = B_0[1 - 2(\arccos\alpha)/\pi]$. We use the variable $\xi = \frac{1}{\delta}[\alpha + \cos(2\pi(x+\beta_0 ct)/\lambda)]$ for the phase of the electromagnetic field. The location of current sheets is determined by setting $\xi = 0$, i.e. $x + \beta_0 ct = (\lambda/(2\pi))\arccos[-\alpha]$. And the location of the transitional field is determined by setting $\xi = \pm 1$, i.e. $x+\beta_0 ct = (\lambda/2\pi)\arccos[-\alpha \pm \delta]$. By subtracting the coordinates we get the half thickness of the current sheet $\Delta = (\lambda\delta)/(2\pi\sqrt{1-\alpha^2})$. To make sure $\Delta \ll \lambda$, we need to have $\delta/\sqrt{1-\alpha^2} \ll 1$. The background cold electron/positron plasma in the wind is uniform, with constant density $n_{e,p}^{\text{cold}} = n_{c0}/2$ and constant temperature $kT_{e,p}^{\text{cold}} = 0.04 m_e c^2$. The hot electron/positron plasma balances the magnetic pressure and keeps the steady profile of the electromagnetic field. Besides frame S which is also the downstream rest, there are several reference frames, including the center-of-mass (CM) frame S' of the wind, the CM frame S_e'' of the hot electrons, and the CM frame S_p'' of the hot positrons. The space-time coordinate transform between S and S' is[27]

$$\begin{cases} x &= \gamma_0(x' - \beta_0 ct') \\ t &= \gamma_0(t' - \beta_0 \frac{x'}{c}) \end{cases} \qquad \begin{cases} x' &= \gamma_0(x + \beta_0 ct) \\ t' &= \gamma_0(t + \beta_0 \frac{x}{c}) \end{cases} \tag{2}$$

where $\gamma_0 = 1/\sqrt{1-\beta_0^2}$ is the Lorentz factor for the transformation between S and S'.

The electromagnetic field in frame S' can be derived by the Lorentz transform of electromagnetic field[27]

$$
\begin{aligned}
&E'_x = E_x = 0 \\
&E'_y = \gamma_0(E_y + \beta_0 B_z) = 0 \\
&E'_z = \gamma_0(E_z - \beta_0 B_y) = 0 \\
&B'_x = B_x = 0 \\
&B'_y = \gamma_0(B_y - \beta_0 E_z) = \frac{B_0}{\gamma_0}\tanh\xi = \frac{B_0}{\gamma_0}\tanh\left\{\frac{1}{\delta}\left[\alpha + \cos\left(\frac{2\pi x'}{\lambda\gamma_0}\right)\right]\right\} \\
&B'_z = \gamma_0(B_z + \beta_0 E_y) = 0
\end{aligned}
\tag{3}
$$

In S' frame, the electric field is zero everywhere, the half thickness of the current sheet is $\Delta' = \gamma_0\Delta$, and the wavelength is $\lambda' = \gamma_0\lambda$.

We assume S''_p is moving at speed $\beta_h\hat{z}$ with respect to S'. The charge density-current tensor $(\rho''_{hp}, 0, 0, 0)$ in S''_p frame can be transformed into the charge density-current tensor $(\rho'_{hp}, j'_{hp,x}, j'_{hp,y}, j'_{hp,z})$ in S' frame and $(\rho_{hp}, j_{hp,x}, j_{hp,y}, j_{hp,z})$ in S frame

$$
\begin{cases}
\rho'_{hp} &= \gamma_h(\rho''_{hp} + \frac{\beta_h}{c} j''_{hp,z}) = \gamma_h\rho''_{hp} \\
j'_{hp,x} &= j''_{hp,x} = 0 \\
j'_{hp,y} &= j''_{hp,y} = 0 \\
j'_{hp,z} &= \gamma_h(j''_{hp,z} + \beta_h c\rho''_{hp}) = \gamma_h\beta_h c\rho''_{hp}
\end{cases}
\qquad
\begin{cases}
\rho_{hp} &= \gamma_0(\rho'_{hp} + \frac{\beta_0}{c} j'_{hp,x}) = \gamma_0\gamma_h\rho''_{hp} \\
j_{hp,x} &= \gamma_0(j'_{hp,x} + \beta_0 c\rho'_{hp}) = \gamma_0\beta_0\gamma_h c\rho''_{hp} \\
j_{hp,y} &= j'_{hp,y} = 0 \\
j_{hp,z} &= j'_{hp,z} = \gamma_h\beta_h c\rho''_{hp}
\end{cases}
\tag{4}
$$

where $\gamma_h = 1/\sqrt{1-\beta_h^2}$. The velocity of S''_p in S is

$$
(v_{hp,x}, v_{hp,y}, v_{hp,z}) = (j_{hp,x}/\rho_{hp}, j_{hp,y}/\rho_{hp}, j_{hp,z}/\rho_{hp}) = c(\beta_0, 0, \beta_h/\gamma_0)
\tag{5}
$$

Assuming the density of the hot electrons/positrons in the current sheet in S frame is $n_{he,hp} = n_{h0}/(2\cosh^2\xi)$, where $n_{h0}/n_{c0} = \eta$ is the overdensity relative to the cold particles outside the layer and is set to be $\eta = 3$[13, 17, 20]. Then in $S''_{he,hp}$ frame we have $n'_{he,hp} = n_{h0}/(\gamma_0\gamma_h)/(2\cosh^2\xi)$.

The temperature of hot component is $kT_h = \eta_T m_e c^2$. In S' frame, the sum magnetic pressure and hot component thermal pressure in x direction is

$$
p'_x = \frac{B'^2_y}{8\pi} + n''_{he}k_B T_h + n''_{pe}k_B T_h = \frac{B_0^2}{8\pi\gamma_0^2}\tanh^2\xi + \frac{n_{c0}\eta\eta_T m_e c^2}{\gamma_0\gamma_h\cosh^2\xi}
\tag{6}
$$

To ensure pressure balance we need $B_0^2/(8\pi\gamma_0^2) = (n_{c0}\eta_n\eta_T m_e c^2)/(\gamma_0\gamma_h)$, thus $\eta_T = \frac{\sigma_0\gamma_h}{2\eta_n}$ where the magnetization parameter is $\sigma_0 = \frac{B_0^2}{4\pi\gamma_0 n_{c0} m_e c^2}$.

The drift velocity $\beta_h\hat{z}$ of the hot particles is determined by keeping the steady profile of electromagnetic field, i.e. in the rest frame of the wind $\nabla\times\boldsymbol{B}' = (4\pi/c)\boldsymbol{J}'$ is satisfied so that the electric field stays zero. Thus

$$
\begin{aligned}
J'_z &= -\rho''_{he}c\beta_h\gamma_h + \rho''_{hp}c\beta_h\gamma_h e = \frac{\eta n_{c0}c\beta_h\cosh^{-2}\xi}{\gamma_0} \\
&= \frac{1}{4\pi}\partial_{x'}B'_y = -\frac{B_0}{2\lambda\gamma_0^2\delta}\cosh^{-2}\xi\sin\left(\frac{2\pi x'}{\lambda\gamma_0}\right)
\end{aligned}
\tag{7}
$$

https://doi.org/10.1051/epjconf/202023507003

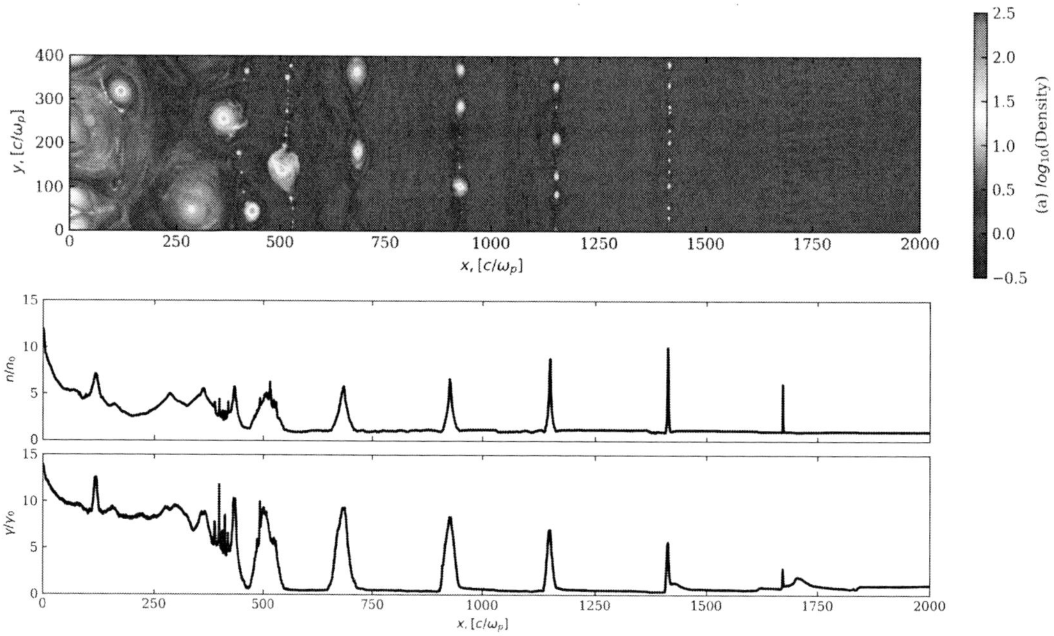

Figure 1. Top: 2D spatial profile of logarithm of particle number density normalized by n_{c0} at time $\omega_p t = 2000$. Middle: 1D spatial profile (averaged over y) of particle number density normalized by n_{c0} at time $\omega_p t = 2000$. Bottom: 1D spatial profile of average γ of all particles normalized by γ_0 at time $\omega_p t = 2000$.

Thus

$$\beta_h = -\frac{B_0}{2\lambda\gamma_0\delta e\eta_n n_{c0}c}\sin\left(\frac{2\pi(x+\beta_0 ct)}{\lambda}\right) = -\frac{\sin\left(\frac{2\pi(x+\beta_0 ct)}{\lambda}\right)}{\sqrt{1-\alpha^2}}\frac{\sqrt{\sigma}(c/\omega_p)}{(\eta_n\gamma_0)\Delta} \tag{8}$$

where $\omega_p = \sqrt{4\pi n_{c0}e^2/(\gamma_0 m_e)}$ is the plasma frequency of the cold background plasma. The time in our simulation is normalized by $1/\omega_p$, and the spatial coordinates in our simulation are normalized by c/ω_p.

The boundary at $x = 0$ is reflecting for particles and conducting for fields. The shock is self-consistently generated by the interaction between the reflected flow and the incoming flow. The simulation is periodic in y direction. In the run we show in this paper, we have $\alpha = 0.1$, $\Delta = d_e$, $\lambda = 640d_e$, $\gamma_0 = 10^4$ and $\sigma_0 = 10$. The length of the simulation box in y direction is $L_y = 400d_e$.

3 Results

As shown in Figure 1, the shock forms self-consistently, and propagates to the right side of the simulation box, compresses and decelerates the current sheets. The results are consistent with the previous study[17]. The shock converts the magnetic energy into the particle kinetic energy, resulting in an average particle Lorentz factor $\langle\gamma\rangle \approx \gamma_0(1 + \sigma_0)$ in the downstream, which is consistent with the jump condition of ultra-relativistic magnetized shock. The location of the shock is approximately $x_{sh} \approx (ct - 1000c/\omega_p)/(2\sqrt{2})$, moving at speed approximately $c/(2\sqrt{2})$. In the upstream, the current

https://doi.org/10.1051/epjconf/202023507003

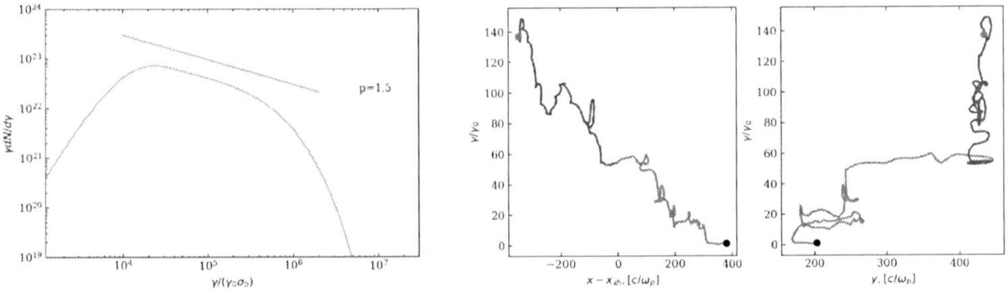

Figure 2. Left: particle energy distribution $\gamma dN/d\gamma$ for all the electrons and positrons in downstream region $0 < x < 354c/\omega_p$ at $\omega_p t = 2000$. Middle and right: particle trajectory for a particle in the simulation. The middle column is particle Lorentz factor (normalized by γ_0) vs. $x - x_{sh}$ coordinate where the location of the shock jump is approximately at $x_{sh} \approx (ct - 1000c/\omega_p)/(2\sqrt{2})$, and the right column is particle Lorentz factor (normalized by γ_0) vs. y coordinate. The red line is the trajectory where the particle is in the upstream, and the red line is the trajectory where the particle is in the downstream. The black dots are for particles at $t = 0$, and the red dots are for particles at $\omega_p t = 2000$.

sheets start to continuously break into a series of magnetic islands separated by X-points. The islands coalesce, grow to larger size, and further grow after passing the shock front.

The energy distribution function at downstream of the shock follows a power law $f(\varepsilon) \propto \varepsilon^{-p+1}$ with $p = 1.5$ as shown in Figure 2. The trajectory for a typical tracer particle that feels the self-consistent electromagnetic field is shown in Figure 2. The trajectories of most high-energy tracer particles are similar to the one in Figure 2. Those tracer particles (i) bounce several times in the upstream and then travel into the downstream, or (ii) travel into the downstream and bounce several times in the downstream, or (iii) do (i) followed by (ii). Few particles get bounced between upstream and downstream, i.e after a particle travels into the downstream of the shock it rarely travels back into the upstream. The fact that we get a $p = 1.5$ power law is consistent with the picture that Fermi mechanism in the reconnection islands is much more efficient than the diffusive shock acceleration.

4 Conclusions and discussions

While a growing body of research using PIC methods focus on the particle acceleration in the spontaneous relativistic magnetic reconnection in the magnetically dominated regime, in this work we extended the study by setting up a simulation with shock driven magnetic reconnection at the termination shock of relativistic striped wind. The analysis shows that many particles are accelerated by Fermi-type mechanism. It is known that PIC simulations only covers a small region in the termination shock, while in reality there is enormous scale separation between the system size and the dynamical scale such as c/ω_p. However, to model the particle acceleration during magnetic reconnection in a macroscopic system, it is important to determine the dominant acceleration mechanism.

Research presented in this paper was supported by the Center for Space and Earth Science (CSES) program, Laboratory Directed Research and Development (LDRD) program 20200367ER of Los Alamos National Laboratory (LANL) and NASA ATP program through grant NNH17AE68I. The research by P. K. was also supported by CSES. CSES is funded by LANL's LDRD program under project number 20180475DR. The simulations were performed with LANL Institutional Computing which is supported by the U.S. Department of Energy National Nuclear Security Administration under Contract No. 89233218CNA000001.

References

[1] H. Yuksel, M.D. Kistler, T. Stanev, Physical Review Letters **103** (2009)
[2] M. Amenomori, Y. Bao, X. Bi, D. Chen, T. Chen, W. Chen, X. Chen, Y. Chen, Cirennima, S. Cui et al., Physical Review Letters **123** (2019)
[3] A.U. Abeysekara, A. Albert, R. Alfaro, C. Alvarez, J.D. Álvarez, J.R.A. Camacho, R. Arceo, J.C. Arteaga-Velázquez, K.P. Arunbabu, D.A. Rojas et al., The Astrophysical Journal **881**, 134 (2019)
[4] A.U. Abeysekara, A. Albert, R. Alfaro, C. Alvarez, J.D. Álvarez, R. Arceo, J.C. Arteaga-Velázquez, D.A. Rojas, H.A.A. Solares, A.S. Barber et al., Science **358**, 911 (2017)
[5] L. Accardo, M. Aguilar, D. Aisa, B. Alpat, A. Alvino, G. Ambrosi, K. Andeen, L. Arruda, N. Attig, P. Azzarello et al., Physical Review Letters **113** (2014)
[6] D. Hooper, I. Cholis, T. Linden, K. Fang, Physical Review D **96** (2017)
[7] A.A. Abdo, B. Allen, D. Berley, S. Casanova, C. Chen, D.G. Coyne, B.L. Dingus, R.W. Ellsworth, L. Fleysher, R. Fleysher et al., The Astrophysical Journal **664**, L91 (2007)
[8] T. Linden, B.J. Buckman, Physical Review Letters **120** (2018)
[9] M.J. Rees, J.E. Gunn, Monthly Notices of the Royal Astronomical Society **167**, 1 (1974)
[10] C.F. Kennel, F.V. Coroniti, The Astrophysical Journal **283**, 694 (1984)
[11] C.F. Kennel, F.V. Coroniti, The Astrophysical Journal **283**, 710 (1984)
[12] F.V. Coroniti, The Astrophysical Journal **349**, 538 (1990)
[13] J.G. Kirk, O. Skjaraasen, The Astrophysical Journal **591**, 366 (2003)
[14] O. Porth, S.S. Komissarov, R. Keppens, Monthly Notices of the Royal Astronomical Society **438**, 278 (2014), `1310.2531`
[15] B. Olmi, L.D. Zanna, E. Amato, N. Bucciantini, Monthly Notices of the Royal Astronomical Society **449**, 3149 (2015)
[16] O. Porth, M.J. Vorster, M. Lyutikov, N.E. Engelbrecht, Monthly Notices of the Royal Astronomical Society **460**, 4135 (2016)
[17] L. Sironi, A. Spitkovsky, The Astrophysical Journal **741**, 39 (2011)
[18] G. Giacinti, J.G. Kirk, The Astrophysical Journal **863**, 18 (2018)
[19] J. Pétri, Y. Lyubarsky, Astronomy & Astrophysics **473**, 683 (2007)
[20] L. Sironi, A. Spitkovsky, The Astrophysical Journal **783**, L21 (2014)
[21] F. Guo, H. Li, W. Daughton, Y.H. Liu, Physical Review Letters **113** (2014)
[22] F. Guo, Y.H. Liu, W. Daughton, H. Li, The Astrophysical Journal **806**, 167 (2015)
[23] G.R. Werner, D.A. Uzdensky, B. Cerutti, K. Nalewajko, M.C. Begelman, The Astrophysical Journal **816**, L8 (2015)
[24] F. Guo, X. Li, W. Daughton, P. Kilian, H. Li, Y.H. Liu, W. Yan, D. Ma, The Astrophysical Journal **879**, L23 (2019)
[25] T.D. Arber, K. Bennett, C.S. Brady, A. Lawrence-Douglas, M.G. Ramsay, N.J. Sircombe, P. Gillies, R.G. Evans, H. Schmitz, A.R. Bell et al., Plasma Physics and Controlled Fusion **57**, 113001 (2015)
[26] Y. Lu, P. Kilian, F. Guo, H. Li, E. Liang (2019), `http://arxiv.org/abs/1909.09613v1`
[27] L.D. Landau, E.M. Lifshitz, C.H. Holbrow, Physics Today **16**, 72 (1963)

Transseries gradient expansion of Yang-Mills plasma

Jakub Jankowski[1,⋆]

[1]*Faculty of Physics, University of Warsaw, ul. Pasteura 5, 02-093 Warsaw, Poland*

Abstract. We discuss a resurgence framework in which both hydrodynamical and transient degrees of freedom of $\mathcal{N} = 4$ SYM plasma are incorporated on an equal footing. In consequence, at late times, all information about initial conditions is encoded in infinitely many exponentially damped modes, controlled by transseries parameters.

1 Introduction

In the last few years the meaning of the hydrodynamic expansion attracted a lot of attention, motivated by applications of hydrodynamics to the phenomenology of Heavy Ion Collisions as well as by various theoretical developments [1]. In particular, much of insight came from the AdS/CFT direction, where it was realised that hydrodynamisation is in general a fast process [2, 3], and that the gradient expansion is generically a divergent series [4]. A tentative reason for both effects is the presence of exponentially damped modes, which are not explicitly captured by the gradient expansion. It turns out that a possible theoretical framework incorporating both sets of degrees of freedom is the theory of resurgence [5], which usually finds its place in theoretical physics in connection with the topic of perturbation theory in quantum mechanics and quantum field theory [5]. Surprisingly it has also been useful in applications to plasma dynamics, as was recently demonstrated in Ref. [6].

2 Hydrodynamization and Bjorken flow

We will start by briefly reviewing the meaning of fast hydrodynamization, which is a process by which system approaches state described by hydrodynamic equations. Physically, fast hydrodynamization is a consequence of exponential damping of initial conditions information, and it is well supported by numerical simulations. To make the problem approachable it is useful to adopt Bjorken symmetry conditions which can be implemented by adopting the coordinate system defined by $ds^2 = -d\tau^2 + \tau^2 dy^2 + dx_\perp^2$ and imposing the requirement of no dependence on the y and $x_\perp$ coordinates [7]. As a consequence, the energy momentum tensor has the following form

$$\langle T_{\mu\nu}(\tau)\rangle = \mathrm{diag}\{\epsilon(\tau), P_L(\tau), P_T(\tau), P_T(\tau)\} . \tag{1}$$

Due to energy-momentum conservation, $\nabla_\mu T^{\mu\nu} = 0$, and conformal invariance, $T^\mu_\mu = 0$, longitudinal and transversal pressures take are expressed as

$$P_L = -\epsilon - \tau\dot{\epsilon} , \qquad P_T = \epsilon + \frac{1}{2}\tau\dot{\epsilon} , \tag{2}$$

⋆e-mail: Jakub.Jankowski@fuw.edu.pl

and the dynamics of the system is determined by a single function of proper-time $\epsilon(\tau)$, *i.e.*, the energy density at mid-rapidity.

A holographic description of boost-invariant dynamics of SYM plasma requires a suitable class of asymptotically-AdS geometries [2, 3]. Once the geometry is numerically determined one can apply the AdS/CFT dictionary to calculate $\epsilon(\tau)$. It is convenient to adopt the notion of effective temperature $T(\tau)$, which is defined by the relation

$$\epsilon(\tau) = \frac{3}{8}\pi^2 N^2 T^4(\tau) \,. \tag{3}$$

It is important to note that $T(\tau)$ is not a temperature in strict thermodynamic sense, yet due to high symmetry constraints, it is a useful physical quantity.

At late time it is possible to calculate $\epsilon(\tau)$ as an asymptotic series which can be expressed as an expansion of the effective temperature

$$T(\tau) = \frac{\Lambda}{(\Lambda\tau)^{1/3}}\left(1 - \frac{1}{6\pi}\frac{1}{(\Lambda\tau)^{2/3}} + \frac{\log 2 - 1}{36\pi^2}\frac{1}{(\Lambda\tau)^{4/3}} + \dots\right). \tag{4}$$

The integration constant Λ appearing here is the only trace of the initial condition, which survives in late time limit. The above expansion is closely related to the gradient expansion of hydrodynamics.

The dependence on Λ suggests that it is useful to introduce the dimensionless pressure anisotropy defined by

$$\mathcal{A}(w) = \frac{P_T(w) - P_L(w)}{P(w)} \,, \tag{5}$$

where $\epsilon(w) = 3P(w)$. Everything is then expressed as a function of the dimensionless quantity $w = \tau T(\tau)$, the proper time measured in units of the relaxation time $\tau_0 \sim 1/T$. The pressure anisotropy is a good measure of the deviation from hydrodynamics, and at late times it is independent of initial conditions. Indeed, numerical simulations show that for generic initial configuration $\mathcal{A}(w)$ follows universal time evolution determined by hydrodynamic behaviour (see Fig. 1). [1] This reflects a universal approach to equilibrium, in which initial state information is dissipated exponentially at rather early time.

3 Transseries gradient expansion

Here we will briefly review applications of resurgence theory to the late time expansion of energy density of $\mathcal{N} = 4$ SYM. For details we refer the reader to Ref. [6].

3.1 Multiparameter transseries

The series of Eq. (4) turns out to be a divergent one [4], and furthermore – as expected based on physical grounds – is supplemented with non-perturbative corrections of exponential nature. Exponentially suppressed terms form an infinite set, and are determined by quasinormal mode (QNM) frequencies of the dual black hole. The asymptotic series of Eq. (4) can be computed to very high order utilizing the AdS/CFT correspondence [4], and takes the following form

$$\varepsilon_{\text{hydro}}(\tau) \sim \frac{\Lambda}{(\Lambda\tau)^{4/3}} \sum_{n=0}^{\infty} \varepsilon_n^{(0)} (\Lambda\tau)^{-2n/3} \,, \tag{6}$$

[1] Different initial conditions are labelled by different entropies as measured by the area of the apparent horizon determined by means of Bekenstein-Hawking relatio [3].

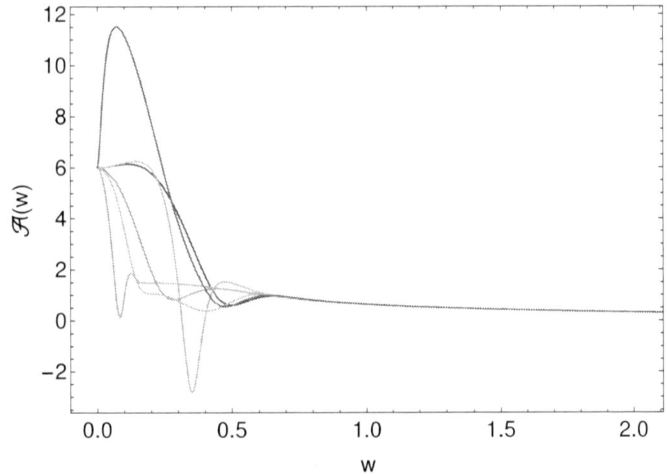

Figure 1. The pressure anisotropy $\mathcal{A}(w)$ for six different initial configurations. The initial conditions all have the same value of the pressure anisotropy, but evolve differently, which reflects the richness of the initial states. However at late times, for $w > 0.7$, all the curves behave identically and match the first order hydro prediction [2, 3].

where coefficients $\varepsilon_n^{(0)}$ diverge factorially, *i.e.*, $\varepsilon_n^{(0)} \sim \Gamma(n+\beta)A_1^{-n-\beta}$for $n \gg 1$. The full expansion of $\epsilon(\tau)$ takes form of multi-parameter transseries with infinitely many transseries parameters [6]. For simplicity let us only show one sample contribution of the least-damped, transient mode determined by quasi-normal mode (QNM) frequency $A_1 \in \mathbb{C}$ [2]

$$\epsilon(\tau) \sim \underbrace{\frac{\Lambda}{(\Lambda\tau)^{4/3}}\sum_{n=0}^{\infty}\varepsilon_n^{(0)}(\Lambda\tau)^{-2n/3}}_{\varepsilon_{\text{hydro}}(\tau)} + \frac{\Lambda\sigma_1}{(\Lambda\tau)^{4/3}}\sum_{n=0}^{\infty}\varepsilon_n^{(1)}(\Lambda\tau)^{-2n/3}e^{-A_1(\Lambda\tau)^{2/3}} + \cdots \tag{7}$$

The transseries parameter associated with the A_1 mode, belonging to an infinite set, $\sigma_1 \in \mathbb{C}$ is free and encodes information about the initial configuration. The original series $\varepsilon_n^{(0)} \sim \Gamma(n+\beta)A_1^{-n-\beta}$ for $n \gg 1$ diverge factorially, and so does the first sector $\varepsilon_n^{(1)} \sim \Gamma(n+\beta_2)A_2^{-n-\beta_2}$ for $n \gg 1$. The transseries expansion of Eq. (7) can be determined using AdS/CFT duality. The parameter $A_2 \in \mathbb{C}$, present in the $\varepsilon_n^{(1)}$ coefficients, is second to most damped QNM mode in the sense that $\text{Im}\,A_1 < \text{Im}\,A_2$.

3.2 Borel transform

In order to make sense of factorially divergent asymptotic series one can perform a Borel transform. In the case at hand it is convenient to adopt a slightly non-standard definition. For example in the original hydrodynamic series it reads

$$\mathcal{B}\left[\varepsilon_{\text{hydro}}\right](\xi) = \xi^{\beta-\frac{1}{2}}\sum_{n=0}^{\infty}\frac{\varepsilon_n^{(0)}}{\Gamma\left(n+\beta+\frac{1}{2}\right)}\xi^n . \tag{8}$$

[2]Imaginary part of A_1 and A_2 is proportional to relaxation time.

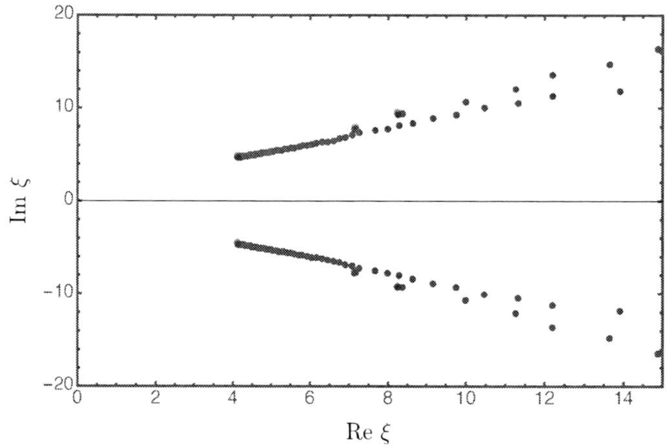

Figure 2. Poles of the Borel-Padé approximant $\mathrm{BP}_{189}\left[\epsilon_{\text{hydro}}\right]$, in the complex ξ–plane. In addition in colour: $\xi = A_1, \overline{A_1}, 2A_1, 2\overline{A_1}$ (blue dots), $\xi = A_2, \overline{A_2}$ (green dots).

The key point now is that series $\mathcal{B}\left[\varepsilon_{\text{hydro}}\right](\xi)$ has a finite radius of convergence determined by the exponent A_1, and can be analytically continued to a function defined in the whole complex plane, called the Borel plane. In the case at hand, due to limited number of terms given, we used diagonal Pade approximant to perform the continuation. To be more specific one approximates the sought after function by a rational function

$$\mathrm{BP}_N\left[\Phi\right](\xi) = \frac{P_N(\xi)}{Q_N(\xi)}, \tag{9}$$

where $P_N(\xi)$ and $Q_N(\xi)$ are polynomials of order N, fitting their coefficients to match a number of consecutive terms in the series of Eq. (8). This ensures the correct radius of convergence of the approximating function. Poles of the Borel-Pade approximant condense on cuts of the underlying function, while numerical artefacts are suppressed by large order of used polynomials. In short, poles of the approximant resemble the singularity structure of the exact Borel transform.

The singularity structure of the analytically-continued Borel transform carries information about non-perturbative degrees of freedom present in the system and is shown in Fig. 2. In particular one can identify there the second least-damped frequency $A_2 \in \mathbb{C}$, corresponding to the second QNM of the dual black hole, as well as non linear terms present in (7), though not shown explicitly.

3.3 Large order relations

One important consequence of the analytic structure mentioned above are the so-called *large order relations* [5], which relate coefficients from the $\varepsilon_n^{(0)}$ sector for $n \gg 1$ with those of $\varepsilon_n^{(1)}$ sector for $n \sim 1$. In our particular example the relation reads

$$\varepsilon_n^{(0)} \sim -\frac{S_{0\to1}}{2\pi\mathrm{i}}\frac{\Gamma(n+\beta)}{A_1^{n+\beta}}\left(\varepsilon_0^{(1)} + \frac{A_1\varepsilon_1^{(1)}}{n+\beta-1} + \frac{A_1^2\varepsilon_2^{(1)}}{(n+\beta-1)(n+\beta-2)} + \cdots\right) + \text{c.c.} + \cdots \tag{10}$$

where $S_{0\to1}$ is the Stokes constant, determined numerically to be $S_{0\to1} = 0.01113\cdots - \mathrm{i}0.03050\cdots$. Each sector comes with its own independent Stokes parameter that characterizes underlying system.

To compare coefficients obtained from holographic computation, and those predicted by large-order relation (10) we make use of the following quantity

$$\Delta_{\mathbf{0}}\varepsilon_m^{(\mathbf{e}_1)} \equiv \frac{\varepsilon_m^{(1)}\big|_{\mathbf{0}-\text{predicted}} - \varepsilon_m^{(1)}\big|_{\text{numerical}}}{\varepsilon_m^{(1)}\big|_{\text{numerical}}}, \qquad m \geq 1\,. \tag{11}$$

Here, coefficients $\varepsilon_m^{(1)}\,|_{\text{numerical}}$ are coefficients of the expansion in the first transsector (corresponding to A_1), computed with AdS/CFT, and $\varepsilon_m^{(1)}\,|_{\mathbf{0}-\text{predicted}}$ are those obtained from $\varepsilon_m^{(0)}$ assuming relation (10) holds. In Fig. 3 we plot $\Delta_{\mathbf{0}}\varepsilon_m^{(\mathbf{e}_1)}$ for a few first values of consecutive number m. The agreement between both computations is of the order of 10^{-50}, which provides strong evidence of the resurgence properties of the gradient expansion in strongly coupled Yang-Mills plasma.

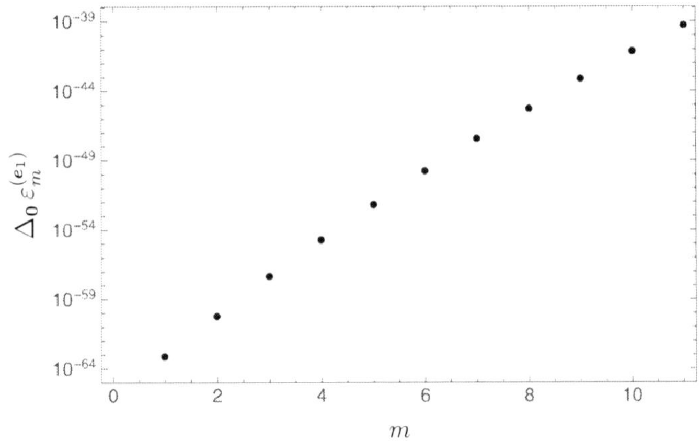

Figure 3. Differences of the values of series coefficients as computed from holography and predicted by large order relations, defined in Eq. (11).

4 Conclusions

In this short note we have reviewed results of Ref. [6] which provide a unified description of hydrodynamic and transient degrees of freedom in strongly coupled $\mathcal{N} = 4$ SYM theory. It is also the first example demonstrating resurgence property in a first principle computation of quantum field theory. The energy density, as a function of proper time, turns out to be a multi-parameter transseries, where the original gradient expansion is supplemented by exponentially damped corrections, each appearing with an independent transseries parameter. Taking into account more and more of those modes allows one to extend ansatz (7) towards earlier times, however never reaching strictly times close to $\tau = 0$. In consequence, information about the initial configuration is present in the system at late times in the form of an infinite number of exponentially damped modes, fast dissipation of which is responsible for short hydrodynamisation time. A few possible extensions of that computation would be to relax restrictive symmetry assumptions or study the Bjorken flow in a non-conformal theory.

Acknowledgements

I would like to thank Michał Spaliński and Mauricio Martinez for useful discussions. This research was supported by the Polish National Science Centre (NCN) grant 2016/23/D/ST2/03125.

References

[1] W. Florkowski, M. P. Heller and M. Spalinski, Rept. Prog. Phys. **81**, no. 4, 046001 (2018)
[2] M. P. Heller, R. A. Janik and P. Witaszczyk, Phys. Rev. Lett. **108**, 201602 (2012)
[3] J. Jankowski, G. Plewa and M. Spaliński, JHEP **1412**, 105 (2014)
[4] M. P. Heller, R. A. Janik and P. Witaszczyk, Phys. Rev. Lett. **110**, no. 21, 211602 (2013)
[5] I. Aniceto, G. Basar and R. Schiappa, Phys. Rept. **809**, 1 (2019)
[6] I. Aniceto, B. Meiring, J. Jankowski and M. Spaliński, JHEP **1902**, 073 (2019)
[7] J. D. Bjorken, Phys. Rev. D **27**, 140 (1983)

Correlations in the Initial Conditions of Heavy-Ion Collisions

Douglas Wertepny[1,*], *Jacquelyn* Noronha-Hostler[2,**], *Matthew* Sievert[2,***], *Skandaprasad* Rao[3,****], and *Noah* Paladino[3,†]

[1]Department of Physics, Ben-Gurion University of the Negev, Beer-Sheva 84105, Israel
[2]University of Illinois at Urbana-Champaign, Urbana, IL 61801, US
[3]Department of Physics and Astronomy, Rutgers University, Piscataway, NJ USA 08854

Abstract.
Ultracentral collisions of heavy nuclei, in which the impact parameter is nearly zero, are especially sensitive to the details of the initial state model and the microscopic mechanism for collective flow. In a hydrodynamic "flow" picture, the final state momentum correlations are a direct response to the fluctuating initial geometry, although models of the initial geometry differ widely. Alternatively, dynamical mechanisms based in the color glass condensate (CGC) formalism can naturally lead to many-body correlations with very different systematics. Here we present a calculation of event-by-event elliptic flow in both the hydrodynamic and CGC paradigms and show that they can be qualitatively distinguished in ultracentral collisions of deformed nuclei. Specifically, the multiplicity dependence in such collisions is qualitatively opposite, with the CGC correlations increasing with multiplicity while the hydrodynamic correlations decrease. The consistency of the latter with experimental data on UU collisions appears to rule out a CGC-mediated explanation. We find that these qualitative features also persist in small deformed systems and can therefore be a valuable test of the microscopic physics in that regime.

1 Introduction

This Proceedings is a brief summary of the important results derived in Ref. [1]; we refer the interested reader there for additional information and details.

While "centrality" in heavy-ion collisions is usually considered to be synonymous with "impact parameter," the physics which drives multiplicity production arises from many factors. In off-central collisions (say, centralities greater than $\sim 10\%$), both the produced multiplicity and the elliptical collision geometry are indeed strongly correlated with the impact parameter b. But in ultracentral collisions (say, centralities less than $\sim 1\%$), where $b \approx 0$ already, the remaining variation in multiplicity production is sensitive to the presence of deformed nuclear geometry. For this reason, ultracentral collisions of deformed ions such as

*e-mail: wertepny@post.bgu.ac.il
**e-mail: jnorhos@illinois.edu
***e-mail: msievert@illinois.edu
****e-mail: svr34@scarletmail.rutgers.edu
†e-mail: noah.paladino@rutgers.edu

uranium have been used to study the impact of nonspherical nuclear geometries [2–4]. This kind of geometry engineering in ultracentral collisions has been used to rule out the two-component Glauber model of initial conditions [5–7] and also extended to other deformed nuclei such as xenon [8–11]. As we demonstrate in this work, such ultracentral collisions of deformed nuclei can also provide a powerful test of the response of the system to initial geometry in various theoretical frameworks. Specifically, we find that two competing models for the microscopic origin of collective flow – hydrodynamics and the color glass condensate – lead to qualitatively opposite predictions for the multiplicity dependence in ultracentral collisions of deformed systems. This is especially important for understanding the role which non-hydrodynamic response may play in small deformed systems [12, 13] and how the transition from large systems with hydrodynamic response may occur [14].

In distinguishing collective flow in hydrodynamics from other multiparticle correlations, it is useful to decompose the two-particle distribution for a given event into "factorized" and "non-factorized" pieces [15]:

$$\frac{dN_2}{d^2p_1dy_1\, d^2p_2dy_2} = \frac{dN_1}{d^2p_1dy_1}\,\frac{dN_1}{d^2p_2dy_2} + \delta_2(p_1, p_2). \tag{1}$$

Here the first term factorizes into two particles produced independently, and the second term denotes the intrinsic, non-factorizing correlations. From here one can define the second cumulant $v_n\{2\}$ as

$$(v_n\{2\})^2 \equiv \frac{\left\langle \int_{p_1p_2} e^{in(\phi_1-\phi_2)} \frac{dN_2}{d^2p_1\,dy_1\,d^2p_2\,dy_2} \right\rangle}{\langle N_{\text{pairs}}\rangle}. = \frac{\left\langle \left| \int_p e^{in\phi} \frac{dN_1}{d^2p\,dy} \right|^2 \right\rangle}{\langle N_{\text{pairs}}\rangle} + \frac{\left\langle \int_{p_1p_2} e^{in(\phi_1-\phi_2)}\,\delta_2(p_1, p_2) \right\rangle}{\langle N_{\text{pairs}}\rangle}. \tag{2}$$

The first term here reflects the indirect correlation of two particles because of their mutual correlation with the event geometry; this is the phenomenon of "collective flow" in hydrodynamic response. The second term represents the dynamical "non-flow" effects which correlate two particles directly with each other. Usually non-flow effects such as jet correlations are suppressed through the use of rapidity cuts, although non-flow correlations which are long-range in rapidity can survive such a procedure.

2 Collective Flow in the Hydrodynamics

In the hydrodynamic paradigm, non-flow effects are suppressed such that all particles are generated independently. In this scenario, $\delta_2 = 0$ and all multiparticle correlations such as $v_n\{2\}$ arise from moments (cumulants) of the event-by-event distribution of the *single*-particle anisotropy

$$V_n \equiv v_n e^{in\psi_n} = \frac{1}{N_{\text{tot}}} \int_p e^{in\phi} \frac{dN_1}{d^2p\,dy}, \tag{3}$$

such that the second-order cumulant

$$v_n\{2\} \overset{\text{flow}}{=} \sqrt{\langle v_n^2\rangle} \tag{4}$$

measures the RMS of the fluctuating anisotropic flow v_n. In turn, the flow vector V_n for a given event arises from the deterministic response of hydrodynamics to the initial conditions. From symmetry considerations and the assumption that long-wavelength modes dominate in hydrodynamics, one can construct a hierarchy of estimators to V_n from the initial state [16]. These estimators are known as eccentricities, defined as

$$\mathcal{E}_n \equiv \varepsilon_n e^{in\Phi_n} = -\frac{\langle \int_r r^n e^{in\phi} s(r,\phi) \rangle}{\langle \int_r r^n s(r,\phi) \rangle} \tag{5}$$

where here the entropy density s is used to characterize the initial state. In central heavy-ion collisions, these estimators are precise enough that simple linear response

$$V_n \approx \kappa_n \mathcal{E}_n \tag{6a}$$

$$v_n\{2\} \approx |\kappa_n| \sqrt{\langle \varepsilon_n^2 \rangle}. \tag{6b}$$

provides a good description of the observed flow harmonics [16].

Within this framework, we can estimate the cumulants $v_n\{2\}$ in ultracentral collisions of deformed nuclei. We generate an initial collision geometry using the Trento code [17], which combines the nuclear profiles of the colliding nuclei into a reduced thickness function T_R. This specifies the initial entropy density $s(x,y) \propto T_R(x,y)$, from which we can determine the event-by-event eccentricities. Then linear response determines the cumulant for sub-bin i relative to the full $0-1\%$ centrality bin as

$$\frac{v_n^i\{2\}}{v_n\{2\}} = \sqrt{\frac{\langle \varepsilon_n^2 \rangle_i}{\langle \varepsilon_n^2 \rangle_{0-1\%}}}. \tag{7}$$

3 Non-Flow Gluon Correlations in the CGC

The most prominent alternative to a hydrodynamic paradigm in certain heavy-ion collisions is an explanation based on multi-gluon correlations arising from the color-glass condensate (CGC) effective theory of high-energy QCD. Because the single-gluon distribution is isotropic,

$$\frac{dN_1}{d^2p\,dy} = \frac{1}{2\pi p_T}\frac{dN_1}{dp_T\,dy} \tag{8}$$

the "collective flow" component of the second-order cumulant $v_n\{2\}$ vanishes. Instead, the cumulant is driven by the explicit two-gluon correlations δ_2, which can be calculated order-by-order in the CGC framework. In the high-p_T "dilute-dilute" limit, the gluon correlation function has an explicit dependence on the the overlapping nuclear profiles T_A, T_B. At leading order, that dependence is quadratic

$$\delta_2(p_1,p_2) \overset{\text{L.O.}}{=} \left(\int d^2x_\perp T_A^2(\vec{x}_\perp)\, T_B^2(\vec{x}_\perp) \right) f(p_1,p_2), \tag{9}$$

but the resulting gluon correlations are explicitly symmetric: $f(p_1,p_2) = f(p_2,p_1)$ [18, 19]. As a result, the quadratic dependence on the nuclear profiles generates the leading contribution to even harmonics n such as $v_2\{2\}$, while the contribution to odd harmonics such as $v_3\{2\}$ vanishes. For these odd harmonics, the leading contribution arises at the next-to-leading order:

$$\delta_2(p_1,p_1) \overset{\text{N.L.O.}}{=} \left(\int d^2x_\perp T_A^3(\vec{x}_\perp)\, T_B^3(\vec{x}_\perp) \right) g(p_1,p_2), \tag{10}$$

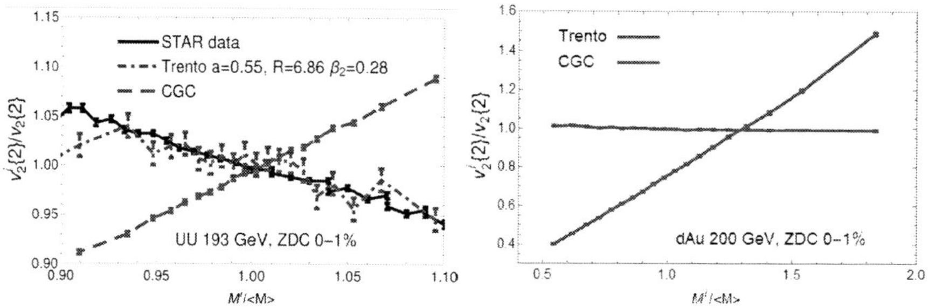

Figure 1. Comparison of the fractional elliptic flow $v_2\{2\}$ fractional versus multiplicity M in ultracentral collisions of deformed ions. Here we show both a large deformed system (UU, left panel) and a small deformed system (dAu, right panel) and normalize the axes by the full $0-1\%$ centrality bin.

where the symmetry is explicitly broken: $g(p_1, p_2) \neq g(p_2, p_1)$ [20]. Here the functions f and g stand for specific integrals of the transverse momenta.

The angular integrals which determine the second cumulant $v_n\{2\}$ then yield

$$(v_n\{2\})^2 = \frac{1}{\langle N_{\text{pairs}} \rangle} \begin{cases} f_n \left\langle \int d^2x_\perp T_A^2(\vec{x}_\perp)\, T_B^2(\vec{x}_\perp) \right\rangle & \text{if n = even} \\ g_n \left\langle \int d^2x_\perp T_A^3(\vec{x}_\perp)\, T_B^3(\vec{x}_\perp) \right\rangle & \text{if n = odd} \end{cases} \tag{11}$$

where the coefficients f_n, g_n are pure numbers which will cancel out in ratios. This allows the cumulants to be calculated, event-by-event, directly from overlap integrals of the nuclear profiles

$$\mathcal{I}_\alpha \equiv \int d^2x_\perp T_A^\alpha(\vec{x}_\perp) T_B^\alpha(\vec{x}_\perp). \tag{12}$$

This includes not only the numerators of the cumulants which invoke $\mathcal{I}_2, \mathcal{I}_3$, but also the denominators $\langle N_{\text{pairs}} \rangle \propto \langle (\mathcal{I}_0)^2 \rangle$ using the phenomenologically-preferred $p = 0$ setting of Trento [21]. Altogether this gives for the CGC scenario

$$\frac{v_n^i\{2\}}{v_n\{2\}} = \sqrt{\frac{\left\langle (\mathcal{I}_0)^2 \right\rangle_{0-1\%}}{\left\langle (\mathcal{I}_0)^2 \right\rangle_i} \frac{\left\langle \mathcal{I}_{2,3} \right\rangle_i}{\left\langle \mathcal{I}_{2,3} \right\rangle_{0-1\%}}}, \tag{13}$$

where one uses $\mathcal{I}_2$ for n = even and $\mathcal{I}_3$ for n = odd. In this way, we can generate an event-by-event comparison between the cumulants $v_n\{2\}$ arising from collective flow in hydrodynamic response and the dynamical gluon correlations arising in the CGC.

4 Results and Conclusions

Results for the second elliptic cumulants $v_2\{2\}$ are shown in Fig. 1 for both a large UU collision and a small dAu collision. In *UU* collisions, the experimental data (black) reveal an anticorrelation between the elliptic flow and the event multiplicity. This decreasing trend is consistent with the expectation of a linear response from hydrodynamics to the initial state (blue). In this scenario, tip-on-tip collisions produce the most multiplicity but the least ellipticity, while side-on-side collisions produce less multiplicity but a greater ellipticity.

These features are qualitatively opposite to the trend shown by the CGC-mediated correlations (red), which increases monotonically as a function of multiplicity. This is because the particle correlations are all proportional to powers of the saturation momentum Q_s^2 which is monotonically correlated with multiplicity. The CGC-mediated mechanism cares only about density, not about geometry. For this reason, the predictions in ultracentral deformed systems are qualitatively opposite.

These features persist generically in smaller systems, as we discuss in detail in Ref. [1]. As an example, we also show the equivalent results for ultracentral dAu collisions. The same qualitative trends persist: the hydrodynamic response is anticorrelated with multiplicity, while the CGC response is positively correlated with it. For dAu the ellipticity in ultracentral collisions is not as great as in UU, so the negative slope persists. Meanwhile, the CGC response is still strongly correlated with multiplicity, even in this small system. Although ample dAu data exists on tape, this analysis of ultracentral events has not yet been performed. We further believe that these results are generic and applicable across the entire spectrum of small and large systems.

References

[1] J. Noronha-Hostler, N. Paladino, S. Rao, M.D. Sievert, D.E. Wertepny (2019), 1905.13323
[2] Y. Pandit (STAR), J. Phys. Conf. Ser. **458**, 012003 (2013), 1305.0173
[3] H. Wang, P. Sorensen (STAR), Nucl. Phys. **A932**, 169 (2014), 1406.7522
[4] L. Adamczyk et al. (STAR), Phys. Rev. Lett. **115**, 222301 (2015), 1505.07812
[5] D. Kharzeev, M. Nardi, Phys. Lett. **B507**, 121 (2001), nucl-th/0012025
[6] A. Goldschmidt, Z. Qiu, C. Shen, U. Heinz, Phys. Rev. **C92**, 044903 (2015), 1507.03910
[7] B. Schenke, C. Shen, P. Tribedy (2019), 1901.04378
[8] G. Giacalone, J. Noronha-Hostler, M. Luzum, J.Y. Ollitrault, Phys. Rev. **C97**, 034904 (2018), 1711.08499
[9] C. Collaboration (CMS) (2018)
[10] S. Acharya et al. (ALICE), Phys. Lett. **B784**, 82 (2018), 1805.01832
[11] T.A. collaboration (ATLAS), *Measurement of the azimuthal anisotropy of charged particle production in Xe+Xe collisions at* $\sqrt{s_{\rm NN}}$*=5.44 TeV with the ATLAS detector*, CERN (CERN, Geneva, 2018)
[12] C. Aidala et al. (PHENIX), Nature Phys. **15**, 214 (2019), 1805.02973
[13] P. Bozek, W. Broniowski, Phys. Rev. Lett. **121**, 202301 (2018), 1808.09840
[14] M.D. Sievert, J. Noronha-Hostler, Phys. Rev. **C100**, 024904 (2019), 1901.01319
[15] M. Luzum, H. Petersen, J. Phys. **G41**, 063102 (2014), 1312.5503
[16] J. Noronha-Hostler, L. Yan, F.G. Gardim, J.Y. Ollitrault, Phys. Rev. **C93**, 014909 (2016), 1511.03896
[17] J.S. Moreland, J.E. Bernhard, S.A. Bass, Phys. Rev. **C92**, 011901 (2015), 1412.4708
[18] Y.V. Kovchegov, D.E. Wertepny, Nucl.Phys. **A906**, 50 (2013), 1212.1195
[19] Y.V. Kovchegov, D.E. Wertepny, Nucl. Phys. **A925**, 254 (2014), 1310.6701
[20] Y.V. Kovchegov, V.V. Skokov, Phys. Rev. **D97**, 094021 (2018), 1802.08166
[21] J.E. Bernhard, J.S. Moreland, S.A. Bass, J. Liu, U. Heinz, Phys. Rev. **C94**, 024907 (2016), 1605.03954

AUTHOR INDEX

EDP Sciences – Web of Conferences
17, Avenue du Hoggar
Parc d'Activité de Courtabouf
BP 112
F-91944 Les Ulis Cedex A
France

ISBN 978-1-7138-1440-5

CURRAN ASSOCIATES INC.
proceedings.com

Tenth International Conference on Mobile Ubiquitous Computing, Systems, Services and Technologies (UBICOMM 2016)

Held at NexTech 2016

Venice, Italy
9—13 October 2016

IARIA